人工智能与
人类未来丛书

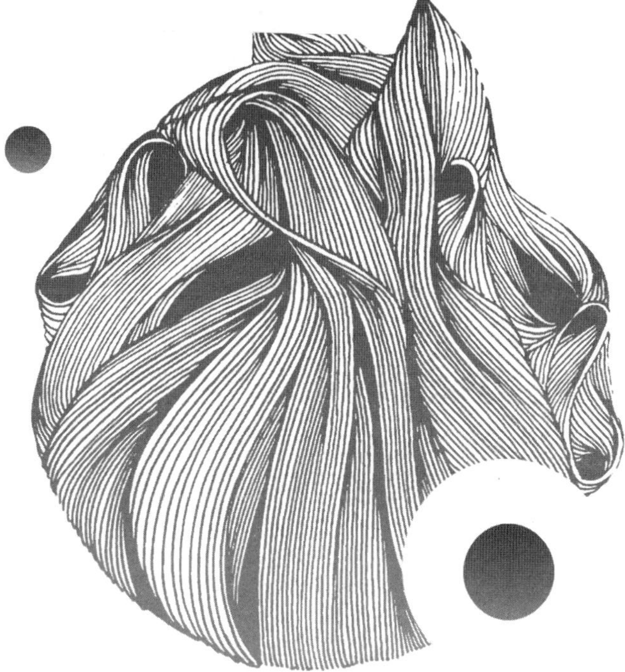

DEEPSEEK
SOURCE CODE
DEEP ANALYSIS

DeepSeek 源码深度解析

薛栋　黄捷　著

北京大学出版社
PEKING UNIVERSITY PRESS

内 容 提 要

本书是一本系统讲解DeepSeek源码及其核心实现原理的技术指南，内容覆盖了从基础概念到高级应用的全流程知识。全书共7章，结构层层递进。第1章对DeepSeek进行了全面概述，帮助读者构建对DeepSeek系统的整体认知。第2章聚焦于环境搭建、代码获取与模型部署接入，为后续深入研究提供基础。第3章深入探讨了MoE（混合专家模型）的基本原理、功能模块与优化技术。第4章详细解析了DeepSeek-V3模型的架构知识，并通过测试验证展示了系统的实际效果。第5章围绕统一多模态大模型展开，介绍了Janus系列架构、核心技术及工具模块。第6章针对高分辨率图像场景，探讨了结合MoE、细粒度特征提取与视觉/语言适配器的多模态模型的知识。第7章聚焦DeepSeek-R1推理大模型，展示了DeepSeek在推理性能与自我进化方面的前沿探索。

本书适合人工智能工程师、深度学习研究者、AI产品开发人员及高校师生阅读。无论您是希望夯实开源模型基础，还是寻找前沿实战案例，本书都将为您提供全面而深入的参考与指导。

图书在版编目(CIP)数据

DeepSeek源码深度解析 / 薛栋，黄捷著. -- 北京 : 北京大学出版社，2025. 5. -- ISBN 978-7-301-36158-0

Ⅰ. TP18

中国国家版本馆CIP数据核字第2025VJ1333号

书　　名	DeepSeek源码深度解析 DeepSeek YUANMA SHENDU JIEXI
著作责任者	薛栋　黄捷　著
责任编辑	刘　云　吴秀川
标准书号	ISBN 978-7-301-36158-0
出版发行	北京大学出版社
地　　址	北京市海淀区成府路205 号　100871
网　　址	http://www.pup.cn　　新浪微博: @ 北京大学出版社
电子邮箱	编辑部 pup7@pup.cn　总编室 zpup@pup.cn
电　　话	邮购部 010-62752015　发行部 010-62750672　编辑部 010-62570390
印 刷 者	北京飞达印刷有限责任公司
经 销 者	新华书店
	787毫米×1092毫米　16开本　27印张　650千字
	2025年5月第1版　2025年5月第1次印刷
印　　数	1-4000册
定　　价	119.00元

推荐序

夯实智能基石，共筑人类未来

人工智能正在改变当今世界。从量子计算到基因编辑，从智慧城市到数字外交，人工智能不仅重塑着产业形态，还改变着人类文明的认知范式。在这场智能革命中，我们既要有仰望星空的战略眼光，也要具备脚踏实地的理论根基。北京大学出版社策划的“人工智能与人类未来丛书”，恰如及时春雨，无论是理论还是实践，都对这次社会变革有着深远影响。

该丛书最鲜明的特色在于其能“追本溯源”。当业界普遍沉迷于模型调参的即时效益时，《人工智能大模型数学基础》等基础著作系统梳理了线性代数、概率统计、微积分等人工智能相关的计算脉络，将卷积核的本质解构为张量空间变换，将损失函数还原为变分法的最优控制原理。这种将技术现象回归数学本质的阐释方式，不仅能让读者的认知框架更完整，还为未来的创新突破提供了可能。书中独创的“数学考古学”视角，能够带读者重走高斯、牛顿等先贤的思维轨迹，在微分流形中理解Transformer模型架构，在泛函空间里参悟大模型的涌现规律。

在实践维度，该丛书开创了“代码即理论”的创作范式。《人工智能大模型：动手训练大模型基础》等实战手册摒弃了概念堆砌，直接使用PyTorch框架下的100多个代码实例，将反向传播算法具象化为矩阵导数运算，使注意力机制可视化为概率图模型。在《DeepSeek源码深度解析》中，作者团队细致剖析了国产大模型的核心架构设计，从分布式训练中的参数同步策略，到混合专家系统的动态路由机制，每个技术细节都配有工业级代码实现。这种“庖丁解牛”式的技术解密，使读者既能把握技术全貌，又能掌握关键模块的实现精髓。

该丛书着眼于中国乃至全世界人类的未来。当全球算力竞赛进入白热化阶段，《Python大模型优化策略：理论与实践》系统梳理了模型压缩、量化训练、稀疏计算等关键技术，为突破“算力围墙”提供了方法论支撑。《DeepSeek图解：大模型是怎样构建的》则使用大量的可视化图表，将万亿参数模型的训练过程转化为可理解的动力学系统，这种知识传播方式极大地降低了技术准入门槛。这些创新不仅呼应了“十四五”规划中关于人工智能底层技术突破的战略部署，还为构建自主可控的技术生态提供了人才储备。

作为人工智能发展的见证者和参与者，我非常高兴看到该丛书的三重突破：在学术层面构建了贯通数学基础与技术前沿的知识体系；在产业层面铺设了从理论创新到工程实践的转化桥梁；在战略层面响应了新时代科技自立自强的国家需求。该丛书既可作为高校培养复合型人工智能人才的立体化教材，又可成为产业界克服人工智能技术瓶颈的参考宝典，此外，还可成为现代公民了解人工智能的必要书目。

站在智能时代的关键路口，我们比任何时候都更需要这种兼具理论深度与实践智慧的启蒙之作。愿该丛书能点燃更多探索者的智慧火花，共同绘制人工智能赋能人类文明的美好蓝图。

于　剑

北京交通大学人工智能研究院院长

交通数据分析与挖掘北京市重点实验室主任

中国人工智能学会副秘书长兼常务理事

中国计算机学会人工智能与模式识别专委会荣誉主任

前言

DeepSeek是杭州深度求索人工智能基础技术研究有限公司推出的AI助手，于2025年1月15日正式上线。

DeepSeek大模型的诞生既是对全球人工智能浪潮的深刻响应，也是中国在大规模语言模型研发领域迈出的坚实步伐。DeepSeek不仅融合了最先进的视觉、语言以及跨模态交互技术，还通过高效的模型训练与推理机制，实现了复杂任务的精准处理和快速响应。DeepSeek为研究人员和开发者搭建了一座连接理论与实践的桥梁，极大地降低了高性能AI技术的应用门槛，推动了前沿技术在各行各业中的广泛应用和产业化进程。

随着人工智能技术朝更高层次、多模态和跨领域应用方向的不断发展，深入理解核心框架和源码实现变得尤为重要。本书正是在这一背景下应运而生，通过系统剖析DeepSeek的设计理念、架构实现及优化策略，为开发者和研究人员提供了翔实的技术指导和实践案例。无论您是初学者还是资深工程师，本书都将帮助您迅速掌握DeepSeek的关键技术，从而在前沿AI领域中获得竞争优势，推动更多创新项目的落地与发展。

本书的特色

- 全面解析：系统梳理DeepSeek的基础概念、源码架构与核心技术，实现从理论到实践的全流程剖析。
- 实战指导：提供详细的环境搭建、源码获取、模型部署与调试流程，帮助读者快速上手并深入理解工程细节。
- 模块架构剖析：详细解析DeepSeek各个模块的源码结构与设计思想，揭示系统整体架构和模块间的调用关系。
- 核心算法解读：逐步解析关键算法与数据流处理过程，从源码层面深入探讨模型训练、推理和优化的实现细节。

● 剖析设计模式与工程实践：剖析源码中采用的设计模式及工程实现策略，展示如何通过优秀的代码组织提高系统的可维护性与扩展性。

● 案例驱动的实战讲解：通过丰富的源码实例和案例分析，带领读者逐步掌握从源码阅读到二次开发的全流程技能。

● 技术前沿：重点解析混合专家模型、DeepSeek-V3与DeepSeek-R1等大规模模型的训练优化、量化计算及权重转换等关键技术。

● 多模态支持：涵盖文本、视觉及跨模态数据处理，展示统一多模态大模型构建与高分辨率图像处理的最新方案。

总之，本书通过理论与实战结合的方式，帮助读者在多模态大模型领域中从零开始、逐步精通，最终使读者具备自主开发与创新的能力。

赠送资源

本书附赠67G学习资源，包括全书案例源代码、PPT课件及教学视频等，读者可以扫描右侧二维码关注“博雅读书社”微信公众号，输入本书77页的资源下载码，即可获得本书的下载学习资源。

博雅读书社

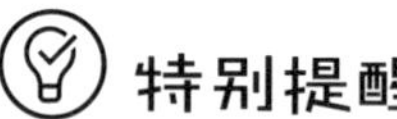

本书从写作到出版，需要一段时间，软件升级可能会有界面变化，读者在阅读本书时，可以根据书中的思路，举一反三地进行学习，不拘泥于细微的变化，掌握使用方法即可。

致谢

在本书的编写过程中，我得到了北京大学出版社各位专业编辑的悉心指导与大力支持。正是他们严谨认真的态度和细致高效的工作，确保了本书能够在有限的时间内顺利出版。对此，我深表感谢。

同时，我也衷心感谢家人在整个写作期间给予的巨大支持与理解，他们的陪伴与鼓励是我坚持完成本书写作的重要动力。

由于个人水平有限，书中难免存在纰漏与不足之处，恳请广大读者不吝赐教，提出宝贵的意见与建议，以便在后续版本中不断完善与改进。

最后，感谢您选择并阅读本书。希望它能成为您编程与技术探索路上的有力向导，并助您在学习与实践中不断进步。祝您阅读愉快！

编者

目录

CONTENTS

第1章 DeepSeek概述

第2章 环境搭建、代码获取与模型部署接入

第3章 混合专家模型（MoE）初探

第4章 基于DeepSeekMoE架构的DeepSeek-V3

第5章 统一多模态大模型

第6章 适用于高分辨率图像的多模态模型

第7章 DeepSeek-R1推理大模型

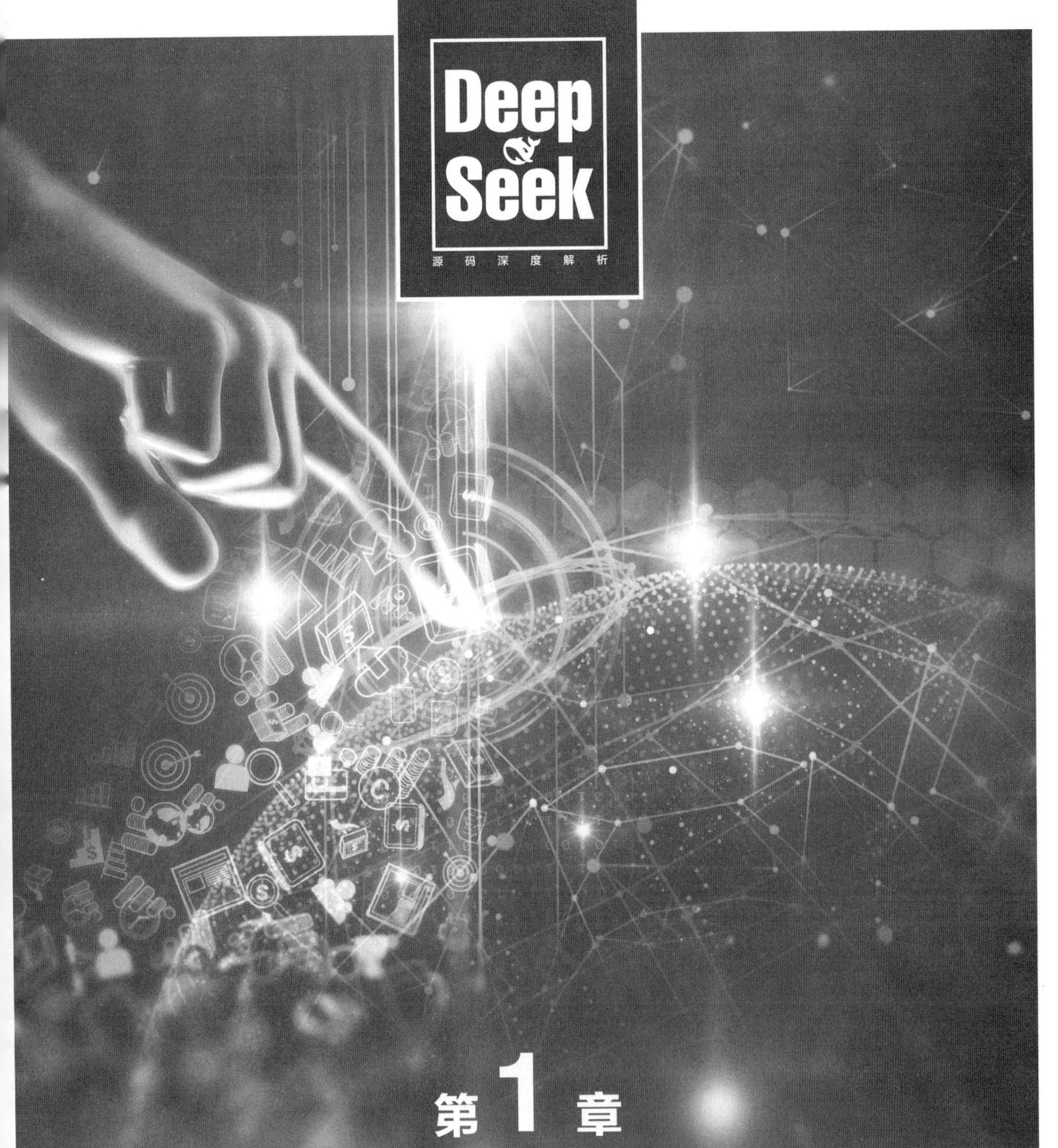

第 1 章 DeepSeek 概述

本章将详细阐述DeepSeek的基本概念、架构、应用与评测信息。首先介绍了DeepSeek的背景、目标及其在各类应用场景中的核心功能，阐明了其研发动机和实际价值；随后，详细描述了DeepSeek的整体架构设计和模块划分，揭示了其在输入嵌入、Transformer模块、优化策略和输出层等方面的创新技术；最后，通过与其他大模型进行对比，展示了DeepSeek在推理速度、计算效率和多任务处理等方面的优势，从而为后续章节对其技术细节的深入探讨奠定了坚实基础。

1.1 DeepSeek 简介

在人工智能的浩瀚星空中，DeepSeek犹如一颗璀璨的科技之星，熠熠生辉，引领着大模型时代的风云变幻。DeepSeek以卓越的创新精神和前沿的技术架构，突破常规极限，将海量知识与智能推理完美融合，展现出惊人的计算力与思维深度。

1.1.1 DeepSeek 介绍

DeepSeek，全称是杭州深度求索人工智能基础技术研究有限公司，总部位于中国杭州。

1. 公司背景

● 成立时间：DeepSeek成立于2023年7月17日，由知名量化资管公司幻方量化创立。

● 公司定位：公司致力于开发先进的LLM（Large Language Model，大语言模型）及相关技术，专注于自然语言处理、机器学习、深度学习等核心技术的研发。

● 核心优势：DeepSeek在硬件资源和技术积累上具备显著优势，具有强大的研发能力和创新精神。

2. 团队构成

● 创始人：梁文锋，浙江大学信息与通信工程专业硕士，曾创立幻方量化，专注于量化投资，后进军通用人工智能领域，创办DeepSeek。

● 核心成员：团队成员多来自国内顶尖高校，如清华大学、北京大学、中山大学、北京邮电大学等，具有深厚的学术背景和丰富的研发经验。

● 团队特点：团队规模较小，不到140人，成员年轻且多为本土培养，注重技术创新和软硬件协同设计。

1.1.2 DeepSeek 的背景与目标

DeepSeek大模型的诞生既是对全球人工智能浪潮的深刻响应，也是中国在大规模语言模型研发领域迈出的坚实步伐。

1. 背景动因

● 全球人工智能变革浪潮：随着深度学习和自然语言处理技术的迅猛发展，全球各大科技公司和研究机构纷纷投入巨资研发大规模预训练模型。这一趋势不仅推动了语义理解、知识获取和多模态交互等前沿领域的突破，也暴露出传统密集模型在参数扩展、推理速度和计算资源消耗上的种种不足。

- 国内需求与挑战：面对日益激烈的国际竞争，中国亟须打造具有自主知识产权的先进大模型，以满足本土语言、文化及特定领域应用（如古文解析、历史研究、编程辅助等）的特殊需求。同时，国内市场对高效、低成本且具备强大泛化能力的智能系统呼声日益高涨，这为DeepSeek的研发提供了充足的市场和技术驱动力。
- 技术瓶颈与突破：在现有大模型的研发过程中，如何在保持大规模参数优势的同时，实现高效推理和低能耗计算成为核心难题。传统密集模型虽然在一定程度上展现出卓越的语言理解能力，但往往难以兼顾实时性和资源利用率，这促使研究者探索如混合专家（Mixture of Experts，MoE）架构、多头潜在注意力（Multi-head Latent Attention，MLA）等新型技术方案，从而催生了DeepSeek系列模型。

2. 目标追求

- 性能与效率的双重突破：DeepSeek大模型旨在通过创新的架构设计（如引入混合专家架构、多头潜在注意力、低精度训练等技术）实现海量参数与高效推理之间的完美平衡，其不仅能大幅提升模型的理解和生成能力，还能在推理速度和资源消耗方面达到前所未有的优化效果。
- 多领域应用与普适性：模型目标不仅用于通用语言处理，更致力于在编程、数学推理、专业知识问答及跨语言应用等多个领域展现出卓越表现。通过大规模多语料预训练和针对性微调，DeepSeek希望成为一个能够服务于学术研究、工业应用和商业创新的全能型人工智能平台。
- 开放共享与生态构建：为推动全球人工智能技术的进步和公平竞争，DeepSeek大模型坚持开放策略，通过开源协议释放模型权重和技术文档，鼓励全球开发者、研究者及企业进行深度合作与持续创新。这不仅有助于形成一个共赢的技术生态，也为后续大模型的标准制定和应用推广提供了宝贵经验。
- 引领未来智能变革：站在技术前沿，DeepSeek致力于突破现有大模型的技术瓶颈，探索更高层次的智能推理与自我反思机制，推动人工智能从“工具”向“智慧体”转变，最终实现对复杂任务的高效解决与决策支持，助力各行各业进入智能化新时代。

总之，DeepSeek大模型既是对当前人工智能发展现状的深刻洞察，也是对未来智能变革的战略布局。它不仅代表着技术创新和效率提升的最新成果，更承载着推动人工智能普惠化、产业化与国际竞争力提升的宏伟目标。

1.1.3 DeepSeek的产品

DeepSeek的核心产品为大语言模型，其整体架构设计经历了多个版本的演进，以下是对其主要产品的介绍。

1. DeepSeek LLM 系列

DeepSeek LLM系列包括 7B和 67B参数的模型，采用了与LLaMA系列相似的架构。这些模型使用了预规范化的解码器（Pre-norm Decoder-only Transformer），结合了均方根（Root Mean Square Layer Nomalization，RMSNorm）作为归一化方法、SwiGLU激活函数、旋转位置嵌入（Rotary Position Embedding，RoPE）和分组查询注意力（Grouped Query Attention，GQA）。词汇表大小为102,400（字节级BPE），上下文长度为 4096。训练数据包含了 2 万亿个英文和中文文本。

2. DeepSeek MoE

在 2024 年 1 月，DeepSeek发布了两款DeepSeek MoE模型（Base和Chat），每个模型拥有16B参数（每个Token激活 2.7B参数），上下文长度为 4K。这些模型采用了稀疏门控混合专家（Sparsely-Gated MoE）架构，包含“共享专家”（始终被查询）和“路由专家”（可能不会被查询）。这种设计有助于平衡专家的使用，避免某些专家被过度使用而其他专家很少被使用的情况。

3. DeepSeek-V2

2024 年 5 月，DeepSeek发布了DeepSeek-V2 系列，包括 4 个模型：DeepSeek-V2、DeepSeek-V2-Lite、DeepSeek-V2-Chat和DeepSeek-V2-Lite-Chat。其中，DeepSeek-V2 模型在预训练阶段使用了 8.1 万亿个Token的数据，扩展了上下文长度（从 4K增加到 128K）。在训练过程中，采用了多头潜在注意力和混合专家架构，以提高模型的性能和效率。

4. DeepSeek-V3

2024 年 12 月，DeepSeek发布了DeepSeek-V3 模型，包括DeepSeek-V3-Base和DeepSeek-V3（聊天模型）。该模型在架构上与V2 类似，但引入了多Token预测机制，以提高解码速度。在训练过程中，模型在 14.8 万亿个多语言语料上进行了预训练，主要包括英文和中文文本。上下文长度从 4K扩展至 128K。此外，模型还进行了监督微调（Supervised Fine-Tuning，SFT）和强化学习（Reinforcement Learn，RL）训练，以提升推理能力。

5. DeepSeek-R1

DeepSeek-R1是DeepSeek于 2025 年 1 月发布的开源LLM Learning，其架构设计在多个方面进行了创新，以提升推理能力和效率。通过强化学习、混合专家架构、多头潜在注意力机制和低精度训练等技术，显著提升了模型的推理能力和效率，为人工智能领域带来了新的突破。概括起来，DeepSeek-R1的主要特征如下。

- 强化学习与混合专家架构：DeepSeek-R1 采用了强化学习技术，摒弃了传统的过程奖励模型（Process-supervised Reward Model，PRM）方法，直接以结果为导向进行奖励，促使AI学会更高效地思考，并展现出初步的反思能力。 此外，DeepSeek-R1采用了MoE架构，每层包含 1 个共享专家和 256 个路由专家，每个专家的中间隐藏维度为 2048。在这些路由专家中，每个Token将激活 8 个专家。
- 多头潜在注意力机制：DeepSeek-R1 引入了多头潜在注意力机制，通过压缩潜在向量来提升性能，并减少推理过程中的内存使用。

- 低精度训练与高效训练框架：在训练过程中，DeepSeek-R1 采用了低精度训练技术，并结合高效的DualPipe训练框架，实现了模型性能的大幅提升与成本的有效控制。
- 开源与社区协作：DeepSeek-R1 作为开源模型，鼓励全球开发者和研究人员进行实验和改进，促进了人工智能领域的协作与创新。

总之，DeepSeek注重模型的扩展性和效率，通过引入混合专家架构、多头潜在注意力机制和多Token预测等技术，不断提升模型在处理复杂任务时的性能。

1.1.4 DeepSeek 的应用场景

DeepSeek大模型凭借其卓越的性能和广泛的应用场景，正在推动人工智能技术在多个领域的创新和发展。

1. 自然语言处理领域

- 智能客服系统开发：DeepSeek-V3能够准确分析并理解用户提问的意图，从而给予高质量的回复，显著提升客户满意度，解决企业客服环节的诸多问题。
- 长文本分析与摘要：DeepSeek-V3具有强大的长文本处理能力，如支持长达128K的输入文本，能有效应对复杂冗长的法律文件，帮助法律从业者快速获取文件的关键信息。
- 文本翻译：利用DeepSeek的多头潜在注意力机制，能够准确理解源语言文本每个词在上下文中的准确含义，从而更精准地将其翻译成目标语言。

2. 代码生成与编程辅助

DeepSeek-V3在代码生成和多语言编程测评中表现优异，其能够理解编程的逻辑需求并生成可用的代码段，适用于初学者进行基础代码编写，以及经验丰富的开发者用于快速生成代码模板等场景。

3. 多模态数据处理

图文内容生成与描述：DeepSeek-V3采用的混合专家架构，支持高效的多模态数据处理，可以融合图像和文本信息进行深入分析，推动多模态AI应用的发展。

4. 金融领域

- 金融舆情分析：DeepSeek与拓尔思联合开发的金融舆情大模型，能够快速准确地分析金融舆情，为投资者提供有价值的参考信息。
- 智能研报生成：中信证券的智能研报系统采用DeepSeek大模型后，错误率降低了90%，大大提高了研报的质量和效率。

5. 教育领域

科大讯飞接入了DeepSeek-Math模型，推出了AI数学辅导应用“星火助学”，其能够根据学生的学习情况，提供个性化的数学学习计划和练习题。

6. 办公领域

金山办公接入了DeepSeek-Writer API，提升了WPS智能写作功能，公文生成效率提升了3倍，错误率下降了90%。

7. 医疗领域

DeepSeek大模型能够输入患者主诉，检索相似病例，生成鉴别诊断列表，通过美国健康保险流通与责任法案（Health Insurance Portability and Accountability Act，HIPAA）认证，支持私有化部署与严格的数据隔离。

8. 法律领域

法律文书处理方面，DeepSeek大模型内置法律条文数据库，且支持实时更新与司法解释无缝对接，能够进行合同条款智能审查、争议焦点精准提取、判决书自动生成。

9. 工业领域

DeepSeek-Max能够通过图像识别（缺陷检测）、文本生成（维修建议）、语音指导（操作辅助）等流程，显著降低漏检率。

总之，DeepSeek大模型凭借其强大的技术架构和广泛的应用场景，正在为各行业提供智能化解决方案，推动了行业的数字化转型和创新发展。

1.1.5 DeepSeek 的核心功能

DeepSeek是一款具有多功能性和技术先进性的人工智能模型，其核心功能涵盖了多个领域，能够为用户提供高效、智能的解决方案。

1. 自然语言处理（Natural Language Processing，NLP）

- 智能问答：DeepSeek能够快速准确地回答各类问题，包括科学知识、历史文化、生活常识和技术问题等。它不仅能提供答案，还能根据用户的追问深入拓展相关内容，就像一位知识渊博的导师。
- 内容生成：DeepSeek可以撰写各种类型的文章，如新闻报道、学术论文、商业文案和小说故事等。它还能生成工作报告、市场调研报告等，帮助用户快速整理和分析数据，提高写作效率。
- 文本翻译：DeepSeek支持多种语言之间的实时翻译，帮助用户扫除语言障碍，轻松融入多语言环境。
- 代码辅助：对程序员而言，DeepSeek是一个强大的代码辅助工具。它可以生成代码、调试代码和优化代码，大大提高编程效率和质量。

2. 计算机视觉

- 图像识别：DeepSeek能够识别图像中的物体、人脸和场景，广泛应用于安防、医疗和零售等领域。
- 图像生成：根据用户输入的描述，DeepSeek可以生成高质量的图像，适用于设计和创意领域。

● 视频分析：DeepSeek能够分析视频内容，提取关键信息，如动作识别、事件检测等。

3. 语音处理

● 语音识别：DeepSeek可以将语音转换为文字，适用于语音助手、会议记录等场景。

● 语音合成：DeepSeek可以将文字转换为语音，生成自然流畅的语音。

● 语音翻译：DeepSeek支持实时语音翻译，帮助用户进行跨语言交流。

4. 数据分析

● 数据处理：DeepSeek能够处理各种类型的数据，包括Excel表格数据、CSV文件数据等。它可以进行数据清洗、数据统计分析和数据分类排序，帮助用户快速整理和分析数据。

● 可视化图表生成：在数据分析的基础上，DeepSeek可以将数据转化为直观的可视化图表，如柱状图、折线图、饼图、散点图等，让数据一目了然。

5. 智能对话与搜索

● 智能对话：DeepSeek能够与用户进行高水平、顺畅的对话，像朋友一样交流，为用户答疑解惑。

● AI搜索：DeepSeek可以进行全网搜索，让用户实时掌握信息，无论是知识查询还是热点追踪，其都能快速搞定。

6. 多语言支持

DeepSeek支持多种语言，包括中文、英文、日文、韩文、法文、德文等常见语言，能够满足全球不同用户的需求，帮助用户轻松实现语言转换和沟通。

7. 个性化定制与扩展

● 自定义知识库：DeepSeek支持上传文件建立自定义知识库，基于这些知识为用户提供更个性化、更具针对性的回答和建议。

● 多模态交互：DeepSeek的部分模型具有图像和文本理解生成能力，用户不仅可以输入文本，还能上传图片进行提问或创作。

DeepSeek凭借其强大的技术架构和丰富的功能，为用户提供了全方位的智能支持，无论是在日常生活还是专业领域，都能展现出卓越的性能和实用性。

1.2 DeepSeek 的架构概览

DeepSeek模型基于经典的Transformer模型架构，并进行了深度优化，采用了混合专家架构，通过稀疏激活提升模型效率。此外，DeepSeek引入了动态路由网络，智能地调配计算资源，以高效处理长文本和复杂逻辑任务。

1.2.1 DeepSeek 的整体架构设计

DeepSeek的整体架构设计以高效处理长文本和提升推理效率为核心目标，主要包含如下所示的几个关键组成部分。

1. MLA

- 低秩联合压缩：MLA通过将Key与Value分解为低秩矩阵并进行联合压缩，减少了需要存储和访问的数据量，从而降低了推理阶段的显存与时间开销。
- 显式位置编码融合：结合位置编码，使得模型能够在压缩后依旧保留序列顺序信息。
- 并行化计算优化：对压缩后的Key-Value进行并行操作，兼顾了注意力的灵活性与推理速度的提升。

2. DeepSeekMoE 架构

- 混合专家系统：DeepSeekMoE架构融合了MoE、MLA和RMSNorm三个核心组件。通过专家共享机制、动态路由算法和潜在变量缓存技术，该模型在保持性能水平的同时，相较传统MoE模型降低了40%的计算开销。
- 动态路由机制：针对输入令牌嵌入，路由器通过门控网络从多个专家中选择最相关的专家。这种机制确保了计算的高效性和模型的稳定性。
- 无辅助损失的负载均衡策略：DeepSeek-V3通过动态调整专家偏置，实现了负载均衡，避免了传统方法中强制负载均衡导致的模型性能下降。

3. 多 Token 预测训练目标（Multi-Token Prediction，MTP）

同时预测多个Token：在训练过程中，模型不仅预测下一个Token，还预测后续多个位置的Token。这种机制增加了训练信号密度，有助于模型学习长期依赖关系，提高生成质量。

4. 层级策略优化

- MoE：内置多个专家子网络，通过精细的门控机制按需激活，扩大模型容量，同时保持计算成本可控。
- 分阶段训练：包括预训练阶段、对齐阶段和领域微调阶段，确保模型在不同任务和领域的表现。

5. 其他优化

- FP8（8-bit Floating Point）混合精度：大幅加快训练速度，在硬件条件支持下可实现更高吞吐量。
- 多语言与多领域数据：模型具备一定的跨语言能力，可在通用场景下保持较佳表现。

总之，DeepSeek的整体架构设计通过这些创新和优化，实现了在超大规模参数与实际推理效率之间的平衡，显著提升了模型的性能和应用价值。

1.2.2 DeepSeek 的模块划分

DeepSeek模型采用了多层次的模块化设计，以提升其性能和效率。

1. 输入嵌入模块

● 功能：将输入文本转化为模型可处理的向量。

● 细节：通过词嵌入和位置嵌入组合，为每个输入Token生成一个固定维度的向量。

2. Transformer 模块

（1）MLA。

● 功能：高效处理序列信息，降低计算和存储需求。

● 细节：通过低秩压缩技术，将Token的特征压缩到较小的潜在空间，再通过上投影矩阵将其恢复到Key、Value空间。

（2）MoE。

● 功能：通过多个专家子网络提高模型容量和计算效率。

● 细节：每个MoE层包含1个共享专家和256个路由专家，每个Token选择8个专家进行处理。

（3）RMSNorm归一化层。

● 功能：稳定训练过程，加速模型收敛。

● 细节：在每个Transformer模块中使用RMSNorm归一化层，对输入数据进行归一化处理。

3. 优化策略模块

（1）MTP。

● 功能：增加训练信号密度，提高生成质量。

● 细节：在训练过程中，模型不仅预测下一个Token，还预测后续多个位置的Token。

（2）负载均衡策略。

● 功能：确保专家负载均衡，提高模型性能。

● 细节：通过动态调整专家偏置项，实现负载均衡，无须额外的辅助损失函数。

4. 输出层

● 功能：将Transformer模块的输出转化为最终的预测结果。

● 细节：通过一个线性层将Transformer的输出映射到词汇表大小的维度，得到每个Token的预测概率分布。

5. 其他辅助模块

（1）FP8混合精度训练模块。

● 功能：降低训练时的GPU内存占用和计算开销。

● 细节：通过精细的量化策略和高精度累加，实现FP8混合精度训练。

（2）残差流分形解码架构。

● 功能：提高推理效率。

● 细节：通过主次双Token预测和动态损失融合，提升单次前向传播的学习效率。

1.2.3 DeepSeek 与其他模型的技术对比

DeepSeek模型在人工智能领域引起了广泛关注，其性能和特点与其他大语言模型相比，展现出独特的优势和差异。

1. 与 GPT 系列对比

● 技术架构：DeepSeek采用混合架构，结合了深度学习与强化学习技术，注重高效性和灵活性，支持快速迭代和定制化开发；GPT系列基于Transformer架构，以强大的语言生成能力和上下文理解能力著称。

● 性能表现：DeepSeek在语言生成任务中表现出色，尤其在中文语境下的表现优于GPT系列，生成的文本更加符合中文表达习惯，且在多轮对话中能够保持较高的连贯性；GPT-4在英文任务中表现优异，但在处理中文时偶尔会出现语义偏差或文化背景理解不足的问题。

● 计算效率与资源消耗：DeepSeek在计算效率上表现优异，其模型设计降低了资源消耗，适合在资源有限的环境中部署；GPT-4由于模型规模较大，对计算资源的需求较高，部署成本较高。

● 应用场景：DeepSeek适用于多种场景，包括智能客服、内容创作、教育辅助和数据分析等，其高效性和灵活性使其在企业级应用中具有较大优势；GPT系列在内容创作、代码生成和学术研究等领域表现优异，但其高昂的部署成本限制了其在中小企业中的应用。

2. 与 Claude 对比

● 技术架构：DeepSeek采用混合架构，注重高效性和灵活性；Claude以“对齐性”为核心设计理念，注重模型的道德和安全性能。

● 性能表现：DeepSeek在语言生成任务中表现出色，尤其在中文语境下的表现优于Claude；Claude在生成内容的安全性上表现优异，但在复杂语言任务上的灵活性和创造力稍显不足。

● 计算效率与资源消耗：DeepSeek在计算效率上表现优异，适合在资源有限的环境中部署；Claude在计算效率上表现较好，但其生成速度略慢于DeepSeek。

● 应用场景：DeepSeek适用于多种场景，包括智能客服、内容创作、教育辅助和数据分析等；Claude在需要高安全性和道德标准的场景（如法律咨询、医疗辅助）中表现优异，但其应用范围相对较窄。

3. 与 Gemini 对比

● 技术架构：DeepSeek采用混合架构，注重高效性和灵活性；Gemini是多模态AI模型，能够同时处理文本、图像和音频等多种数据类型，其架构设计注重多模态融合。

● 性能表现：DeepSeek在语言生成任务中表现出色，尤其在中文语境下的表现优于Gemini；Gemini在多模态任务中表现突出，但在纯文本生成任务上略逊一筹。

● 计算效率与资源消耗：DeepSeek在计算效率上表现优异，适合在资源有限的环境中部署；Gemini由于模型规模较大，对计算资源的需求较高，部署成本较高。

● 应用场景：DeepSeek适用于多种场景，包括智能客服、内容创作、教育辅助和数据分析等；Gemini在多模态任务（如图像描述、视频分析）中表现突出，适合用于多媒体内容生成和分析。

4. 与 Switch Transformer 对比

● 参数效率：在配置64个专家（其中8个共享）的情况下，DeepSeekMoE较Switch Transformer

（64个专家）实现了1.8倍的吞吐量提升，同时参数量降低了30%。

● 训练效率：相比参数规模相当（13B）的密集Transformer，DeepSeekMoE训练速度提升了2.1倍。

● 推理性能：MLA缓存机制使自回归任务的延迟降低35%。

● 模型性能：在WikiText-103测试集上，DeepSeekMoE的困惑度达到12.3，优于Switch Transformer的14.1；在WMT 14 EN-DE测试集上，DeepSeekMoE的BLEU得分达44.7，较Transformer++提升2.1分。

5. 与LLaMA对比

● 训练成本：DeepSeek-V3的训练费用相比LLaMA等大模型要少得多，据外媒估计，Meta的大模型LLaMA-3.1的训练投资超过了5亿美元。

● 性能表现：DeepSeek-V3在多项评测中表现优异，表现比LLaMA更好，甚至直逼世界顶尖的闭源模型GPT-4o和Claude-3.5-Sonnet。

总体而言，DeepSeek模型在性能、成本效益、开源策略、技术架构和应用领域等方面，与其他大语言模型相比，展现出独特的优势和差异。DeepSeek官网展示了其与其他大模型的对比数据，如下图所示。

deepseek.com

DeepSeek-V3 的综合能力

DeepSeek-V3 在推理速度上相较历史模型有了大幅提升。
在目前大模型主流榜单中，DeepSeek-V3 在开源模型中位列榜首，与世界上最先进的闭源模型不分伯仲。

	Benchmark (Metric)	DeepSeek V3	DeepSeek V2.5 0905	Qwen2.5 72B-Inst	Llama3.1 405B-Inst	Claude-3.5 Sonnet-1022	GPT-4o 0513
	Architecture	MoE	MoE	Dense	Dense	-	-
	# Activated Params	37B	21B	72B	405B	-	-
	# Total Params	671B	236B	72B	405B	-	-
	MMLU (EM)	88.5	80.6	85.3	**88.6**	88.3	87.2
	MMLU-Redux (EM)	**89.1**	80.3	85.6	86.2	88.9	88.0
	MMLU-Pro (EM)	75.9	66.2	71.6	73.3	**78.0**	72.6
	DROP (3-shot F1)	**91.6**	87.8	76.7	88.7	88.3	83.7
English	IF-Eval (Prompt Strict)	86.1	80.6	84.1	86.0	**86.5**	84.3

DeepSeek与其他大模型的对比数据

根据上图中的数据，可以总结出以下信息。

（1）综合性能与推理能力。

● 综合能力出众：在大规模多任务语言理解（Massive Multitask Language Understanding，MMLU）的英语评测中，DeepSeek-V3的表现处于高水平，甚至与部分闭源模型（如Claude-3.5和

GPT-4o）相当。在中文评测（CLUEWSC、C-Eval和C-SimpleQA）上，DeepSeek-V3 同样取得了最高分数，显示出其跨语言综合能力的均衡性。

- 推理速度和效率提升明显：DeepSeek-V3 相较于历史模型（如DeepSeek-V2.5、Qwen2.5 和 LLaMA3.1）在推理速度上有大幅提升，这表现在DROP、IF-Eval、LiveCodeBench等多项指标上，其 3-shot F1 分数、Prompt Strict模式下的表现及代码生成任务均领先于其他开源模型。

（2）参数架构与效率。

MoE架构优势：DeepSeek-V3 采用MoE架构，使得其在总参数量（671B）远高于某些密集模型（如Qwen2.5的72B、LLaMA3.1的405B）的同时，通过仅激活部分参数（37B）实现高效计算。这种设计不仅提升了模型容量，也保证了推理的高效能。

（3）代码（Code）与数学（Math）能力。

- 代码生成任务：在HumanEval-Mul、LiveCodeBench及Codeforces等代码任务上，DeepSeek-V3 的表现均优于同类开源模型，显示出其在复杂编程和逻辑推理任务上的能力。
- 数学题解能力：在DeepSeek官网展示的对比数据中，AIME 2024、MATH-500 和 CNMO 2024 等数学评测数据表明，DeepSeek-V3在数学推理和问题解决上有明显优势，其Pass@1及EM分数均高于其他模型，体现了很强的逻辑和数学处理能力。

（4）与闭源模型旗鼓相当。

虽然在部分指标如GPQA-Diamond和SimpleQA上，闭源模型（如Claude-3.5 和GPT-4o）仍有一定优势，但从整体来看，DeepSeek-V3在大多数评测中都处于领先地位或与顶尖闭源模型不相上下，成为开源模型中的佼佼者。

（5）对比结论。

- DeepSeek-V3在多个领域和任务中表现出色，尤其是在English、Code和Math等领域的任务中，其表现与世界上最先进的闭源模型不分伯仲。
- DeepSeek-V3在开源模型中位列榜首，显示出其强大的竞争力。
- DeepSeek-V3在多个指标上表现优异，显示出其在技术架构和训练方法上的优化效果。

综上所述，DeepSeek-V3在推理速度、综合语言理解、代码生成及数学推理等多个维度上均展现出显著的优势。其采用的MoE架构和高效的参数激活机制使其在保持大规模模型容量的同时，实现了高效计算，已成为目前大模型主流榜单中开源模型的领跑者，并与世界上最先进的闭源模型比肩。

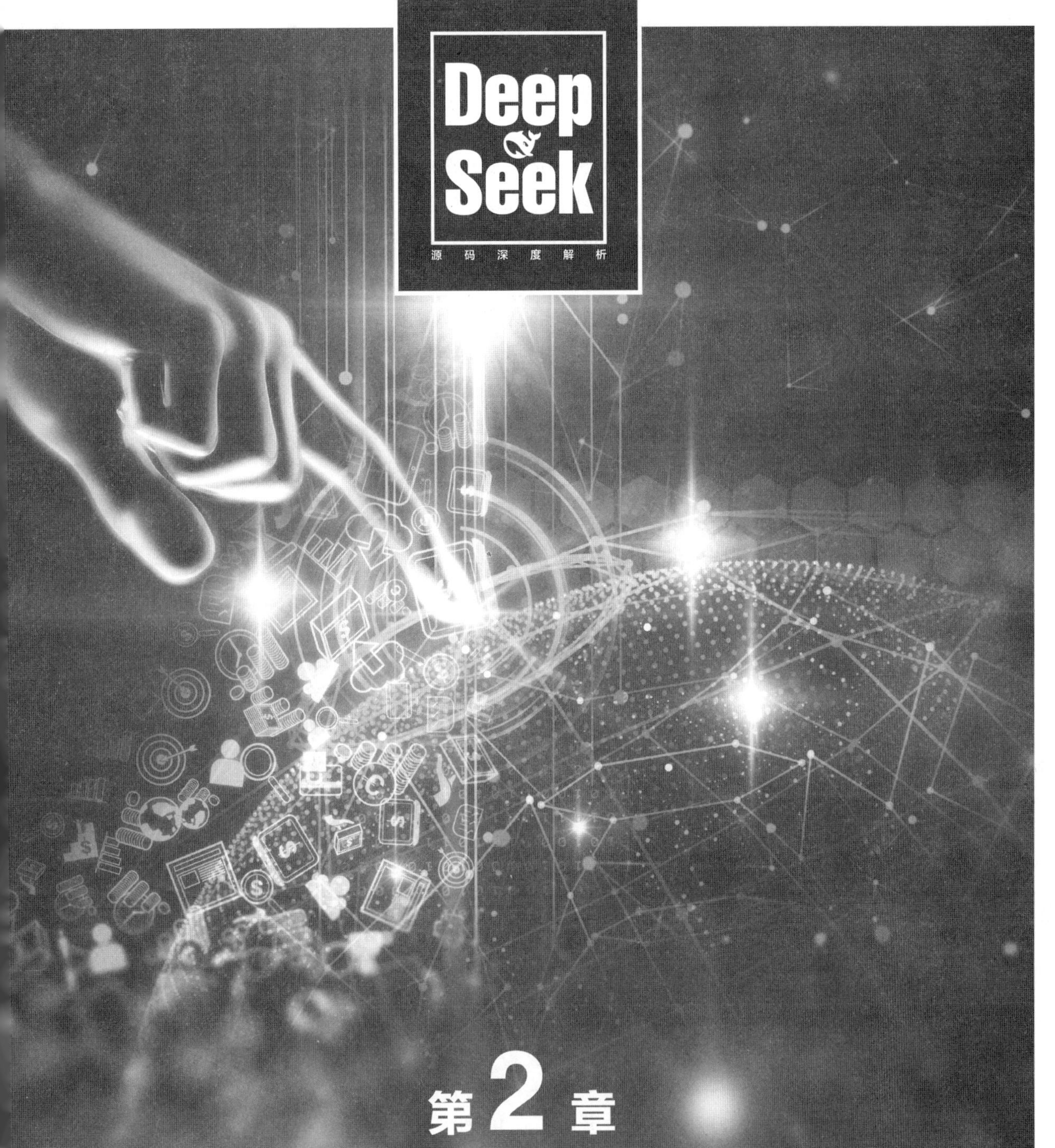

第2章 环境搭建、代码获取与模型部署接入

环境搭建与代码获取是学习DeepSeek源码的基础步骤，本章将详细讲解相关内容。首先，在环境准备方面，需要满足一定的硬件环境要求，如具备合适的GPU等计算资源，同时要进行软件环境配置，包括安装操作系统、搭建Python环境等，并安装所需的依赖库，以确保代码能够正常运行。其次，在代码获取与管理方面，可以从GitHub克隆代码到本地，通过合理使用代码分支管理来组织和开发不同的功能或修复不同的问题，同时要进行代码更新与同步，以保持本地代码与远程仓库代码的一致性。最后，以DeepSeek-R1模型为例，讲解了本地部署DeepSeek大模型的知识。

2.1 环境准备

要研究 、分析DeepSeek的开源代码，首先需要搭建适当的硬件环境，确保计算资源满足模型运行需求。接着配置软件环境，包括安装特定版本的操作系统、Python、CUDA等。最后按照项目提供的指南，安装所需的依赖库，以确保代码能够正确运行。

2.1.1 硬件环境要求

要深入分析和学习DeepSeek的源代码，不同的DeepSeek开源模型所需的硬件环境也不同。硬件环境的具体要求主要取决于所选模型的参数规模，不同DeepSeek模型版本对硬件的配置建议如下表所示。

不同DeepSeek模型版本对硬件的配置建议

模型类型	最小GPU配置	CPU配置	内存要求	磁盘空间
R1-Zero全量版	RTX 4090(24GB)	Xeon 8核+128GB	128GB	500GB
R1蒸馏版-70B	RTX 3090(24GB)	i9-13900K+64GB	64GB	320GB
R1蒸馏版-14B	RTX 3060(12GB)	Ryzen 7+32GB	32GB	80GB
R1蒸馏版-1.5B	不需要GPU	任意4核处理器+8GB	8GB	12GB

DeepSeek-V3源码采用FP8 训练，并开源了原生FP8 权重。由于模型参数量巨大（约 700B），对显存的要求极高，可能需要数百GB的显存。因此，个人用户难以在普通硬件上运行该模型，建议使用高性能计算集群或云服务。

注意

模型参数量越大，对硬件配置的要求越高。如果硬件配置不足，模型将使用CPU进行计算，这会显著延长推理时间。建议在选择和部署DeepSeek模型时，根据模型的参数规模和您的硬件配置进行匹配。确保您的硬件环境满足模型运行的最低要求，以获得最佳的性能和体验。

2.1.2 软件环境配置

要深入分析和学习DeepSeek的源代码，除硬件配置外，软件环境的配置也至关重要。

1. 操作系统

DeepSeek的源码分析和开发环境对操作系统没有严格限制，但建议使用以下系统以获得更好的兼容性和支持。

- Linux：如Ubuntu 20.04 或更高版本。
- Windows：Windows 10 或更高版本。
- macOS：macOS 11.0（Big Sur）或更高版本。

2. Python 环境

DeepSeek的大部分代码是用Python编写的，因此需要配置合适的Python环境。

- Python版本：建议使用Python 3.8 或更高版本。
- 虚拟环境：为了避免与其他项目的依赖冲突，建议使用venv或conda创建虚拟环境。

3. 深度学习框架

DeepSeek模型基于PyTorch框架开发，因此需要安装PyTorch。根据计算机的GPU情况选择合适的PyTorch版本，例如，使用的是NVIDIA GPU且CUDA版本为11.7，可以使用以下命令安装：

```
pip install torch torchvision torchaudio --extra-index-url https://download.
pytorch.org/whl/cu117
```

4. 必要的依赖库

在分析源码前，需要安装项目依赖的库。通常，这些依赖项会列在项目的requirements.txt或environment.yml文件中。例如DeepSeek-V3项目中的依赖文件requirements.txt如图2-1所示，内容如下。

```
torch==2.4.1
triton==3.0.0
transformers==4.46.3
safetensors==0.4.5
```

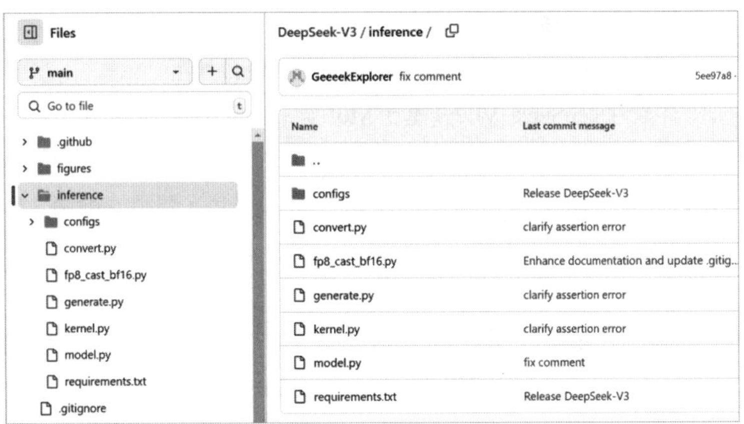

图2-1 DeepSeek-V3项目中的依赖文件requirements.txt

可以使用pip命令安装文件requirements.txt中列出的依赖库，例如：

```
pip install transformers==4.46.3
```

为了避免环境被污染，建议使用虚拟环境安装，在创建并激活虚拟环境后，运行以下命令安装依赖库：

```
pip install -r requirements.txt
```

如果使用conda，则运行以下命令安装依赖库：

```
conda env create -f environment.yml
```

5. 开发工具

选择合适的集成开发环境（Integrated Development Environment，IDE）或代码编辑器，有助于提高代码阅读和分析的效率。在此建议选择如下工具。

- PyCharm：专为Python开发设计的IDE，提供强大的调试和代码分析功能。
- Visual Studio Code：一款流行的免费代码编辑器，支持丰富的插件。

通过上面的软件环境配置，将为深入分析和学习DeepSeek的源代码做好充分的准备。建议根据项目的具体要求和您的研究方向，选择和配置适合的开发环境。

2.2 源码获取与管理

代码获取与管理是项目开发的基础环节，首先通过从GitHub克隆代码到本地，为开发工作做好准备。在开发过程中，合理运用代码分支管理，能够有效地组织和开发不同的功能或修复不同的问题，确保代码的稳定性和可维护性。同时，定期进行代码更新与同步，可保持本地代码与远程仓库的一致性。

2.2.1 开源项目简介

目前，DeepSeek在GitHub中的开源项目如下。

（1）DeepSeek-VL2：基于MoE架构的多模态大模型，支持高级视觉—语言理解。

（2）awesome-Deepseek-Integration：DeepSeek模型的应用集成示例库，提供了应用程序编程接口（Application Programming Interface，API）调用、微调、部署的代码示例，包含了与常见框架（如LangChain、Hugging Face等）的整合案例。

（3）DeepSeek-V3：第三代通用大语言模型，平衡性能与推理效率。

（4）DeepSeek-R1：强化学习驱动的AI智能体框架，支持多工具调用与环境交互。

（5）Janus：统一多模态理解与生成的基座模型，支持任意模态输入任意模态输出（文本↔图像↔视频）。

（6）DeepSeek-V2：高效经济的MoE语言模型，在通用NLP任务中接近GPT-4水平。

（7）DeepSeek-Coder-V2：代码智能模型，超越闭源模型性能，支持300+编程语言与复杂代码推理。

（8）ESFT（Expert Specialized Fine-Tuning）：专家专用微调框架，针对垂直领域的小样本高效微调。

（9）DreamCraft3D：基于扩散模型的3D内容生成工具，可以从单张图像生成高质量3D网格。

（10）DeepSeek-Prover-V1.5：数学定理自动证明模型，支持形式化验证与自然语言推理。

（11）DeepSeek-Coder：初代代码生成模型，支持主流编程语言。

（12）DeepSeek-VL：第一代视觉—语言多模态模型，支持图像描述、视觉定位、跨模态检索。

（13）DeepSeek-Math：数学推理专用语言模型，具备从小学数学到高等数学的解题能力。

（14）Awesome-Deepseek-Coder：DeepSeek-Coder生态资源合集，集成了第三方插件、教程、微调指南。

（15）DeepSeek-LLM：初代通用大语言模型（7B/13B/67B），中英双语优化，支持知识密集型任务。

（16）DeepSeek-MoE：高效混合专家模型架构，提升了模型的性能和效率。

（17）3FS（Fire-Flyer File System）：是一款高性能的分布式文件系统，专为AI训练和推理工作负载设计。能够充分利用现代固态硬盘（SSD）和远程直接内存访问（RDMA）网络的全部带宽，加速数据访问操作，从而显著提升AI模型的训练和推理效率。

（18）DualPipe：一种双向流水线并行算法，用于在V3/R1训练中实现计算与通信的重叠。通过允许不同部分并行工作，消除了传统流水线并行中的低效"流水线气泡"，从而最大限度地减少了GPU的空闲时间，提高了训练效率。

（19）DeepGEMM：一个高效的FP8 GEMM（矩阵乘法）内核，采用细粒度缩放技术，旨在优化矩阵计算的效率。通过低精度计算提升速度，同时利用英伟达CUDA技术修正误差，确保计算的准确性。

（20）DeepEP：一个高效的专家并行通信库，专为MoE模型训练和推理设计。它能够智能地分配专家负载，确保不同GPU之间的负载平衡，从而提高GPU利用率并减少通信开销。

（21）open-infra-index：这是一个包含生产级AI基础设施工具的项目，旨在支持高效的AGI（通用人工智能）开发和社区驱动的创新。提供了多种工具和框架，帮助开发者更好地利用DeepSeek的技术。

（22）profile-data：该数据集包含了DeepSeek在训练和推理框架中的性能分析数据，帮助社区更好地理解通信与计算重叠策略及底层实现细节。

（23）smallpond：一个基于DuckDB和3FS构建的轻量级数据处理框架，适用于高效的数据预处理和分析。

（24）FlashMLA：针对Hopper GPU的高效MLA解码内核，专门用于优化显卡的计算效率。它能够动态分配算力，避免资源浪费，从而让AI模型的处理速度接近硬件极限。

（25）EPLB（Expert Parallel Load Balancer）：它是一个专家并行负载均衡器，用于解决MoE架构中专家负载不平衡的问题。它通过智能分配专家，确保不同GPU之间的负载平衡，从而提高整体效率。

基于篇幅的限制，本书只分析具有代表性的开源项目，即最新、最通用的开源项目。

2.2.2 获取源码

在GitHub上获取源代码有多种方法，大家可以根据自己的需求选择适合的方式。

1. 下载 ZIP 压缩包

如果只想获取项目的当前版本代码，而不需要进行版本控制，可以直接下载ZIP压缩包。以获取DeepSeek-R1源码为例，具体步骤如下。

（1）在目标仓库的页面，单击右上角绿色的“Code”按钮。在下拉菜单中，选择“Download ZIP”，如图2-2所示。

（2）下载完成后，解压缩文件即可获得项目的源代码。

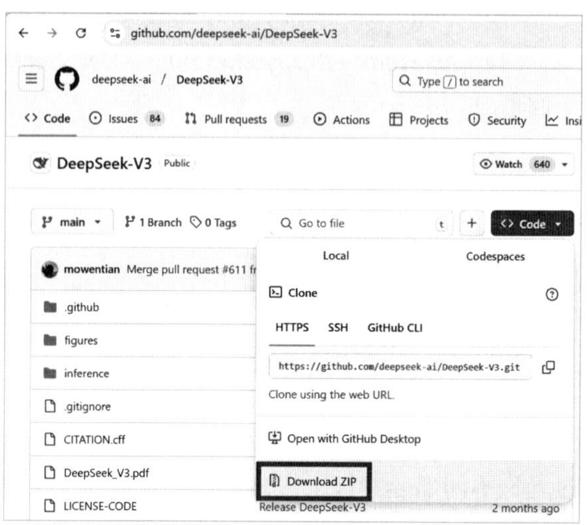

图2-2 选择“Download ZIP”

2. 克隆（Clone）仓库到本地

如果希望在本地对项目进行版本控制和开发，使用Git克隆仓库是最佳选择。以获取DeepSeek-R1源码为例，具体步骤如下。

（1）在目标仓库的页面，单击绿色的“Code”按钮。选择HTTPS或SSH链接，复制链接https://github.com/deepseek-ai/DeepSeek-R1.git，如图2-3所示。

（2）在本地打开终端或命令提示符，导航到目标目录，然后运行以下命令：

```
https://github.com/deepseek-ai/
DeepSeek-R1.git
```

这将在当前目录下创建一个与仓库同名的文件夹，并将仓库的所有内容下载到该文件夹中。

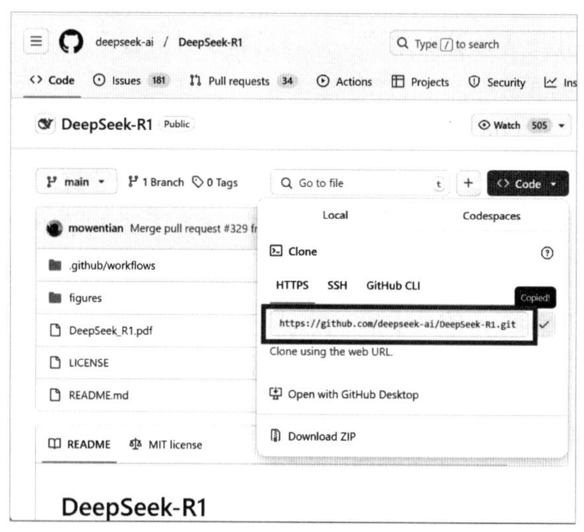

图2-3 复制链接

3. 直接在线浏览和复制代码

如果您只需要查看或复制特定的代码片段，可以直接在GitHub上浏览代码。具体步骤如下：

（1）导航到目标仓库的页面，浏览到您感兴趣的文件或目录。

（2）单击文件名以查看其内容，然后复制所需的代码片段。

4. Fork 仓库

如果希望对项目进行修改并可能贡献回原始项目，可以选择Fork仓库。具体步骤如下：

（1）在目标仓库的页面，单击右上角的“Fork”按钮，如图2-4所示。

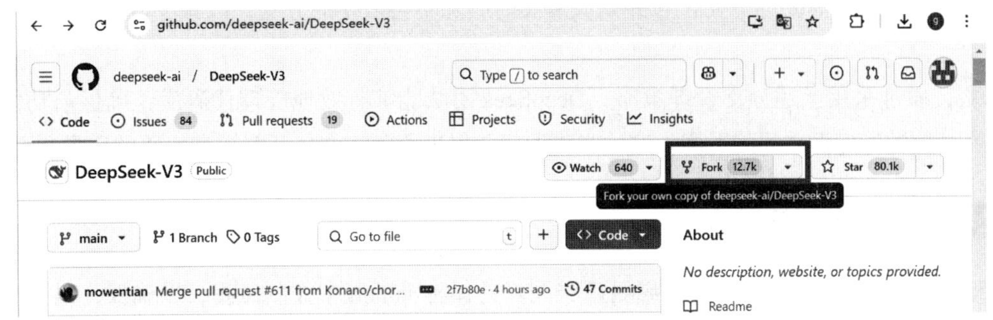

图2-4　单击右上角的“Fork”按钮

（2）这将在您的GitHub账户下创建该仓库的副本。

（3）然后，可以按照前面介绍的克隆仓库的步骤，将Fork的仓库克隆到本地进行开发。

总之，根据自己的具体需求，选择上述方法之一即可从GitHub获取源代码。

2.2.3 代码分支管理

DeepSeek源码在GitHub上的分支管理主要遵循Git分支管理的通用规则和流程，下面是关于DeepSeek源码分支管理的一些信息。

1. 分支类型

- 主分支（Main Branch）：通常用于发布稳定版本，是整个项目的主要开发线路。
- 功能分支（Feature Branch）：用于开发新功能的分支，开发者可以在不影响主分支的情况下进行实验和添加新功能，开发完成后可以合并回主分支。
- 修复分支（Bugfix Branch）：专门用于修复bug的分支，修复完成后可以合并回主分支。

2. 分支管理操作

- 创建分支：在GitHub上，可以在仓库页面的分支选择框中输入新的分支名称，并单击“Create branch”按钮来创建新分支。也可以在本地使用命令行创建分支，如“git checkout -b new-feature-branch”，该命令会创建一个名为“new-feature-branch”的新分支并自动切换到该分支。
- 切换分支：使用git checkout branch-name命令可以切换到不同的分支。
- 合并分支：当一个功能开发完成后，通常需要将其合并回主分支。可以在本地使用git merge feature-branch命令进行合并，也可以在GitHub上通过创建Pull Request来合并分支。创建Pull Request时，GitHub会自动比较两个分支之间的差异，并提供在线代码审查功能。
- 删除分支：如果某个分支的工作已经完成，可以使用git branch -d branch-name命令删除它。如果分支尚未合并，需要使用强制删除命令git branch -D branch-name。

3. 分支管理策略

- 定期合并主分支：在开发新功能时，应定期合并主分支到功能分支，以保持功能分支的更新，避免其与主分支的差异过大，降低合并冲突的可能性。

- 及时删除不再需要的分支：合并完成后及时删除不再需要的分支，以减少混乱，保持仓库的整洁。

总之，通过采用上述分支管理策略，DeepSeek团队能够有效地管理代码库，确保新功能的开发与现有功能的稳定性之间的平衡。这种方法不仅能促进团队协作，还能提高代码的质量和可靠性。

2.2.4 代码更新与同步

DeepSeek项目在GitHub上的源码更新与同步遵循标准的版本控制流程，以确保代码的稳定性和持续改进。

1. 代码更新机制

功能开发：开发者在功能分支上进行新功能的开发和测试。在功能完成并经过初步测试后，提交代码并发起合并请求（Pull Request，PR）。

代码审查：项目维护者对合并请求进行代码审查，确保代码质量和功能的正确性。审查通过后，代码被合并到开发分支（Development Branch）。

集成测试：在开发分支上进行集成测试，确保新功能与现有功能的兼容性。如果测试通过，且开发分支达到稳定状态，代码将被合并到主分支，并准备发布新的版本。

2. 代码同步机制

- 本地同步：开发者在本地仓库中，通过执行git pull命令，从远程仓库获取最新的代码更新，确保本地代码与远程仓库保持同步。
- 远程同步：开发者在本地完成功能开发并通过测试后，使用git push命令，将本地功能分支的更新推送至远程仓库。随后，发起合并请求，等待项目维护者的审查与合并。
- 版本发布：主分支积累了足够的稳定功能后，项目会创建新的版本标签（Tag），并发布新版本的源码。例如，DeepSeek-V3 的首个版本已上线并同步开源，用户可以在GitHub上获取最新的源码。

3. 代码更新流程

（1）提交更新：开发者在本地完成代码修改后，通过git add、git commit等命令将更改提交到本地仓库。

（2）推送更新：使用git push命令将本地提交的更改推送到远程GitHub仓库，通常推送到对应的分支。

（3）创建PR：如果开发者需要将功能分支或修复分支的更改合并到主分支，可以在GitHub上创建PR。创建PR后，其他开发者可以对代码进行审查和讨论。

（4）合并PR：经过审查和讨论后，如果代码没有问题，仓库管理员或具有相应权限的开发者可以合并PR，将更改合并到主分支。

4. 代码同步方法

- 拉取远程更新：开发者在本地仓库中使用git pull命令拉取远程仓库的最新更改，以保持本地代码与远程代码的一致性。
- 推送本地更新：如前文所述，使用git push命令将本地的代码更新推送到远程仓库。
- 同步特定分支：如果需要同步特定分支的代码，可以在创建PR时选择对应的源分支和目标分支进行合并。

5. 代码更新策略

- 定期更新：DeepSeek项目会定期发布更新，包括新功能的添加、性能优化、bug修复等。开发者可以根据项目的需求和计划，定期检查并更新代码。
- 版本控制：通过Git的版本控制功能，DeepSeek项目可以方便地管理不同版本的代码。开发者可以使用git tag命令创建标签来标记特定的版本。
- 分支管理：DeepSeek项目使用分支管理来组织代码更新。主分支通常用于发布稳定版本，功能分支用于开发新功能，修复分支用于修复bug。合理的分支管理，可以确保代码的稳定性和可维护性。

总之，通过上述代码更新与同步机制，DeepSeek项目确保了开发过程的有序进行，维护了代码的高质量和稳定性。开发者可以通过定期同步远程仓库的更新，保持本地代码的最新状态，并积极参与项目的开发和改进。

2.3 DeepSeek模型的本地部署与接入

本地部署DeepSeek是指将DeepSeek大语言模型及相关运行环境安装在本地服务器或个人计算机上，使其能够在本地环境中独立运行，无须依赖外部网络或云服务。这种部署方式能够确保数据的隐私性和安全性，既能避免数据在传输过程中泄露，也能降低对外部网络的依赖，提高系统的稳定性和响应速度。此外，本地部署还允许用户根据自身需求对模型进行定制化优化，以更好地满足特定业务场景或个性化需求。本节将详细讲解本地部署DeepSeek模型的基本步骤。

2.3.1 安装Ollama

（1）登录Ollama官网，根据您计算机的操作系统类型下载对应的Ollama版本，目前Ollama支持macOS、Linux和Windows主流操作系统。笔者选择的是Windows系统版本，如图2-5所示。

（2）单击“Download for Windows”按钮开始下载，下载完成之后是一个“.exe”格式的安装包，以管理员身份打开它。

（3）在弹出界面中点击“Install”按钮开始安装，如图2-6所示。这里需要注意的是，需要将Ollama安装在C盘的，不支持更改路径，因此C盘的空余空间必须大于5GB。

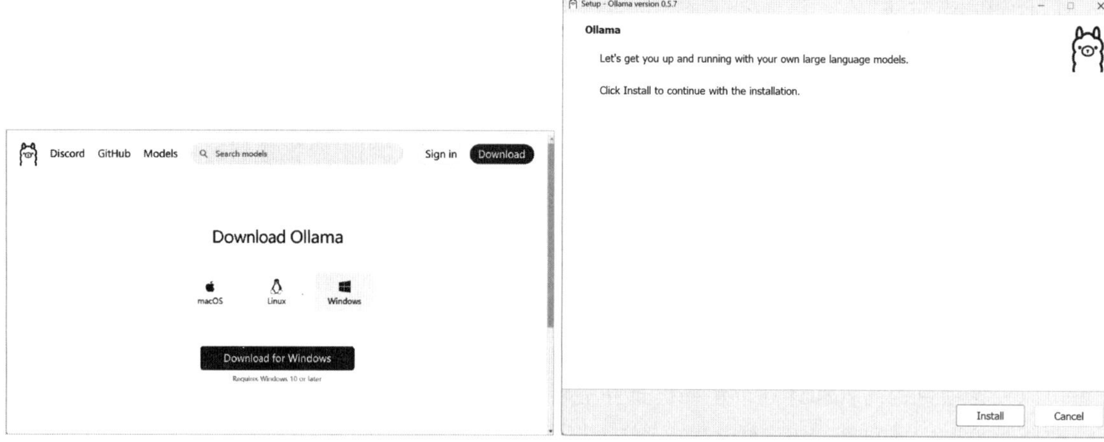

图2-5　Windows系统版本的Ollama　　图2-6　单击“Install”按钮

（4）在弹出的“Installing”界面中等待进度条完成，如图2-7所示。

（5）完成安装后，为了确保Ollama服务已启动，在终端中输入下面的ollama命令进行验证，若弹出如图2-8所示的界面则表示成功安装。

```
ollama -h
```

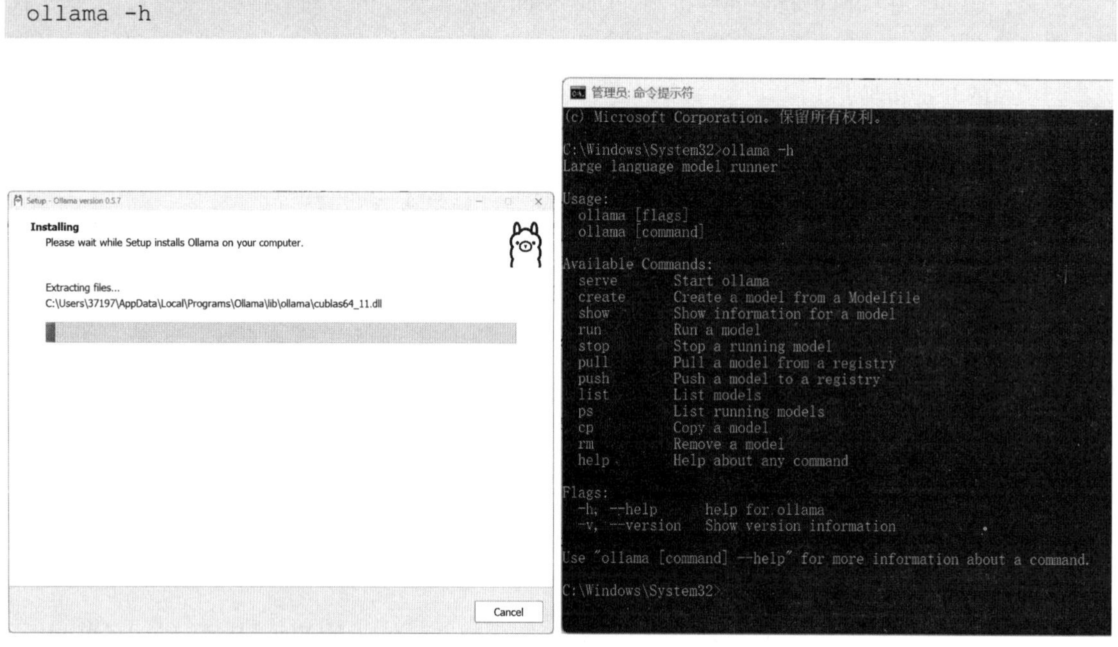

图2-7　“Installing”界面　　图2-8　成功安装并启动Ollama

2.3.2 部署DeepSeek模型

（1）登录Ollama官网，单击顶部的“Models”来到模型界面，如图2-9所示。

（2）单击“deepseek-r1”链接来到DeepSeek的模型界面，在下拉列表中有多个不同大小的模型版本，如1.5b、7b、8b、14b、32b、70b、671b。模型大小不同，可适配的计算机显存、显卡及内存配置也不同，如图2-10所示。

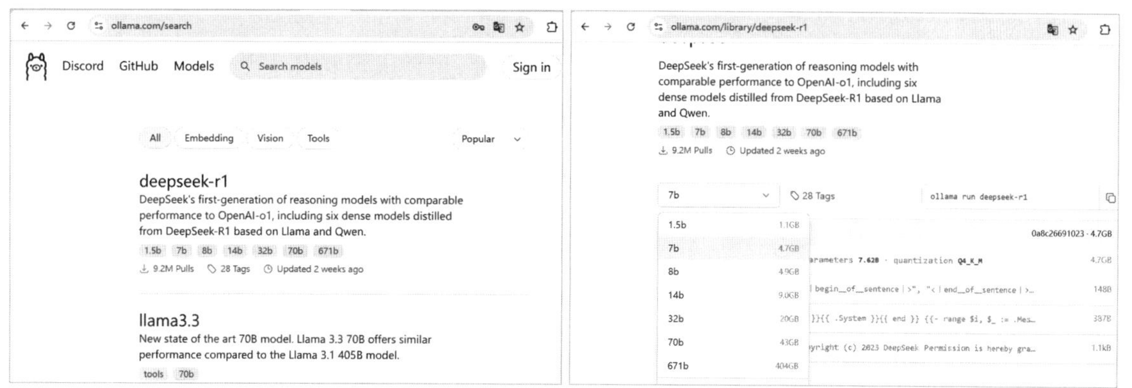

图2-9　模型界面　　　图2-10　下拉列表中的DeepSeek模型

（3）大家可以根据硬件配置选择合适的模型版本，假设要安装70b版本，在下拉列表中选择70b，然后复制对应的命令“ollama run deepseek-r1:70b”，如图2-11所示。

（4）在命令行界面输入刚刚复制的命令，按下回车键后开始安装70b版本的DeepSeek模型。安装时间可能会有点长，需要耐心等待（模型越大等待时间越长），安装成功的界面如图2-12所示。

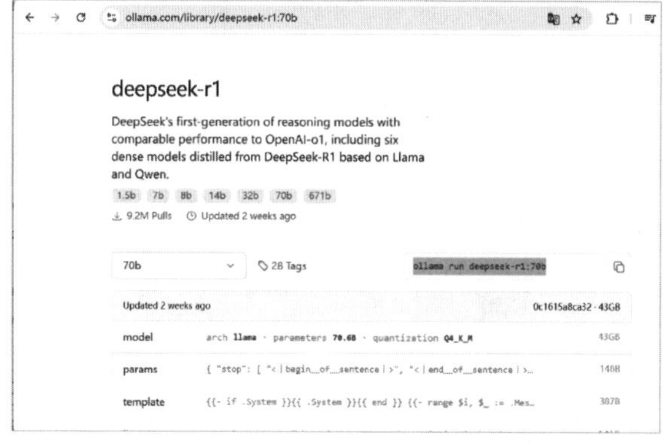

图2-11　安装70b版本

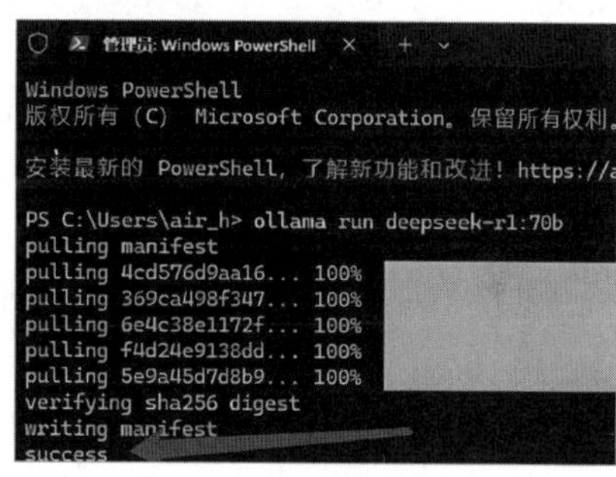

图2-12　安装成功界面

（5）安装成功后，在终端界面输入如下命令即可启动DeepSeek进行聊天，输入“/bye”命令可退出模型。

```
ollama run deepseek-r1:70b
```

2.3.3 Chatbox部署可视化

Chatbox是一款开源免费的AI客户端工具，专为本地部署的AI模型（如DeepSeek）而设计，提供简洁美观的界面，让用户能够轻松与AI模型进行交互。

使用Chatbox部署DeepSeek模型的步骤如下：

（1）登录Chatbox官网下载安装包，单击“免费下载 (for Windows)”按钮下载Windows版本的安装包，如图2-13所示。

（2）下载完成后得到一个“.exe”格式的可安装文件，鼠标左键双击开始安装。在弹出的“安装选项”界面中勾选“仅为我安装”，然后单击“下一步”按钮，如图2-14所示。

图2-13　Chatbox官网

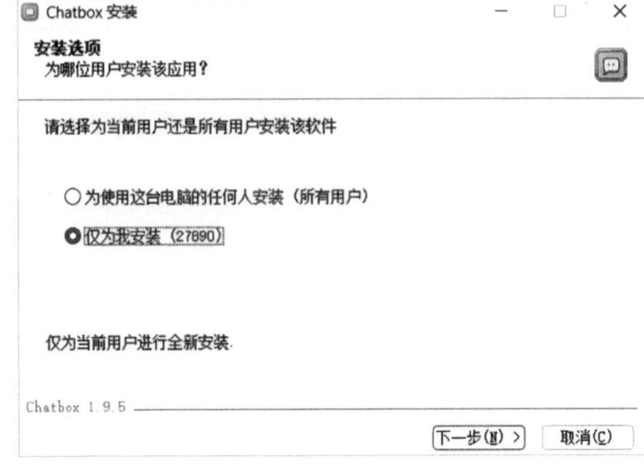

图2-14　勾选“仅为我安装”

（3）在弹出的“选定安装位置”界面中选择安装位置，然后单击“下一步”按钮，如图2-15所示。

（4）安装好之后自动运行Chatbox，单击“使用自己的API Key或本地模型”按钮，配置刚刚部署的模型，如图2-16所示。

（5）在弹出的“选择并配置AI模型提供方”界面中选择“Ollama API”，如图2-17所示。

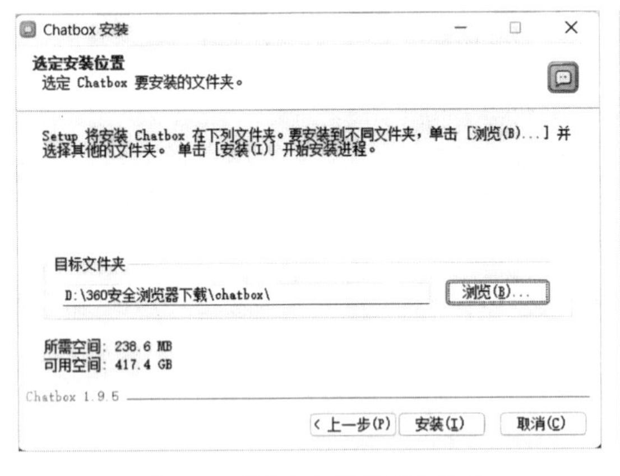

图2-15　“选定安装位置”界面

图2-16　启动界面

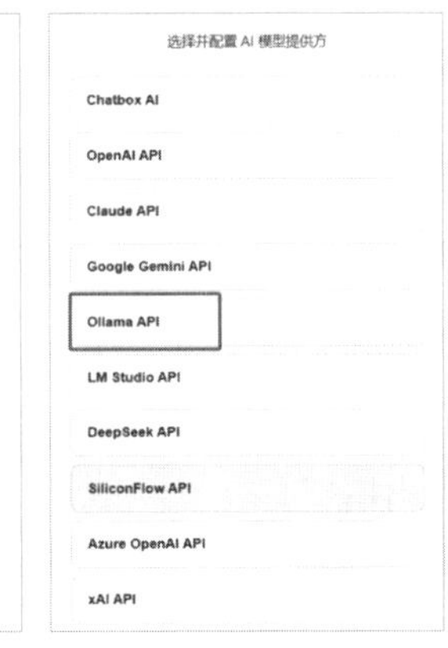

图2-17　选择“Ollama API”

（6）然后选择在本地已经部署好的模型，如前面部署的70b版本的DeepSeek模型。当然也可以选择其他已经部署好的模型，例如，作者还部署了14b版本的DeepSeek模型，如图2-18所示。

这样就把DeepSeek模型部署到本地了，并且能可视化使用DeepSeek进行对话，如图2-19所示。

图2-18　选择部署好的DeepSeek模型

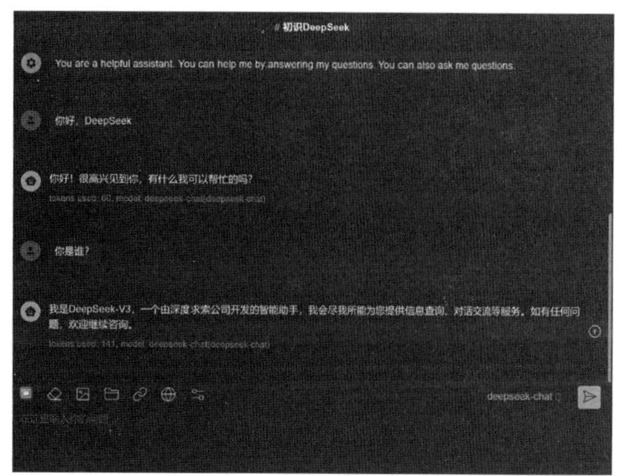

图2-19　可视化使用DeepSeek

2.3.4 DeepSeek接入整合

除了前面介绍的Chatbox，还可以使用其他工具接入DeepSeek服务。DeepSeek提供了开源项目Awesome DeepSeek Integrations，如图2-20所示。Awesome DeepSeek Integrations旨在整合DeepSeek API到各种流行的软件和工具，帮助开发者和用户更便捷地利用DeepSeek的强大功能。

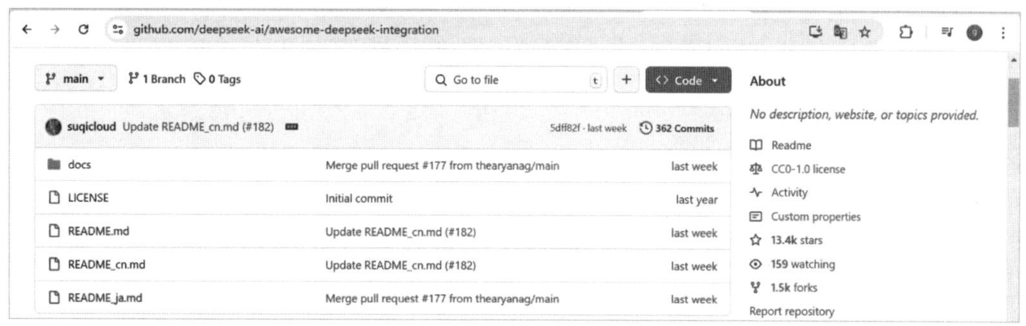

图2-20　开源项目Awesome DeepSeek Integrations

Awesome DeepSeek Integrations项目通过整合DeepSeek API到各种工具和平台，极大地拓展了DeepSeek的应用场景，为开发者和用户提供了丰富的选择。在说明文档README_cn.md中列出了可以接入DeepSeek的内容，具体说明如下所示。

1. 多平台集成

- Chatbox：一个支持Windows、Mac和Linux的桌面客户端，集成了多种前沿的LLM模型。
- LibreChat：一个可定制的开源应用，无缝集成DeepSeek，提供增强的AI交互体验。
- Pal – AI Chat Client：为iOS和iPadOS设计的定制化聊天工具。
- Video Subtitle Master：支持批量生成视频字幕，并可将字幕翻译为多种语言，支持Mac和Windows平台。

2. 开发框架和工具

- PHP Client：为DeepSeek API提供的PHP客户端库，方便开发者在PHP应用中集成。
- Laravel Integration：为Laravel应用提供的DeepSeek集成工具。
- Deepseek Tokenizer：一个高效且轻量级的DeepSeek模型分词库。

3. 浏览器扩展和插件

- Immersive Translate：支持网页翻译的浏览器插件。
- Lulu Translate：提供鼠标选择翻译、段落对比翻译和PDF翻译功能，支持DeepSeek AI。
- RssFlow：智能RSS阅读器扩展，支持通过DeepSeek模型进行内容理解。

4. IDE 和编辑器插件

- VS Code Extensions：如Continue和Cline，为VS Code提供AI编程辅助。
- JetBrains Extensions：为JetBrains IDE提供的AI编程助手。
- Neovim Extensions：如avante.nvim和llm.nvim，支持在Neovim中使用DeepSeek。

5. 其他工具和应用

- Story-Flicks：通过一句话快速生成高清故事短视频。
- 16x Prompt：AI编码工具，支持上下文管理和复杂编码任务的提示生成。
- Portkey AI：统一API，支持与 1600+ LLM模型交互，提供Python和Node SDK。

在Awesome DeepSeek Integrations项目中，为上面的每种工具提供了简单的接入教程，例如，为VS Code插件Continue接入DeepSeek的教程在awesome-deepseek-integration/docs/continue/README_cn.md中，如图2-21所示。

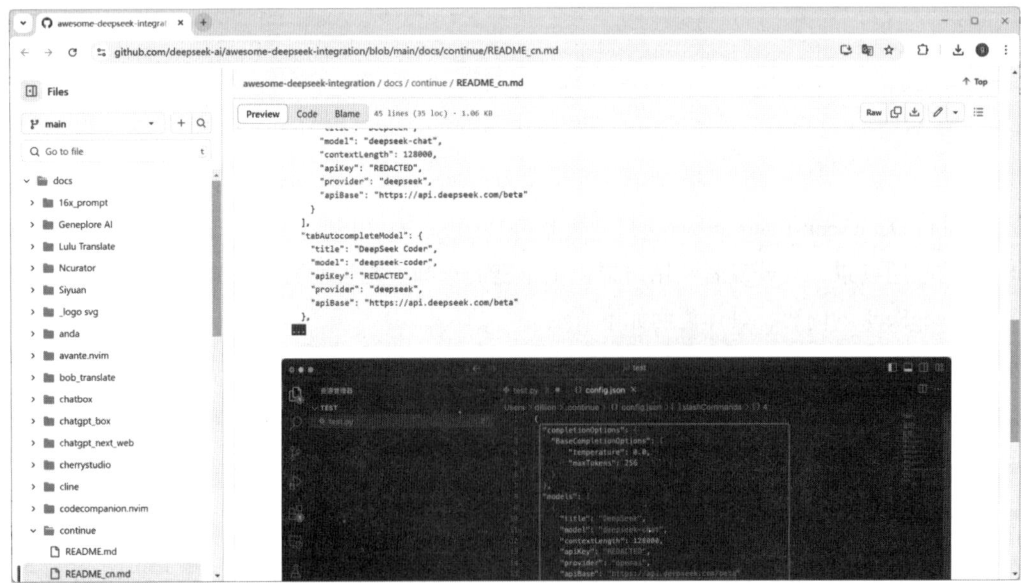

图2-21　为VS Code插件Continue接入DeepSeek

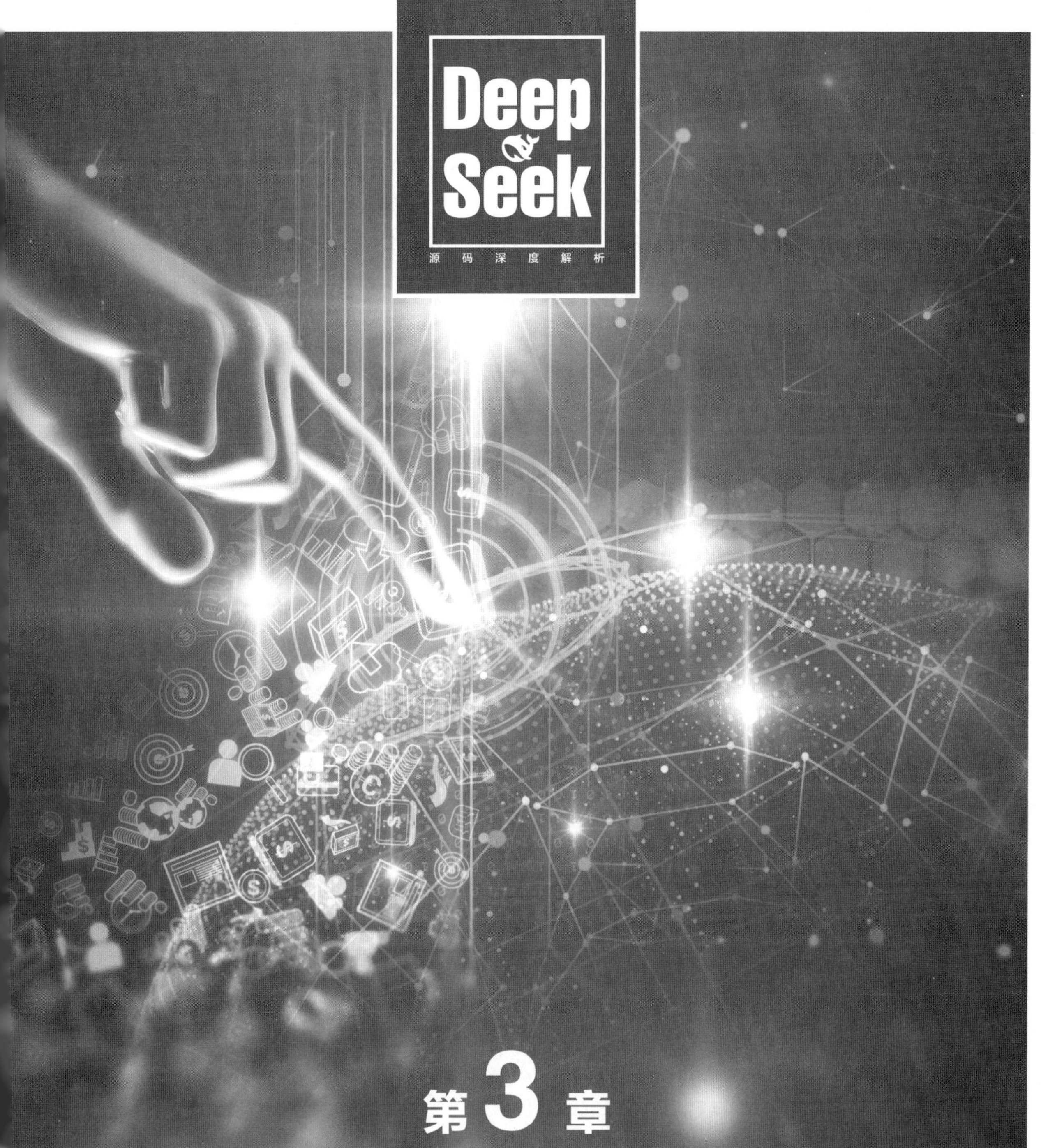

混合专家模型（MoE）初探

混合专家模型（Minture of Experts，MoE）是由DeepSeek AI开发的一种创新的语言模型，旨在实现专家的终极专业化。该模型通过细粒度专家分割和共享专家隔离两种主要策略，提高了专家的专业化程度和模型性能。本章将详细剖析DeepSeek的开源项目DeepSeekMoE，初步探索MoE语言模型的精髓。

3.1 项目介绍

MoE语言模型最早见于DeepSeek开源项目DeepSeekMoE（https://github.com/deepseek-ai/DeepSeek-MoE），旨在通过混合专家技术实现高效的模型训练和推理。

3.1.1 基本特点

DeepSeekMoE是一个16.4B参数的语言模型，采用了创新的MoE架构。MoE架构通过将模型划分为多个专家网络，每个专家网络专注于处理特定的任务或领域，从而在保持高性能的同时显著降低计算成本。DeepSeekMoE在2T的英文和中文数据上进行训练，表现出与DeepSeek 7B和LLaMA2 7B相当的性能，但仅需后面二者40%左右的计算资源。

MoE架构的核心优势如下。

1. 创新的 MoE 架构

MoE架构通过动态路由机制实现高效的多模态信息处理，其核心组成如下。

（1）专家模型：包含多个专家子模型，每个专家专注于处理特定类型的输入（如图像切片、文本片段等）。

（2）门控网络：根据输入特征动态选择最相关的专家模型，分配计算资源。其公式为

$$y=\sum_{i=1}^{n} G(x)_i \cdot E_i \mid (x)$$

其中：

- $G(x)_i$是门控网络对第i个专家的权重。
- $E_i \mid (x)$是第i个专家的输出。

（3）负载均衡：通过负载均衡策略，确保每个专家模型的计算负载均匀分布，避免资源浪费。

注意

在实际应用中，通过使用上述MoE架构得到了如下好处。

- 高效的计算资源利用：DeepSeekMoE 16B模型在推理时仅激活约2B参数，显著减少了计算资源的消耗。
- 细粒度的专家划分：模型将专家网络细粒度划分，并通过动态路由机制将输入Token分配到1～2个专家子网络，提升了模型的效率和专业化能力。

2. 开放性

（1）项目提供了DeepSeekMoE 16B Base和DeepSeekMoE 16B Chat的模型检查点，可以在单个40GB内存的GPU上部署，无须量化。

（1）项目基于MIT许可证开源，支持商业使用。

3. 评估结果

（1）基准测试表现。

- Open LLM Leaderboard：DeepSeekMoE 16B大幅超越具有相似激活参数数量的开源模型，在计算量仅为LLaMA2 7B的40%左右时，性能与之相当。
- 内部基准测试：DeepSeekMoE 16B在多个任务上表现出色，如语言建模、知识密集型任务（Pile、HellaSwag、TriviaQA、NaturalQuestions）、代码生成（HumanEval和MBPP）、数学推理（GSM8K和MATH）等。
- 聊天模型评估：DeepSeekMoE 16B Chat在语言理解、推理（PIQA、ARC、BBH）、机器阅读理解（RACE）、知识密集型任务（TriviaQA、NaturalQuestions）等方面表现出色。

（2）计算量对比。

- 与DeepSeek 7B相比：DeepSeekMoE 16B仅使用40.5%的计算量，性能相当。
- 与LLaMA2 7B相比：DeepSeekMoE 16B使用39.6%的计算量，性能优于LLaMA2 7B。

总之，DeepSeekMoE是一个高效、灵活且支持商业用途的开源语言模型项目，适用于各种自然语言处理任务。通过MoE架构，DeepSeekMoE在保持高性能的同时显著降低了计算成本，适合在资源受限的环境中使用。

3.1.2 开源内容

目前，DeepSeek MoE项目在其GitHub仓库中开源了以下内容。

- 模型检查点：提供了DeepSeekMoE 16B Base和DeepSeekMoE 16B Chat模型的检查点，供研究和应用使用。
- 微调脚本：在finetune目录下，包含了用于在下游任务上微调模型的脚本，支持使用DeepSpeed（训练模型）进行高效训练。
- 依赖项说明：提供了requirements.txt文件，列出了项目所需的Python依赖库，方便用户快速配置环境。
- 使用指南：README.md文件详细介绍了模型的特点、评估结果、快速开始指南、许可证信息及引用方式，帮助用户了解和使用该模型。
- 许可证文件：包含LICENSE-CODE和LICENSE-MODEL文件，分别说明了代码和模型的使用许可，明确了使用条款和条件。

这些开源内容为研究人员和开发者提供了全面的资源，支持他们在不同的应用场景中使用和微调DeepSeekMoE模型。

3.2 功能模块

DeepSeek MoE项目是一种基于MoE架构的大规模语言模型，旨在通过动态路由和专家共享机制，实现高效的模型训练和推理。DeepSeek MoE项目的主要功能模块包括混合专家模块和微调模块，具体说明如下所示。

1. MoE 模块

（1）多头潜在注意力层：引入潜在向量用于缓存自回归推理过程中的中间计算结果，优化键值缓存，降低生成任务中的计算量。

（2）混合专家系统层：采用动态路由机制，从多个专家中选择最相关的专家进行计算。同时，创新性地引入专家共享设计，部分专家在不同令牌或层间共享参数，优化模型性能。

（3）RMSNorm归一化层：用于对模型进行归一化处理，提升训练稳定性和模型性能。

2. 微调模块

本项目开源了模型微调代码，支持通过低秩适应（Low-Rank Adaptation，LoRA）和量化技术（4/8-bit）高效微调MoE模型。该模型微调代码的主要功能如下。

- 数据预处理：构建指令模板、文本分词和标签掩码处理。
- 模型加载配置：支持FlashAttention和量化参数的配置。
- 分布式训练管理：基于DeepSpeed框架，实现高效的分布式训练。
- 模型保存回调：提供自定义的模型保存机制，确保训练过程中的模型检查点被正确保存。
- 混合精度训练：支持混合精度训练，优化显存使用，提升训练效率。

通过上述功能模块，DeepSeek MoE项目实现了对MoE架构的优化，支持在单张40GB显存GPU上运行具有16B参数的模型，显著提升了大规模语言模型的训练和推理效率。

3.3 ZeRO 配置

ZeRO是一种显存优化技术，其核心原理在于通过将模型的不同部分（包括模型参数、梯度和优化器状态）划分到多个GPU上，以减少每个GPU上的内存占用，从而支持更大规模的深度学习模型训练。

3.3.1 ZeRO 优化器介绍

ZeRO技术的实现分为如下所示的三个阶段，每个阶段在显存占用和通信开销之间找到不同的平衡。

- ZeRO Stage 1：仅对优化器状态进行分片存储，每个GPU保留完整的梯度和模型参数，易于实现，且通信量相对较小，适用于中等规模的模型训练，但显存节省有限。

● ZeRO Stage 2：在第一阶段的基础上，进一步对梯度进行分片，每个GPU只存储自己负责的部分梯度和优化器状态，而模型参数仍然完整存储在每个GPU上，较大幅度降低显存需求，同时通信开销适中，是大多数大规模模型训练的理想选择，高效且节省资源。

● ZeRO Stage 3：对所有相关数据（包括模型参数、优化器状态和梯度）进行分片存储，每个GPU只存储自己负责的一部分数据，显存需求最小，适合超大规模模型（如GPT-3）的训练，但通信开销高，对网络带宽要求高。

ZeRO技术特别适用于训练大规模深度学习模型，尤其是在显存资源有限的情况下，其优势如下。

● 显存节省：通过分片存储，ZeRO能够显著降低每个GPU上的内存占用，使得在相同的显存条件下能够训练更大的模型。

● 计算效率提升：虽然在高阶段（如Stage 3）可能会增加通信开销，但总体而言，ZeRO技术通过优化显存使用，可以提高整体计算效率。

● 灵活性：ZeRO技术提供了不同的阶段选择，用户可以根据具体需求（如模型规模、显存资源和网络带宽等）选择合适的优化策略。

3.3.2 第2阶段优化配置

文件ds_config_zero2_no_offload.json是DeepSpeed的配置文件，用于在训练过程中启用BF16精度并采用ZeRO优化的第二阶段（不使用Offload方案）。在配置中设定了自动调整梯度累积步数、梯度裁剪、训练批量大小等参数，同时通过设置如allgather、reduce_scatter、overlap_comm等选项来优化内存和通信效率，从而在保持高性能的同时降低计算资源的消耗。

```
{
    "bf16": {
        "enabled": true
    },

    "zero_optimization": {
        "stage": 2,
        "allgather_partitions": true,
        "allgather_bucket_size": 1e8,
        "overlap_comm": true,
        "reduce_scatter": true,
        "reduce_bucket_size": 1e8,
        "contiguous_gradients": true
    },

    "gradient_accumulation_steps": "auto",
    "gradient_clipping": "auto",
    "steps_per_print": 2000,
    "train_batch_size": "auto",
```

```
    "train_micro_batch_size_per_gpu": "auto",
    "wall_clock_breakdown": false
}
```

对上述各个参数的具体说明如下。

（1）bf16：配置BF16（Brain Floating Point 16）精度。设置为true，表示启用BF16精度训练。

（2）zero_optimization：配置ZeRO优化器的相关参数，具体说明如下所示。

- stage：设置为2，表示使用ZeRO优化的第二阶段，主要优化模型状态的内存占用。
- allgather_partitions：设置为true，表示在需要时收集分散的模型参数分区，以减少内存占用。
- allgather_bucket_size：设置为 1e8（1 亿），指定allgather操作的桶大小，以控制通信开销和平衡性能。
- overlap_comm：设置为true，表示在反向传播过程中重叠通信和计算，以提高效率。
- reduce_scatter：设置为true，表示在反向传播时使用reduce-scatter操作，以优化梯度的聚合。
- reduce_bucket_size：设置为 1e8（1 亿），指定reduce-scatter操作的桶大小，以控制通信开销和平衡性能。
- contiguous_gradients：设置为true，表示将梯度存储为连续的内存块，以提高内存访问效率。

（3）gradient_accumulation_steps：设置为 "auto"，表示自动确定梯度累积的步数，以实现所需的全局批量大小。

（4）gradient_clipping：设置为 "auto"，表示自动确定梯度裁剪的阈值，以防止梯度爆炸。

（5）steps_per_print：设置为 2000，表示每训练 2000 步打印一次日志信息。

（6）train_batch_size：设置为 "auto"，表示自动确定全局训练批量大小。

（7）train_micro_batch_size_per_gpu：设置为 "auto"，表示自动确定每个GPU的微批量大小。

（8）wall_clock_breakdown：设置为false，表示不输出各部分训练时间的详细分解。

上述参数共同配置了DeepSpeed的训练行为，特别是在使用ZeRO优化和BF16 精度时，以提高训练效率和性能。

3.3.3 第 3 阶段优化配置

文件ds_config_zero3.json也是一个JSON配置文件，用于DeepSpeed库的配置。它启用了BF16混合精度训练，设置了AdamW优化器和WarmupLR学习率调度器，并配置了ZeRO优化的第三阶段参数，包括优化器和参数的卸载、通信重叠、梯度连续性等，还设置了梯度累积步数、梯度裁剪、打印步数、训练批次大小等参数。

```
{
    "bf16": {
        "enabled": "auto"
    },
    "optimizer": {
```

```
        "type": "AdamW",
        "params": {
            "lr": "auto",
            "betas": "auto",
            "eps": "auto",
            "weight_decay": "auto"
        }
    },

    "scheduler": {
        "type": "WarmupLR",
        "params": {
            "warmup_min_lr": "auto",
            "warmup_max_lr": "auto",
            "warmup_num_steps": "auto"
        }
    },

    "zero_optimization": {
        "stage": 3,
        "offload_optimizer": {
            "device": "cpu",
            "pin_memory": true
        },
        "offload_param": {
            "device": "cpu",
            "pin_memory": true
        },
        "overlap_comm": true,
        "contiguous_gradients": true,
        "sub_group_size": 1e9,
        "reduce_bucket_size": "auto",
        "stage3_prefetch_bucket_size": "auto",
        "stage3_param_persistence_threshold": "auto",
        "stage3_max_live_parameters": 1e9,
        "stage3_max_reuse_distance": 1e9,
        "stage3_gather_16bit_weights_on_model_save": true
    },

    "gradient_accumulation_steps": "auto",
    "gradient_clipping": "auto",
    "steps_per_print": 20,
    "train_batch_size": "auto",
    "train_micro_batch_size_per_gpu": "auto",
    "wall_clock_breakdown": false
}
```

3.3.4 优化总结

在DeepSeek MoE项目中，ds_config_zero2_no_offload.json和ds_config_zero3.json是DeepSpeed配置文件，分别对应ZeRO优化器的第二阶段（ZeRO Stage 2）和第三阶段（ZeRO Stage 3）。这两个配置文件的主要区别在于内存优化的程度和策略。

1. ZeRO Stage 2（ds_config_zero2_no_offload.json）

- 优化范围：对优化器状态和梯度进行分片处理，实现了内存减少到原来的1/8，同时通信开销和数据并行度维持不变。
- 配置特点：在配置文件中，zero_optimization的stage设置为2，且未启用参数或优化器状态的卸载（offload），所有数据均保留在GPU显存中。

2. ZeRO Stage 3（ds_config_zero3.json）

- 优化范围：分片优化器状态、梯度和参数，内存减少的幅度和数据并行度成线性关系。举例来说，在64个GPU（Nd等于64）的情况下，内存消耗可以降低为原来的1/64，通信量相对会增加50%。
- 配置特点：在配置文件中，zero_optimization的stage设置为3，并启用了参数和优化器状态的卸载（offload），将其存储在CPU内存中，以进一步减少GPU显存占用。

总的来说，相比ds_config_zero2_no_offload.json，ds_config_zero3.json通过更深入的内存优化策略，能够支持训练更大规模的模型，但也带来了更多的通信开销和复杂性。

3.4 模型微调

在本项目中，微调脚本文件finetune.py提供了一套全面的工具，用于对DeepSeek MoE预训练语言模型进行微调。支持加载特定任务的数据、对数据进行预处理和编码，以及通过多种配置选项（如LoRA量化、分布式训练等）对模型进行高效训练。用户可以根据自己的需求，通过命令行参数或配置文件调整微调策略，以优化模型在特定任务或数据集上的性能。

3.4.1 微调原理

在DeepSeek MoE项目中，文件finetune.py通过以下技术对预训练模型进行微调。

- LoRA技术：通过在MoE模型的特定层（如q_proj, v_proj, k_proj, o_proj等）插入低秩矩阵，以少量的可训练参数来调整预训练模型的行为，避免直接微调大量参数导致的灾难性遗忘和高计算成本。
- 量化技术（4/8-bit）：采用量化的方式将模型的权重从浮点型（如 32-bit或 16-bit）转换为更低精度的格式（4-bit或 8-bit），从而显著减少模型在GPU或CPU上的内存占用，使得在资源有限的设备上也能运行大规模的后续训练。

文件finetune.py的具体实现流程如下。

（1）模型加载与配置。

- 根据配置参数加载预训练的MoE模型，并根据bits参数（如4或8）决定是否进行量化。
- 启用flash_attention_2或optimized的注意力实现方式，以提高计算效率。
- 如果启用LoRA，则根据配置初始化LoRA模块，并对目标模块（如trainable.split(',')）应用LoRA。

（2）数据加载和预处理。

- 使用load_dataset加载训练数据。
- 通过train_tokenize_function对数据进行预处理，包括构建指令提示、分词和预处理标签。

（3）训练准备。

- 将预处理后的数据集传递给Trainer，并配置学习率、批量大小、梯度累积步数等训练参数。
- 使用DataCollatorForSupervisedDataset对输入数据和标签进行收集和格式化。

（4）训练过程。

- 利用Hugging Face的Trainer进行模型训练，支持从检查点恢复训练。
- 在训练过程中使用混合精度（如BF16或FP16）来加速计算并减少内存使用。

（5）模型保存与推理。

- 在训练过程中，通过SavePeftModelCallback回调函数定期保存LoRA适配器的权重。
- 训练结束后，保存模型状态或使用safe_save_model_for_hf_trainer函数将模型权重安全地保存到指定目录。

通过上述流程，finetune.py能够在高效利用计算资源的同时，实现对MoE模型的快速微调，使其适应特定的下游任务。

3.4.2 生成提示文本

定义函数build_instruction_prompt()，用于生成包含指令和响应格式的提示文本。它接受一个字符串参数instruction，并返回一个多行字符串，其中包含预定义的上下文说明、指令部分及响应部分的模板。在生成的字符串中，instruction的内容被插入指定位置，以构建完整的提示文本。

```
# 定义常量
IGNORE_INDEX = -100                        # 忽略的索引值
EOT_TOKEN = "<|EOT|>"                      # 响应结束标记
logger = logging.getLogger(__name__)       # 创建日志记录器

# 构建指令提示的函数
def build_instruction_prompt(instruction: str):
    return f'''
You are an AI assistant, developed by DeepSeek Company. For politically
sensitive questions, security and privacy issues, you will refuse to answer.
### Instruction:
```

```
{instruction.strip()}
### Response:
'''.strip()
```

3.4.3 配置模型微调参数

定义一个名为ModelArguments的数据类，用于配置模型微调的参数。其中包括可训练的模型参数列表、LoRA（低秩适应）相关设置、需要保存的模块、是否使用LoRA、预训练模型的路径、注意力机制的实现方式，以及量化相关的配置，如是否使用双量化、量化数据类型和位数等。这些配置有助于在微调过程中灵活地调整模型的各项参数，以满足不同的训练需求。

```
# 定义模型参数类
@dataclass
class ModelArguments:
    trainable: Optional[str] = field(
        default="q_proj,v_proj,k_proj,o_proj,gate_proj,down_proj,up_proj",
# 默认可训练的模型参数
        metadata={"help": "Comma-separated list of model parameters to train."}
    )
    lora_rank: Optional[int] = field(
        default=8,     # LoRA 的秩
        metadata={"help": "Rank of LoRA"}
    )
    lora_dropout: Optional[float] = field(
        default=0.1,   # LoRA 的 dropout 率
        metadata={"help": "Dropout rate for LoRA."}
    )
    lora_alpha: Optional[float] = field(
        default=32.0,                                  # LoRA 的 alpha 值
        metadata={"help": "Alpha value for LoRA."}
    )
    modules_to_save: Optional[str] = field(
        default="embed_tokens,lm_head",                # 需要保存的模块
        metadata={"help": "Comma-separated list of modules to save."}
    )
    use_lora: Optional[bool] = field(
        default=False,                                 # 是否使用 LoRA
        metadata={"help": "Whether to use LoRA."}
    )
    model_name_or_path: Optional[str] = field(
        default="deepseek-ai/deepseek-moe-16b",    # 模型路径
        metadata={"help": "Path to pretrained model or model identifier from
                  huggingface.co/models."}
    )
```

```
    attn_implementation: Optional[str] = field(
        default="flash_attention_2",                    # 注意力实现方式
        metadata={"help": "Attention implementation to use."}
    )
    double_quant: bool = field(
        default=True,                                   # 是否使用双量化
        metadata={"help": "Compress the quantization statistics through
                  double quantization."}
    )
    quant_type: str = field(
        default="nf4",                                  # 量化数据类型
        metadata={"help": "Quantization data type to use. Should be one of
`fp4` or `nf4`."}
    )
    bits: int = field(
        default=16,                                     # 使用的位数
        metadata={"help": "How many bits to use."}
    )
```

3.4.4 设置训练数据

定义数据类DataArguments，用于指定训练数据的路径。它包含一个名为data_path的字段，用于存储训练数据的位置。

```
# 定义数据参数类
@dataclass
class DataArguments:
    data_path: str = field(
        default=None,  # 数据路径
        metadata={"help": "Path to the training data."}
    )
```

3.4.5 配置超参数

TrainingArguments类继承自Hugging Face的transformers.TrainingArguments，用于配置训练过程中的超参数，如缓存目录、优化器类型和模型最大序列长度等。这些设置有助于控制训练过程的各个方面，确保模型以最佳方式进行微调。

```
# 定义训练参数类
@dataclass
class TrainingArguments(transformers.TrainingArguments):
    cache_dir: Optional[str] = field(default=None)    # 缓存目录
    optim: str = field(
        default="adamw_torch",                        # 优化器类型
```

```
        metadata={"help": "Optimizer to use."}
    )
    model_max_length: int = field(
        default=512,                                        # 模型最大序列长度
        metadata={"help": "Maximum sequence length. Sequences will be right
                  padded (and possibly truncated)."}
    )
```

3.4.6 保存模型

SavePeftModelCallback是一个自定义的回调类，旨在与Hugging Face的Trainer一起使用，以在训练过程中正确保存PEFT（Paramefer-Efficient Fine-Tuning，参数高效微调）模型。该回调在模型保存时，仅保存PEFT模型的适配器权重，而不是整个基础模型，从而节省存储空间并提高加载效率。此外，在训练结束时，它会创建一个名为completed的文件，指示训练已顺利完成。

```
# 定义保存 LoRA 模型回调类
class SavePeftModelCallback(transformers.TrainerCallback):
    def save_model(self, args, state, kwargs):
        logger.info('Saving PEFT checkpoint...')  # 保存 LoRA 模型
        if state.best_model_checkpoint is not None:
            checkpoint_folder = os.path.join(state.best_model_checkpoint,
                                             "adapter_model")
        else:
            checkpoint_folder = os.path.join(args.output_dir,
                                             f"{PREFIX_CHECKPOINT_DIR}-
                                             {state.global_step}")
        peft_model_path = os.path.join(checkpoint_folder, "adapter_model")
        kwargs["model"].save_pretrained(peft_model_path)
        kwargs["tokenizer"].save_pretrained(peft_model_path)

    def on_save(self, args, state, control, **kwargs):
        self.save_model(args, state, kwargs)
        return control

    def on_train_end(self, args, state, control, **kwargs):
        def touch(fname, times=None):
            with open(fname, 'a'):
                os.utime(fname, times)
        touch(os.path.join(args.output_dir, 'completed'))
        self.save_model(args, state, kwargs)
```

在使用PEFT进行模型微调时，正确保存和加载适配器权重至关重要。根据Hugging Face的文档，建议在保存模型时，仅保存适配器权重，不必保存整个基础模型。

另外，Hugging Face的官方文档提供了有关如何使用回调自定义训练过程的详细信息。通过实

现SavePeftModelCallback，您可以确保在训练过程中正确保存PEFT模型的适配器权重，并在训练结束时生成一个指示训练顺利完成的文件。

3.4.7 获取最新检查点

函数get_last_checkpoint()用于获取指定目录中最新的检查点。首先，检查目录是否存在一个名为“completed”的文件，如果存在，则表示训练已完成，返回None；否则，会遍历目录中以“checkpoint-”开头的子目录，找到其中编号最大的一个，认为其是最新的检查点，并返回其路径。如果未找到任何符合条件的子目录，同样返回None。

```
# 获取最新的检查点
def get_last_checkpoint(checkpoint_dir):
    if os.path.isdir(checkpoint_dir):
        is_completed = os.path.exists(os.path.join(checkpoint_dir, 'completed'))
        if is_completed:
            return None
        max_step = 0
        for filename in os.listdir(checkpoint_dir):
            if os.path.isdir(os.path.join(checkpoint_dir, filename)) and
                          filename.startswith(PREFIX_CHECKPOINT_DIR):
                max_step = max(max_step, int(filename.replace(f'{PREFIX_
                               CHECKPOINT_DIR}-', '')))
        if max_step == 0:
            return None
        latest_ckpt_dir = os.path.join(checkpoint_dir, f'{PREFIX_CHECKPOINT_
                                       DIR}-{max_step}')
        logger.info(f"Found a previous checkpoint at: {checkpoint_dir}")
        return latest_ckpt_dir
    return None
```

3.4.8 安全保存模型

函数safe_save_model_for_hf_trainer()的主要功能是将Hugging Face的Trainer对象中的模型安全地保存到指定的目录中。在保存模型时，该函数首先将模型的状态字典（state_dict）从GPU内存转移到CPU内存，以减少GPU内存的占用。然后，它调用Trainer对象的_save方法，将模型的状态字典保存到指定的输出目录。这种方法确保了在使用分布式训练或混合精度训练时，模型能够被正确地保存，避免了可能出现的内存问题。

```
# 安全保存模型以供 HF Trainer 使用
def safe_save_model_for_hf_trainer(trainer: transformers.Trainer, output_
dir: str):
    state_dict = trainer.model.state_dict()
```

```
    if trainer.args.should_save:
        cpu_state_dict = {key: value.cpu() for key, value in state_dict.
                          items()}
        del state_dict
        trainer._save(output_dir, state_dict=cpu_state_dict)
```

3.4.9 分词处理

函数 _tokenize_fn()的作用是对输入的一批字符串进行分词处理，并返回分词后的ID序列及相关的长度信息。该函数通过tokenizer将文本转换为模型可理解的ID序列，并返回这些序列及其长度信息，为后续的模型训练和推理提供数据支持。

```
# 分词函数
def _tokenize_fn(strings: Sequence[str], tokenizer: transformers.
PreTrainedTokenizer) -> Dict:
    tokenized_list = [
        tokenizer(
            text,
            max_length=tokenizer.model_max_length,
            truncation=True,
        )
        for text in strings
    ]
    input_ids = labels = [np.array(tokenized.input_ids) for tokenized in
                          tokenized_list]
    input_ids_lens = labels_lens = [len(tokenized.input_ids) for tokenized
                                    in tokenized_list]
    return {
        "input_ids": input_ids,
        "labels": labels,
        "input_ids_lens": input_ids_lens,
        "labels_lens": labels_lens,
    }
```

3.4.10 文本预处理

函数preprocess()实现了对输入源文本和目标文本的预处理，主要用于为模型训练准备结构化的输入和标签数据。其核心功能是将源文本和目标文本拼接后进行分词，并通过标记源文本部分的标签为 -100（IGNORE_INDEX），使得模型在训练时仅对目标文本部分进行预测。

```
# 预处理数据
def preprocess(sources: Sequence[str], targets: Sequence[str], tokenizer:
transformers.PreTrainedTokenizer) -> Dict:
```

```
    examples = [s + t for s, t in zip(sources, targets)]
    examples_tokenized, sources_tokenized = [_tokenize_fn(strings, tokenizer)
                                             for strings in (examples,
                                             sources)]
    input_ids = examples_tokenized["input_ids"]
    labels = copy.deepcopy(input_ids)
    for label, source_len in zip(labels,
                                 sources_tokenized["input_ids_lens"]):
        label[:source_len] = IGNORE_INDEX
    return {
        "input_ids": input_ids,
        "labels": labels,
    }
```

3.4.11 数据收集器

下列代码定义了一个名为DataCollatorForSupervisedDataset的数据收集器类，用于在监督微调过程中将多个数据实例整理成模型训练所需的批量数据格式。其主要功能是将输入的input_ids和labels转换为张量并进行填充（padding），以确保批量数据的维度一致，同时生成attention_mask，用于指示模型哪些位置是有效的输入数据。

```
# 定义监督微调的数据收集器类
@dataclass
class DataCollatorForSupervisedDataset(object):
    tokenizer: transformers.PreTrainedTokenizer

    def __call__(self, instances: Sequence[Dict]) -> Dict[str, torch.Tensor]:
        input_ids, labels = tuple([instance[key] for instance in instances]
                                  for key in ("input_ids", "labels"))
        input_ids = [torch.tensor(x) for x in input_ids]
        input_ids = torch.nn.utils.rnn.pad_sequence(
            input_ids, batch_first=True,
            padding_value=self.tokenizer.pad_token_id
        )
        labels = [torch.tensor(x) for x in labels]
        labels = torch.nn.utils.rnn.pad_sequence(labels, batch_first=True,
                                                 padding_value=IGNORE_INDEX)
        return {
            "input_ids": input_ids,
            "labels": labels,
            "attention_mask": input_ids.ne(self.tokenizer.pad_token_id),
        }
```

3.4.12 训练数据的分词和预处理

下列代码定义了一个名为train_tokenize_function()的函数，用于对训练数据进行分词和预处理。函数train_tokenize_function()接收examples和tokenizer作为输入，examples包含用户提供的指令和对应的输出。函数首先调用build_instruction_prompt为每个指令构建提示文本，然后将输出拼接上结束标记Token（EOT_TOKEN）。接着，函数对这些提示和输出进行分词，并通过preprocess函数将分词后的ID序列转换为适合模型训练的格式。最后，函数返回一个包含输入ID和标签的数据字典，用于后续的训练过程。

```
# 定义训练分词函数
def train_tokenize_function(examples, tokenizer):
    sources = [build_instruction_prompt(instruction) for instruction in
               examples['instruction']]
    targets = [f"{output}\n{EOT_TOKEN}" for output in examples['output']]
    data_dict = preprocess(sources, targets, tokenizer)
    return data_dict
```

3.4.13 构建和配置模型

函数build_model()用于根据提供的参数构建和配置模型，支持量化和LoRA（Low-Rank Adaptation）配置。该函数的主要功能和实现细节如下。

- 确定计算数据类型：根据training_args.bf16的值，确定计算数据类型为torch.bfloat16或torch.float16。
- 加载预训练模型：使用transformers.AutoModelForCausalLM.from_pretrained方法加载预训练模型。根据model_args.bits的值，决定是否进行4位或8位量化。如果启用了LoRA，则需配置量化参数，包括量化类型、是否使用双量化等。
- 设置模型属性：设置模型的model_parallel和is_parallelizable属性为True，以支持模型并行化。根据training_args.bf16的值，设置模型的torch_dtype。
- LoRA配置：如果启用了LoRA且量化位数小于16，则调用prepare_model_for_kbit_training函数准备模型进行k-bit训练。如果提供了检查点目录，则从检查点加载LoRA适配器。否则，初始化LoRA模块，配置目标模块、LoRA的秩、dropout率、alpha值等参数。
- 模型层的数据类型转换：遍历模型的所有层，根据需要将特定层的数据类型转换为torch.bfloat16或torch.float32。
- 返回模型：返回配置好的模型。

```
# 构建模型
def build_model(model_args, training_args, checkpoint_dir):
    compute_dtype = torch.bfloat16 if training_args.bf16 else torch.float16
                                                          # 确定计算数据类型
```

```
    model = transformers.AutoModelForCausalLM.from_pretrained(
        model_args.model_name_or_path,
        load_in_4bit=model_args.bits == 4,
        load_in_8bit=model_args.bits == 8,
        quantization_config=BitsAndBytesConfig(
            load_in_4bit=model_args.bits == 4,
            load_in_8bit=model_args.bits == 8,
            llm_int8_threshold=6.0,
            llm_int8_has_fp16_weight=False,
            bnb_4bit_compute_dtype=compute_dtype,
            bnb_4bit_use_double_quant=model_args.double_quant,
            bnb_4bit_quant_type=model_args.quant_type,
        ) if model_args.use_lora else None,
        torch_dtype=compute_dtype,
        trust_remote_code=True,
    )

    if compute_dtype == torch.float16 and model_args.bits == 4:
        if torch.cuda.is_bf16_supported():
            logger.warning("=" * 80)
            logger.warning("您的 GPU 支持 bfloat16，您可以通过添加参数 --bf16 来加
                            速训练！")
            logger.warning("=" * 80)
    setattr(model, 'model_parallel', True)
    setattr(model, 'is_parallelizable', True)
    model.config.torch_dtype = torch.bfloat16 if training_args.bf16 else
                               torch.float32

    if model_args.use_lora and model_args.bits < 16:
        model = prepare_model_for_kbit_training(model, use_gradient_
checkpointing=training_args.gradient_checkpointing)

    if model_args.use_lora:
        if checkpoint_dir is not None:
            logger.info(f"从 {checkpoint_dir} 加载适配器。")
            model = PeftModel.from_pretrained(model, checkpoint_dir,
                                              is_trainable=True)
        else:
            logger.info("初始化 LoRA 模块 ...")
            target_modules = model_args.trainable.split(',')
            modules_to_save = model_args.modules_to_save
            if modules_to_save is not None:
                modules_to_save = modules_to_save.split(',')
            lora_rank = model_args.lora_rank
            lora_dropout = model_args.lora_dropout
            lora_alpha = model_args.lora_alpha
```

```
            peft_config = LoraConfig(
                task_type=TaskType.CAUSAL_LM,
                target_modules=target_modules,
                inference_mode=False,
                r=lora_rank,
                lora_alpha=lora_alpha,
                lora_dropout=lora_dropout,
                modules_to_save=modules_to_save
            )
            model = get_peft_model(model, peft_config)

    for name, module in model.named_modules():
        if isinstance(module, LoraLayer):
            if training_args.bf16:
                module = module.to(torch.bfloat16)
        if 'norm' in name or 'gate' in name:
            module = module.to(torch.float32)
        if 'lm_head' in name or 'embed_tokens' in name:
            if hasattr(module, 'weight'):
                if training_args.bf16 and module.weight.dtype == torch.float32:
                    module = module.to(torch.bfloat16)
    return model
```

3.4.14 训练模型

函数train()是DeepSeek MoE项目中的主训练函数，负责整个模型的训练流程。函数train()实现了一个完整的训练流程，用于对预训练的MoE语言模型进行微调。首先加载用户提供的数据，通过分词和预处理将其转化为模型可接受的输入格式。构建模型时应用了LoRA（低秩适应）和量化技术，以优化显存占用和训练效率。在训练过程中，使用了梯度累积、动态损失缩放等技术，并通过检查点机制支持从断点恢复训练。训练结束后，模型的权重会被保存下来，以便后续的评估或推理使用。

```
# 定义训练函数
def train():
    parser = transformers.HfArgumentParser((ModelArguments, DataArguments,
                                            TrainingArguments))
    model_args, data_args, training_args = parser.parse_args_into_
                                            dataclasses()
    log_level = training_args.get_process_log_level()
    logger.setLevel(log_level)
    datasets.utils.logging.set_verbosity(log_level)
    transformers.utils.logging.set_verbosity(log_level)
    transformers.utils.logging.enable_default_handler()
    transformers.utils.logging.enable_explicit_format()
```

```
if training_args.local_rank == 0:
    logger.info("=" * 100)
    logger.info(training_args)

tokenizer = transformers.AutoTokenizer.from_pretrained(
    model_args.model_name_or_path,
    model_max_length=training_args.model_max_length,
    padding_side="right",
    use_fast=True,
    trust_remote_code=True
)

logger.info(f"填充标记：{tokenizer.pad_token} ({tokenizer.pad_token_id})")
logger.info(f"开始标记：{tokenizer.bos_token} ({tokenizer.bos_token_id})")
logger.info(f"结束标记：{tokenizer.eos_token} ({tokenizer.eos_token_id})")

resume_from_checkpoint_dir = get_last_checkpoint(training_args.output_dir)
model = build_model(model_args, training_args, resume_from_checkpoint_dir)

raw_train_datasets = load_dataset(
    'parquet',
    data_files=data_args.data_path,
    split="train",
    cache_dir=training_args.cache_dir
)
if training_args.local_rank > 0:
    torch.distributed.barrier()

train_dataset = raw_train_datasets.map(
    train_tokenize_function,
    batched=True,
    batch_size=3000,
    num_proc=32,
    remove_columns=raw_train_datasets.column_names,
    load_from_cache_file=True,
    desc="运行编码",
    fn_kwargs={"tokenizer": tokenizer}
)

if training_args.local_rank == 0:
    torch.distributed.barrier()

if training_args.local_rank == 0:
    logger.info(f"训练数据集样本数量：{len(train_dataset)}")
    for index in random.sample(range(len(train_dataset)), 3):
        logger.info(f"训练集样本 {index}: 输入 ID - {train_dataset[index]
```

```
                        ['input_ids']}, 标签 - {train_dataset[index]
                        ['labels']}.")
            logger.info(f"训练集样本 {index}: 问题 : {train_dataset[index]
                        ['instruction']}\n 回答 : {train_dataset[index]
                        ['output']}")

    data_collator = DataCollatorForSupervisedDataset(tokenizer=tokenizer)
    data_module = dict(train_dataset=train_dataset, eval_dataset=None, data_
                       collator=data_collator)

    trainer = Trainer(model=model, tokenizer=tokenizer, args=training_args,
                      **data_module)
    if model_args.use_lora:
        trainer.add_callback(SavePeftModelCallback)
    trainer.train(resume_from_checkpoint=resume_from_checkpoint_dir)
    trainer.save_state()
    if not model_args.use_lora:
        safe_save_model_for_hf_trainer(trainer=trainer,
                                       output_dir=training_args.output_dir)

# 主程序入口
if __name__ == "__main__":
    train()
```

函数train()的具体实现流程如下。

（1）解析命令行参数：使用transformers.HfArgumentParser解析命令行参数，将模型参数、数据参数和训练参数分别解析为ModelArguments、DataArguments和TrainingArguments类的实例。

（2）设置日志记录：根据训练参数中的日志级别设置日志记录器的日志级别，并启用默认的日志处理器和显示格式。

（3）加载分词器：使用transformers.AutoTokenizer.from_pretrained()方法加载预训练的分词器，并设置分词器的最大序列长度、填充方向、是否使用快速分词器等参数。

（4）加载模型：调用build_model函数构建和配置模型，包括加载预训练模型、设置量化参数、初始化LoRA模块等。

（5）加载训练数据：使用datasets.load_dataset()方法加载训练数据集，并根据数据路径、缓存目录等参数进行配置。

（6）数据预处理：使用train_tokenize_function()对训练数据进行预处理，包括构建指令提示、分词和标签处理等。

（7）数据收集器：创建DataCollatorForSupervisedDataset实例，用于在训练过程中将数据整理成模型所需的格式。

（8）初始化训练器：使用Hugging Face的Trainer类初始化训练器，传入模型、分词器、训练参数和数据模块等。

（9）添加回调函数：如果启用了LoRA，则添加SavePeftModelCallback()回调函数，用于在训练过程中保存LoRA适配器的权重。

（10）开始训练：调用trainer.train方法开始训练模型，并根据需要从检查点恢复训练。

（11）保存训练状态：训练结束后，调用trainer.save_state()方法保存训练状态，包括模型权重、优化器状态等。

（12）保存模型：如果未启用LoRA，则调用safe_save_model_for_hf_trainer()函数将模型权重安全地保存到指定目录。

3.4.15 微调模型

运行下面的命令，文件finetune.py将启动训练，加载指定的模型和数据，并按照配置进行微调。在训练过程中，检查点和日志将被保存到指定的输出目录，并且可以通过TensorBoard查看训练指标。

```
DATA_PATH="<your_data_path>"
OUTPUT_PATH="<your_output_path>"
MODEL_PATH="<your_model_path>"

cd finetune
deepspeed finetune.py \
    --model_name_or_path $MODEL_PATH \
    --data_path $DATA_PATH \
    --output_dir $OUTPUT_PATH \
    --num_train_epochs 3 \
    --model_max_length 1024 \
    --per_device_train_batch_size 16 \
    --per_device_eval_batch_size 1 \
    --gradient_accumulation_steps 4 \
    --evaluation_strategy "no" \
    --save_strategy "steps" \
    --save_steps 100 \
    --save_total_limit 100 \
    --learning_rate 2e-5 \
    --warmup_steps 10 \
    --logging_steps 1 \
    --lr_scheduler_type "cosine" \
    --gradient_checkpointing True \
    --report_to "tensorboard" \
    --deepspeed configs/ds_config_zero3.json \
    --bf16 True \
    --use_lora False
```

对上述各个命令参数的具体说明如下所示。

- --model_name_or_path：指定预训练模型的路径或名称。
- --data_path：训练数据的路径。

- --output_dir：模型和其他输出文件的保存目录。
- --num_train_epochs：训练的轮数。
- --model_max_length：模型输入的最大序列长度。
- --per_device_train_batch_size：每个设备上的训练批次大小。
- --per_device_eval_batch_size：每个设备上的评估批次大小。
- --gradient_accumulation_steps：梯度累积的步骤数，即在反向传播前累积多少个步骤的梯度。
- --evaluation_strategy：评估策略，此处设置为不进行评估。
- --save_strategy：模型保存策略，此处设置为按步骤保存。
- --save_steps：每隔多少步骤保存一次模型。
- --save_total_limit：保存的模型检查点的总数限制。
- --learning_rate：学习率。
- --warmup_steps：学习率预热的步骤数。
- --logging_steps：日志记录的步骤间隔。
- --lr_scheduler_type：学习率调度器类型，此处为余弦退火。
- --gradient_checkpointing：是否启用梯度检查点，以节省内存。
- --report_to：指定日志报告的工具，此处为TensorBoard。
- --deepspeed：指定DeepSpeed配置文件的路径。
- --bf16：是否使用bfloat16 精度进行训练。
- --use_lora：是否使用LoRA（低秩适应）技术进行模型微调。

上述命令使用DeepSpeed来加速和优化模型的微调过程，通过指定DeepSpeed配置文件（如configs/ds_config_zero3.json），可以利用ZeRO Stage 3来有效地管理内存和计算资源，从而支持大规模模型的训练。

此外，在上述命令中还设置了梯度累积、学习率调度、日志记录等参数，以控制训练过程的各个方面。这些参数的配置需要根据具体的硬件资源和任务需求进行调整，以获得最佳的训练效果。

3.5 调用模型

DeepSeekMoE项目的完整训练流程和训练代码并未开源，但预训练后的模型已经通过Hugging Face平台公开发布。本节将详细讲解使用预训练模型的知识。

3.5.1 下载模型

DeepSeekMoE项目提供了两个版本的模型供用户下载，具体说明如下所示。

1. DeepSeekMoE 16B Base（基础模型）

DeepSeekMoE 16B Base是一个开源的MoE语言模型，具有 16.4B参数。该模型采用创新的

MoE架构，通过细粒度专家分割和共享专家隔离两种主要策略，实现了专家的终极专业化。它在2T英文和中文Token上从零开始训练，性能与DeepSeek 7B和LLaMA2 7B相当，但所需计算资源仅为后面二者的 40% 左右。该模型可在单个 40GB显存的GPU上部署，无须量化。DeepSeekMoE 16B Base模型如图3-1所示。

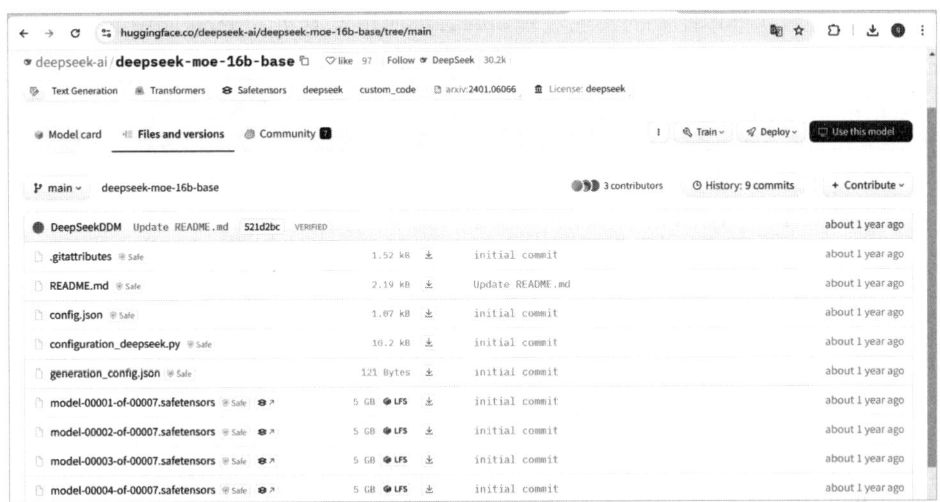

图3-1　DeepSeekMoE 16B Base模型

2. DeepSeekMoE 16B Chat（聊天模型）

DeepSeekMoE 16B Chat是基于DeepSeekMoE 16B Base微调得到的聊天模型。它在多个基准测试中表现出色，尤其是在语言理解、推理和代码生成方面。与DeepSeek 7B Chat和LLaMA2 7B SFT相比，DeepSeekMoE 16B Chat在计算量仅为二者 40% 左右的情况下，仍能取得与二者相同甚至更好的性能。该模型同样可在单个 40GB显存的GPU上部署，无须量化。DeepSeekMoE 16B Chat模型如图3-2所示。

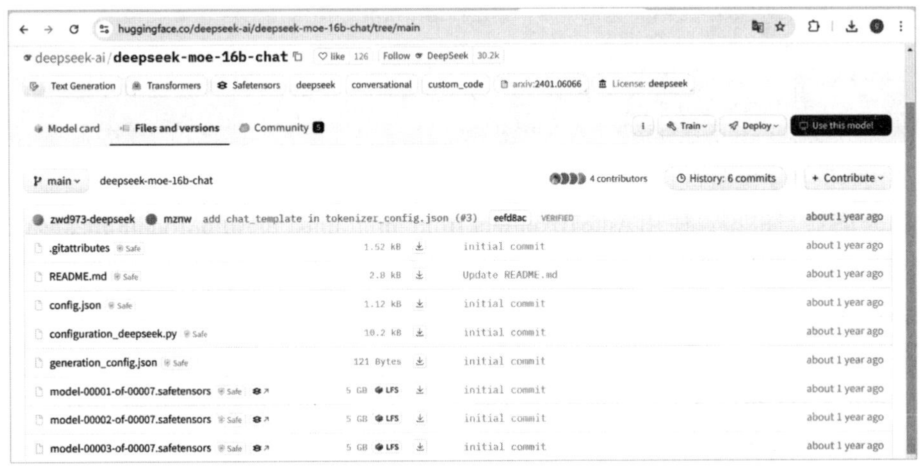

图3-2　DeepSeekMoE 16B Chat模型

我们可以直接从Hugging Face上获取训练好的模型权重，而无须重新训练模型。

3.5.2 调用模型

DeepSeek MoE项目的开源文件，介绍了利用Hugging Face Transformers实现两种不同场景（文本补全和对话生成）推理任务的实现过程。两者都使用了Transformers库中的AutoTokenizer和AutoModelForCausalLM类，同时加载了各自对应的生成配置（GenerationConfig）并设置了填充标记（pad_token_id）为结束标记（eos_token_id），以确保生成过程正确终止。

1. 文本补全

文本补全（Text Completion）是指给定一段输入文本，模型生成后续内容，实现对连续文本的自动扩展。例如下面的演示代码：

```
import torch
from transformers import AutoTokenizer, AutoModelForCausalLM,
GenerationConfig

model_name = "deepseek-ai/deepseek-ai/deepseek-moe-16b-base"
tokenizer = AutoTokenizer.from_pretrained(model_name)
model = AutoModelForCausalLM.from_pretrained(
    model_name, torch_dtype=torch.bfloat16, device_map="auto"
)
model.generation_config = GenerationConfig.from_pretrained(model_name)
model.generation_config.pad_token_id = model.generation_config.eos_token_id

text = "An attention function can be described as mapping a query and a set
of key-value pairs to an output, where the query, keys, values, and output
are all vectors. The output is"
inputs = tokenizer(text, return_tensors="pt")
outputs = model.generate(**inputs.to(model.device), max_new_tokens=100)

result = tokenizer.decode(outputs[0], skip_special_tokens=True)
print(result)
```

对上述代码的具体说明如下所示。

- 加载分词器与模型：使用AutoTokenizer.from_pretrained和AutoModelForCausalLM.from_pretrained根据模型名称加载预训练模型。这里指定了torch_dtype=torch.bfloat16 以降低内存占用，并采用device_map="auto" 自动将模型加载到可用的设备（如GPU）。
- 生成配置（GenerationConfig）：通过GenerationConfig.from_pretrained加载与模型对应的生成配置，并将pad_token_id设置为模型的eos_token_id。这一操作能确保在生成过程中，当生成结束标记时可以正确识别并停止生成。
- 输入文本处理：使用分词器将输入文本转换为PyTorch张量（return_tensors="pt"），为后续模型推理提供标准格式的输入。

- 文本生成：调用model.generate方法进行推理，设置max_new_tokens=100 表示生成最多 100 个新Token。注意，这里输入被自动移动到模型所在设备（inputs.to(model.device)）。
- 输出解码：使用tokenizer.decode将生成的Token序列转换为人类可读文本，参数skip_special_tokens=True用于跳过特殊控制标记。

上述代码适用于需要对给定文本续写（或填充）的场景，常用于创作、问答、代码生成等任务中。

2. 对话生成

对话生成（Chat Completion）是指模拟对话场景，使用专门为聊天任务微调的模型，通过“聊天模板”将用户消息格式化后生成回复。例如下面的演示代码：

```
import torch
from transformers import AutoTokenizer, AutoModelForCausalLM,
GenerationConfig

model_name = "deepseek-ai/deepseek-moe-16b-chat"
tokenizer = AutoTokenizer.from_pretrained(model_name)
model = AutoModelForCausalLM.from_pretrained(
    model_name, torch_dtype=torch.bfloat16, device_map="auto"
)
model.generation_config = GenerationConfig.from_pretrained(model_name)
model.generation_config.pad_token_id = model.generation_config.eos_token_id

messages = [
    {"role": "user", "content": "Who are you?"}
]
input_tensor = tokenizer.apply_chat_template(
    messages, add_generation_prompt=True, return_tensors="pt"
)
outputs = model.generate(input_tensor.to(model.device), max_new_tokens=100)

result = tokenizer.decode(outputs[0][input_tensor.shape[1]:], skip_special_
tokens=True)
print(result)
```

对上述代码的具体说明如下所示。

- 加载聊天模型与分词器：与文本补全类似，但这里加载的是为对话任务微调的模型"deepseek-ai/deepseek-moe-16b-chat"，该模型经过专门的聊天数据（对话语料）训练，更适合生成对话回复。
- 聊天输入模板处理：这里与单纯文本补全不同，输入不再是普通文本，而是以消息列表（含角色字段）形式给出。使用tokenizer.apply_chat_template方法对消息进行格式化处理，自动添加生成提示（generation prompt），使模型能够理解对话结构。例如，在聊天模板中通常会自动加入“User:”和“Assistant:”等提示符，帮助模型区分对话双方。

- 生成过程与输出解码：生成过程调用方式与文本补全类似，不过在解码时，需要剔除输入部分（通过 outputs[0][input_tensor.shape[1]:] 切片），仅保留模型生成的回复部分。

上述代码适用于构建聊天机器人或对话系统，通过对消息列表进行模板格式化，实现对话生成。

注意

关键技术点

- **设备和精度优化：通过设置 torch_dtype=torch.bfloat16 和 device_map="auto"，能够在推理时降低内存占用并自动选取最优设备（通常是 GPU），提高计算效率。**
- **生成配置管理：使用 GenerationConfig 能够灵活控制生成过程中的参数（如采样策略、最大生成 token 数量等），同时通过设置 pad_token_id 确保特殊标记的正确处理。**
- **输入格式化与聊天模板：文本补全直接对普通字符串进行编码，而聊天生成则需要对消息列表进行特殊格式化处理（如 apply_chat_template 方法），使得模型能区分不同对话角色并在生成时输出合适的回复格式。**
- **输出解码与后处理：生成结果需要经过分词器解码为字符串，同时要在聊天场景中截取掉原始对话部分，仅保留模型生成的新回复。**

总之，上述两个调用示例展示了如何利用 Hugging Face Transformers 进行两种不同场景的推理任务（文本补全和对话生成）。其中，前者侧重于连续文本的生成，后者则通过聊天模板处理对话上下文，实现更加人性化的交互。无论是哪种任务，核心流程均包括模型和分词器的加载、输入编码、生成推理及输出解码，同时配合生成配置进行灵活参数调整。通过这些技术，开发者可以快速构建高效且准确的文本生成或聊天系统。

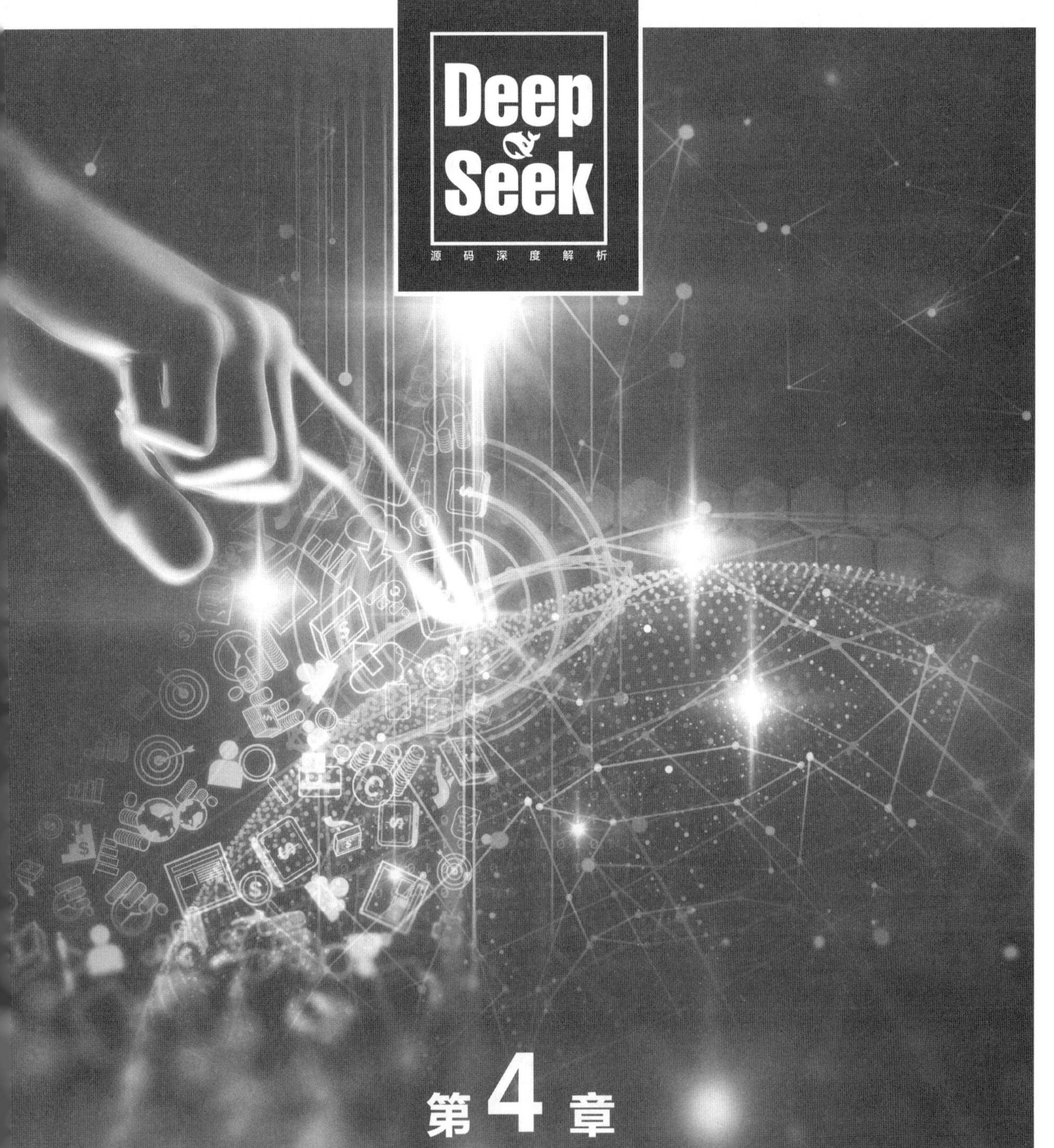

第 4 章

基于 DeepSeekMoE 架构的 DeepSeek-V3

DeepSeek-V3是由DeepSeek团队开发的一款开源大型语言模型，基于Mixture-of-Experts（MoE）架构，拥有6710亿个参数，每个token激活370亿个参数。DeepSeek-V3的目标是提供高性能、低成本、易用且支持商业使用的开源语言模型。

4.1 项目介绍

DeepSeek-V3致力于实现高效推理与成本效益的训练，其核心在于采用了MLA及DeepSeekMoE架构。这些技术均在DeepSeek-V2中得到了充分验证，而DeepSeek-V3在此基础上实现了更多创新，如无辅助损失的负载平衡策略和多token预测训练目标，从而大幅提升模型性能。

4.1.1 核心特点

DeepSeek-V3项目通过算法创新与工程优化的深度结合，为开源社区提供了当前最强的MoE模型之一，特别适合需要复杂推理能力的应用场景。其高效的训练方案也为大模型研发提供了重要参考。

- 创新的架构与训练策略：采用MLA和DeepSeekMoE架构，基于DeepSeek-V2的高效架构，引入无辅助损失的负载平衡策略和多token预测训练目标，以降低性能损耗，提升模型性能。
- 高效的预训练与知识蒸馏：使用FP8混合精度训练框架，优化算法、框架和硬件，实现计算与通信的重叠，提高训练效率，降低训练成本。DeepSeek-V3通过知识蒸馏，从DeepSeek-R1系列模型中训练和提高推理性能。
- 强大的多语言支持：DeepSeek-V3在多样性高、质量好的token上进行预训练，支持多种语言任务和应用，例如英语、中文等。
- 高性能的推理能力：针对不同硬件平台（如NVIDIA GPU、AMD GPU和华为Ascend NPU等）提供多种部署方式，以实现高性能的推理。例如，使用SGLang可以在NVIDIA和AMD GPU上运行DeepSeek-V3，并支持MLA优化、DP Attention、FP8 (W8A8) 和FP8 KV Cache等技术，从而实现业界领先的推理延迟和吞吐量。

4.1.2 训练流程

DeepSeek-V3的训练流程涵盖数据构建、模型架构设计、训练策略和精度优化等多个方面。首先，模型在经过严格筛选和清洗的14.8万亿（14.8T）高质量、多样化的token上进行预训练。在架构设计上，采用了MLA和DeepSeekMoE架构，以实现高效推理和成本效益的训练。在训练过程中，引入了无须辅助损失的负载平衡策略和多token预测训练目标，以提升模型性能。此外，模型支持FP8混合精度训练，并对训练框架进行了全面优化，以加速训练并减少GPU内存占用。综合应用这些策略，使得DeepSeek-V3在保持强大性能的同时，实现了高效训练。

1. 预训练阶段

● 使用FP8混合精度训练框架，首次在超大规模模型上验证了FP8的有效性。

● 在14.8万亿高质量、多样化的token上进行预训练，耗时仅需266.4万个H800 GPU小时（如果使用一个H800GPU，则需要266.4万小时来完成训练。如果使用多个H800 GPU，则实际训练时间会更短，因为计算资源是并行的）。

● 利用算法、框架和硬件的协同设计，成功突破了跨节点MoE训练中的通信瓶颈，实现了几乎完全的计算与通信重叠，从而进一步降低了训练成本。

2. 后续训练阶段

● 通过监督微调和强化学习进一步优化模型性能。

● 引入知识蒸馏技术，从DeepSeek-R1系列中提取长链式思考（Chain of Thought，CoT）能力，使得DeepSeek-V3在推理时表现出更强的逻辑和推理能力，同时保持输出风格和长度的控制。

注意，尽管DeepSeek-V3的完整训练流程和训练代码并未开源，但预训练后的模型权重已经通过Hugging Face平台公开发布。根据模型下载部分的说明，用户可以直接下载两种模型文件。

（1）DeepSeek-V4-Base：包含671B总参数和37B激活参数，支持128K上下文长度，如图4-1所示。

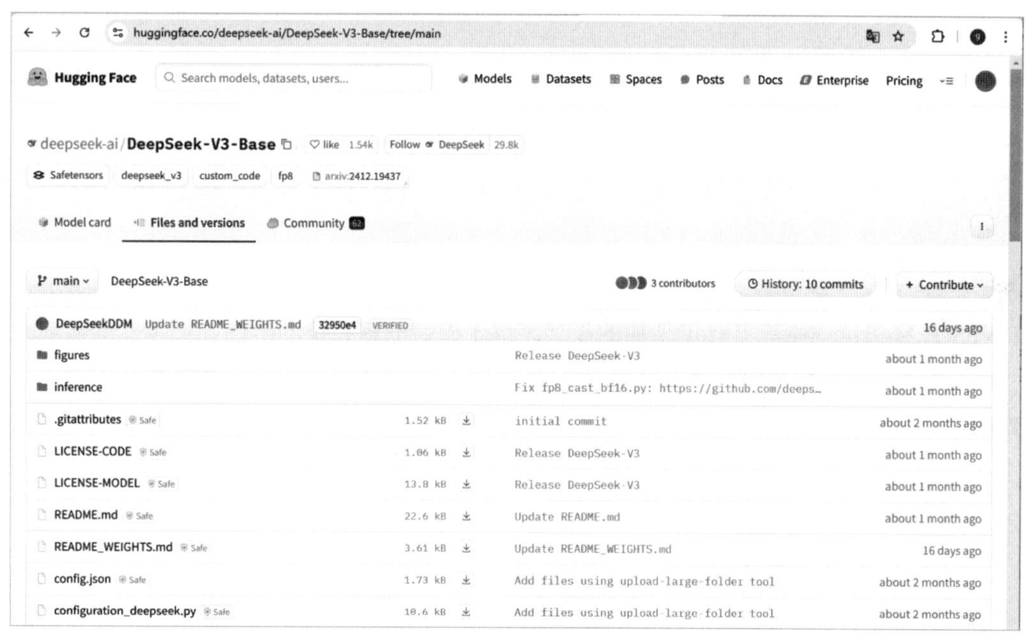

图4-1 DeepSeek-V4-Base模型

（2）DeepSeek-V3：同样包含671B总参数和37B激活参数，支持128K上下文长度，如图4-2所示。

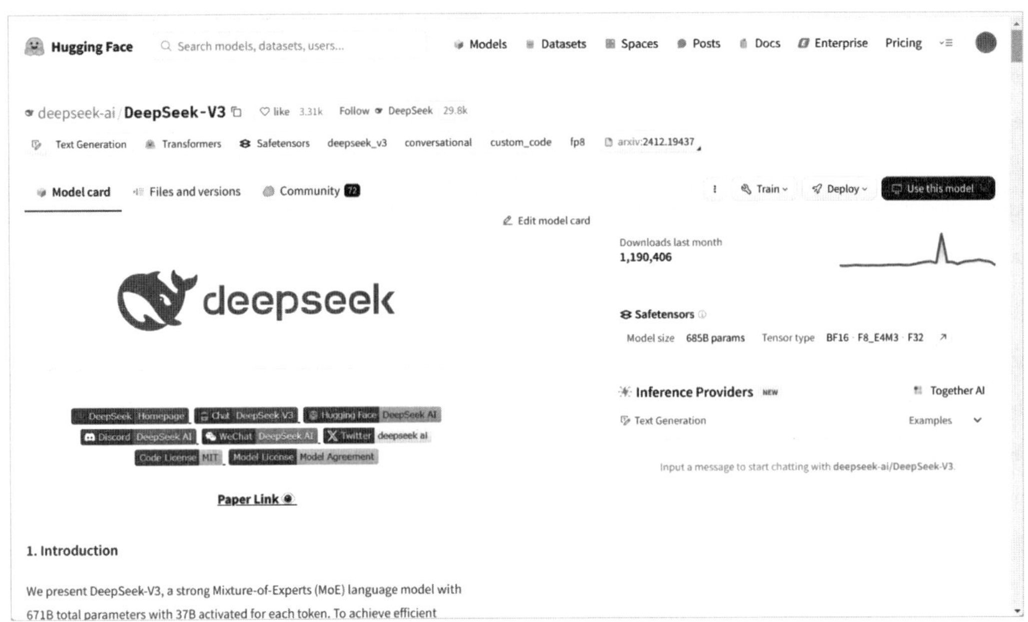

图4-2　DeepSeek-V3模型

我们可以直接从Hugging Face上获取训练好的模型权重，而无须重新训练模型。

4.1.3 与DeepSeekMoE项目的区别

DeepSeekMoE与DeepSeek-V3是DeepSeek公司在AI大模型领域推出的两项重要技术，本书前面已经介绍了DeepSeekMoE项目的源码，两者的定位、架构和应用场景存在显著区别。以下是两者的核心差异分析。

1. 定位与范畴

DeepSeekMoE：一种基于MoE架构的设计，属于模型的核心技术组件。它通过整合专家共享机制、动态路由算法和优化的注意力机制（如MLA），旨在提升模型的训练效率和推理性能。

DeepSeek-V3：一个完整的开源大语言模型，基于DeepSeekMoE架构构建，并结合了多项创新技术（如多token预测、FP8混合精度训练等），是面向实际应用的综合模型。

2. 技术特点

（1）DeepSeekMoE的核心创新。

- 动态路由与专家共享：引入共享专家参数设计，减少冗余计算，同时通过动态路由选择最相关的专家，提升模型效率。
- MLA：优化键值缓存机制，减少推理阶段的浮点运算量（减少25%），并支持长文本处理。
- RMSNorm归一化：替代传统的LayerNorm，简化计算并提升训练稳定性。

（2）DeepSeek-V3的扩展技术。

- 无辅助损失负载均衡策略：避免传统MoE模型依赖辅助损失导致的性能下降，通过动态调整专家偏置实现负载均衡。

- 多token预测：同时预测多个后续token，提升训练数据利用率和推理速度。
- FP8混合精度训练：首次在超大规模模型上验证FP8训练的有效性，显著降低GPU内存占用并加速训练。

3. 性能与效率

（1）DeepSeekMoE。

- 在相同参数规模下，训练速度比密集模型快2.1倍，推理延迟降低35%。
- 在长文本处理任务（如10k token文档问答）中准确率显著高于标准Transformer。

（2）DeepSeek-V3。

- 综合性能超越主流开源模型（如Llama-3.1-405B、Qwen2.5-72B），接近闭源模型GPT-4o和Claude-3.5。
- 训练成本仅为558万美元，远低于同类模型（如Llama-3.1的5亿美元）。

4. 应用场景

- DeepSeekMoE：适用于需要高效计算资源的场景，如实时对话系统（810 token/秒）、轻量级部署（内存占用降低40%）和长文本处理。
- DeepSeek-V3：作为完整模型，更侧重通用任务，尤其在代码生成、数学推理和中文理解领域表现突出，并能通过API提供商业服务。

总之，DeepSeekMoE是DeepSeek-V3的核心架构之一，前者聚焦于底层技术优化（如计算效率、动态路由），后者则是基于该架构构建的完整模型，结合了更多创新训练策略和工程优化。两者共同体现了DeepSeek在“高性能+低成本”大模型研发上的技术路线。

4.2 开源信息介绍

目前（作者写本书时），DeepSeek-V3 模型源码的开源信息如下。

1. 开源内容

- 模型权重：官方通过Hugging Face平台开源了FP8 格式的模型权重（671B参数），并提供了从FP8 到BF16 的转换脚本，方便社区适配不同硬件环境。
- 推理支持：开源社区已适配多种推理框架，包括SGLang（原生FP8 推理）、LMDeploy（BF16/FP8）、TensorRT-LLM（BF16）和MindIE（华为昇腾支持）等。
- 技术文档：公开了技术报告（DeepSeek_V3.pdf），详细描述了模型架构、训练策略（如FP8混合精度训练、无辅助损失负载均衡等）和性能评测。

2. 训练代码暂时未开源的原因推测

- 技术保密性：训练代码可能涉及DeepSeek自研的FP8 混合精度框架、DualPipe流水线并行算法等核心技术，这些是其在高效训练（总成本仅 557 万美元）和性能优化上的关键优势。

● 工程复杂性：MoE模型的分布式训练需要复杂的硬件协同设计（如跨节点全对全通信优化），公开代码可能涉及大量定制化基础设施支持，难以直接复现。

● 商业化考量：DeepSeek同时提供API服务，若开源训练代码可能影响其商业模式的竞争力。

3. 社区替代方案与未来可能性

● 基于现有架构复现：技术报告中已公开关键架构设计（如MLA注意力、DeepSeekMoE），社区可参考这些信息尝试复现部分训练流程。

● 潜在开源计划：DeepSeek在技术报告中强调“以开源精神追求普惠AGI”，未来可能会逐步开放更多训练细节或工具链。

4.3 模型权重

文件README_WEIGHTS.md中详细描述了DeepSeek-V3 模型权重文件的结构和加载规则，以及FP8 权重的量化配置和反量化方法。这些信息对于理解模型的权重组织方式和量化策略非常重要，有助于开发者在实际应用中正确加载和使用模型权重。

4.3.1 权重结构

DeepSeek-V3 的权重结构由主模型权重和MTP模块组成，其中主模型权重包含输入/输出嵌入层及 61 个Transformer隐藏层，参数总量达 671B，而MTP模块则附加了额外的预测层，用于提升模型的多token预测能力，其参数独立于主模型，以支持更高效的推理和性能优化。

1. 主模型权重

（1）组成：输入/输出嵌入层和完整的 61 个Transformer隐藏层。

（2）参数数量。

● 总参数：671B。

● 激活参数：36.7B（包括 0.9B的嵌入层和 0.9B的输出头）。

（3）结构细节。

● 嵌入层：model.embed_tokens.weight。

● Transformer隐藏层：model.layers.0 到model.layers.60，共num_hidden_layers层。

● 输出层：model.norm.weight和lm_head.weight。

2. MTP 模块

（1）组成：由num_nextn_predict_layers字段定义的额外MTP模块，在此模型中设置为 1。

（2）参数数量。

● 参数：11.5B独立参数（不包括共享的 0.9B嵌入层和 0.9B输出头）。

- 激活参数：2.4B（包括共享的 0.9B 嵌入层和 0.9B 输出头）。

（3）结构细节。

- embed_tokens：与主模型权重的嵌入层共享参数。
- enorm & hnorm：用于推测解码的RMSNorm参数。
- eh_proj：对范数结果进行降维投影的参数。
- 额外的Transformer隐藏层：model.layers.61.self_attn和mlp（结构与主模型隐藏层相同）。
- shared_head：与主模型权重的输出头共享参数。

4.3.2 加载规则

DeepSeek-V3 模型的权重加载过程遵循特定的规则，以确保主模型和MTP模块的正确加载和初始化。

1. 主模型权重加载

主模型权重的加载依赖config.json文件中的num_hidden_layers参数。该参数指定了主模型中Transformer隐藏层的数量。加载过程如下。

（1）读取配置文件：系统首先读取config.json文件，获取num_hidden_layers的值。

（2）确定隐藏层数量：根据num_hidden_layers的值，确定主模型中Transformer隐藏层的数量。例如，如果num_hidden_layers为 61，则主模型包含 61 个Transformer隐藏层。

（3）加载权重：系统根据上述确定的层数，从权重文件中加载相应的主模型权重，包括输入/输出嵌入层和所有Transformer隐藏层的权重。

2. MTP 模块加载

MTP模块的加载依赖num_nextn_predict_layers参数，该参数指定了MTP模块的数量。加载过程如下。

（1）读取配置文件：系统读取config.json文件，获取num_nextn_predict_layers的值。

（2）确定MTP模块数量：根据num_nextn_predict_layers的值，确定MTP模块的数量。例如，如果num_nextn_predict_layers为 1，则表示有 1 个MTP模块。

（3）计算层ID：MTP模块的层ID紧接在主模型隐藏层之后。具体计算方式为，MTP模块的起始层ID = num_hidden_layers。例如，如果num_hidden_layers为 61 且num_nextn_predict_layers为 1，则MTP模块的层ID为 61。

（4）加载权重：系统根据计算得到的层ID，从权重文件中加载相应的MTP模块权重，包括额外的Transformer隐藏层和相关参数。

假设文件config.json中的配置信息如下：

```
{
  "num_hidden_layers": 61,
  "num_nextn_predict_layers": 1
```

```
}
```

主模型权重的加载过程如下：

（1）系统读取num_hidden_layers的值为61。

（2）加载主模型61个Transformer隐藏层的权重以及输入/输出嵌入层的权重。

MTP模块的加载过程如下：

（1）系统读取num_nextn_predict_layers的值为1。

（2）计算MTP模块的层ID为61。

（3）加载MTP模块的权重，包括额外的Transformer隐藏层（model.layers.61）和相关参数。

总的来说，通过配置文件中的num_hidden_layers和num_nextn_predict_layers两个参数，DeepSeek-V3能够灵活地加载主模型和MTP模块权重，使得模型结构在硬件部署和后续优化时保持统一且易于扩展，确保了模型的完整性和功能性。

4.3.3 FP8权重

FP8权重是一种采用8位浮点数格式的模型权重，用于在深度学习模型中实现高效的存储和计算。FP8格式通过降低权重的存储精度，将模型的内存需求降低一半，同时显著提高吞吐量，对模型准确性的负面影响极小。

DeepSeek-V3模型原生支持FP8权重格式，采用128×128块缩放，以优化模型的存储，提高计算效率。FP8权重文件通过特定的量化配置来描述量化方法，确保模型在低精度计算下的性能和精度。

1. FP8配置

FP8权重文件引入了quantization_config字段，用于详细描述量化方法。下面是一个配置示例：

```
"quantization_config": {
  "activation_scheme": "dynamic",
  "fmt": "e4m3",
  "quant_method": "fp8",
  "weight_block_size": [128, 128]
}
```

对上述配置的具体说明如下所示。

- 量化格式：quant_method设置为fp8，fmt设置为e4m3，这对应于FP8的e4m3格式（4位指数，3位尾数），在PyTorch中对应于torch.float8_e4m3fn。
- 权重块大小：weight_block_size设置为[128, 128]，表示权重以128×128的块进行处理。
- 激活量化方案：activation_scheme设置为dynamic，表示使用动态激活量化方法。

2. 反量化方法

FP8权重文件包含weight_scale_inv字段，用于存储每个权重块的反量化尺度。以下是对反量化操作的详细说明。

- 存储格式：weight_scale_inv 以 float32 张量的形式存储，与权重数据一起存储在权重文件中。这种存储方式确保了反量化过程的精度和效率。
- 反量化公式：如果权重块未对齐到 128，则在计算尺度之前将其零填充到 128。量化后，移除填充部分，以确保权重块的尺寸正确。反量化过程如下。

```
反量化权重 = FP8 权重块 * weight_scale_inv
```

通过上述反量化过程，模型在运行时能够实现每个token、每 128 个通道的在线量化。这种方法不仅提高了模型的存储效率，还在一定程度上加速了计算过程。

- 运行时操作：通过反量化FP8 权重，运行时操作可以实现每个token、每 128 个通道的在线量化。这种细粒度的量化方法确保了模型在低精度计算下的性能和精度，同时提高了计算效率和内存利用率。

注意

FP8 量化虽然在存储和计算效率上具有优势，但也可能带来量化误差。因此，在实际应用中，需要根据具体任务和模型要求，权衡精度和效率之间的平衡。

总之，DeepSeek-V3 模型通过原生支持FP8 权重格式，采用 128×128 块缩放和动态激活量化方案，实现了高效的存储和计算。反量化方法通过weight_scale_inv字段存储每个权重块的反量化尺度，确保了模型在低精度计算下的精度和性能。结合使用这些技术，使得DeepSeek-V3 模型在保持高性能的同时，显著降低了内存占用和计算成本。

4.4 超参数配置

DeepSeek-V3 模型采用了MoE架构，拥有总计 6710 亿个参数，每个token激活约 370 亿个参数。DeepSeek-V3模型包含 61 层Transformer层，隐藏层维度为 7168。在MLA机制中，设置了 128 个注意力头，每个头的维度为 128。所有可学习参数均以标准差 0.006 进行随机初始化。此外，DeepSeek-V3模型采用了MTP训练目标，以提高性能。

4.4.1 小规模版本（16B）的配置

文件config_16B.json是DeepSeek-V3 模型的 16B版本配置文件，配置信息如下：

```
{
    "vocab_size": 102400,
    "dim": 2048,
    "inter_dim": 10944,
    "moe_inter_dim": 1408,
```

```
    "n_layers": 27,
    "n_dense_layers": 1,
    "n_heads": 16,
    "n_routed_experts": 64,
    "n_shared_experts": 2,
    "n_activated_experts": 6,
    "route_scale": 1.0,
    "q_lora_rank": 0,
    "kv_lora_rank": 512,
    "qk_nope_head_dim": 128,
    "qk_rope_head_dim": 64,
    "v_head_dim": 128,
    "mscale": 0.707
}
```

对上述配置信息的具体说明如下。

- vocab_size：表示词汇表的大小，即模型能够识别和处理的独立token数量。这里设置为102400，说明模型具有的词汇覆盖能力，适用于多语言及大规模文本的处理。
- dim：指模型的隐层维度，也就是每个token的嵌入向量或隐藏状态的维度。2048 的维度保证了模型在捕捉语义和上下文信息时有足够的表达能力，同时也控制了计算资源的需求。
- inter_dim：这是Transformer模块中前馈神经网络（Feed-Forward Network，FFN）的中间层维度。通常，FFN的中间维度会大于dim，以提供足够的非线性变换能力，10944 的设置能够让模型在每一层中进行更复杂的特征提取和转换。
- moe_inter_dim: 1408：针对MoE模块的中间层维度。与常规FFN相比，MoE模块中每个专家的内部维度可能较小，以实现参数共享和高效计算，1408 的设置即为专家网络量身定制的中间维度。
- n_layers：表示Transformer隐藏层的层数，即模型的深度。27 层的深度在保证足够表示能力的同时，也有助于控制模型的训练难度和计算开销。
- n_dense_layers：指在某些模块中额外使用的密集层（Dense Layer）的数量。这里设置为 1，说明在特定部分有一个额外的全连接层，用于进一步特征融合或变换。
- n_heads：表示MHA机制中注意力头的数量。16 个头可以让模型在不同子空间中同时捕获多种语义关系，提高注意力机制的表达能力。
- n_routed_experts：在MoE架构中，表示用于路由（routing）的专家数量。64 个专家为模型提供了丰富的专家资源，以便在不同任务或不同输入场景下动态选择最合适的专家进行处理。
- n_shared_experts：表示在不同层之间共享的专家数量。共享专家能够降低整体参数量，同时利用跨层信息共享，提高模型的泛化能力。
- n_activated_experts：指在每个前向传播过程中，针对一个token或一批数据，实际激活（参与计算）的专家数量。6 个激活专家的设计使得模型在保持高效计算的同时，能够灵活地利用多个专家的优势。
- route_scale：为路由过程引入的缩放因子。设置为 1.0 表示在路由时不进行额外的缩放操作，但该参数为未来可能的调整提供了灵活性，便于在不同任务中微调专家选择的力度。

● q_lora_rank：这是针对query投影使用的LoRA低秩矩阵分解的秩。0 表示在query投影中未采用LoRA技术，这是因为该部分已达到理想效果或出于计算效率考虑。

● kv_lora_rank：针对key和value投影采用LoRA技术时，设置在低秩近似中所用的矩阵秩为 512。较高的LoRA秩能够在保证低秩近似效果的同时，提供足够的自由度来适应复杂的数据分布，改善模型在微调时的表现。

● qk_nope_head_dim：这一参数与不使用位置编码的query/key部分相关，指定了对应注意力头的维度为 128，有助于区分带与不带位置编码的注意力计算方式。

● qk_rope_head_dim：指定了采用RoPE进行位置编码的query/key部分的维度为 64。这种设计允许模型在处理长距离依赖关系时更加灵活，同时减少因位置编码引入的参数量。

● v_head_dim：指value投影中每个注意力头的维度，设置为 128。与query和key的维度设置相辅相成，共同决定了注意力机制的计算精度和表示能力。

● mscale：这是一个缩放因子，用于归一化或初始化时对部分参数进行缩放处理。0.707 接近于 $\frac{1}{\sqrt{2}}$ 的数值，在深度学习中常用于保持梯度稳定和输出均值适中。

4.4.2 中规模版本（236B）的配置

文件config_236B.json是DeepSeek-V3模型的236B版本配置文件，与前面介绍的小规模版本（16B）配置文件config_16B.json相比，新增了n_expert_groups参数，用于定义MoE结构中专家的分组数量。配置信息如下：

```
{
    "vocab_size": 102400,
    "dim": 5120,
    "inter_dim": 12288,
    "moe_inter_dim": 1536,
    "n_layers": 60,
    "n_dense_layers": 1,
    "n_heads": 128,
    "n_routed_experts": 160,
    "n_shared_experts": 2,
    "n_activated_experts": 6,
    "n_expert_groups": 8,
    "n_limited_groups": 3,
    "route_scale": 16.0,
    "q_lora_rank": 1536,
    "kv_lora_rank": 512,
    "qk_nope_head_dim": 128,
    "qk_rope_head_dim": 64,
```

```
    "v_head_dim": 128
}
```

config_236B.json文件定义了DeepSeek-V3 模型 236B 版本的配置参数，这些参数与 16B 版本相比有显著的调整，特别是在模型的隐藏层维度、层数、注意力头数和MoE结构等方面。这些调整使得 236B 版本的模型具有更高的复杂度和更强的表达能力，能够处理更复杂的任务和数据。

4.4.3 大规模版本（671B）的配置

文件config_671B.json是 DeepSeek-V3 模型的 671B 版本配置文件，与前面介绍的中规模版本（236B）配置文件config_236B.json相比，新增了score_func参数，在MoE模型中用于路由机制的评分函数。配置信息如下：

```
{
    "vocab_size": 129280,
    "dim": 7168,
    "inter_dim": 18432,
    "moe_inter_dim": 2048,
    "n_layers": 61,
    "n_dense_layers": 3,
    "n_heads": 128,
    "n_routed_experts": 256,
    "n_shared_experts": 1,
    "n_activated_experts": 8,
    "n_expert_groups": 8,
    "n_limited_groups": 4,
    "route_scale": 2.5,
    "score_func": "sigmoid",
    "q_lora_rank": 1536,
    "kv_lora_rank": 512,
    "qk_nope_head_dim": 128,
    "qk_rope_head_dim": 64,
    "v_head_dim": 128,
    "dtype": "fp8"
}
```

4.5 模型架构

在本开源项目中，文件model.py实现了DeepSeek-V3模型，此模型基于Transformer架构实现，采用了MLA和MoE等创新技术，并引入了MTP。

4.5.1 DeepSeek-V3 模型架构介绍

DeepSeek-V3 不仅整合了传统 Transformer 的核心机制，还通过 MoE 模块和定制的注意力层扩展了模型的容量，提高了灵活性，同时支持分布式训练和低精度计算，为大规模语言模型的高效训练与推理提供了有力保障。DeepSeek-V3 模型的结构和主要组成如下。

1. 基础架构

- Transformer 编码器：DeepSeek-V3 采用多层 Transformer 编码器结构，每层包含自注意力机制和前馈神经网络，用于对输入序列进行深度特征提取。
- 层数和隐藏维度：模型使用 61 层 Transformer，隐藏维度为 7168，提供了强大的表达能力和模型容量。

2. MLA

- 低秩压缩：MLA 通过将注意力键和值压缩成低秩的潜在向量，显著降低了推理过程中的内存占用。
- 查询压缩：同样对注意力查询进行低秩压缩，减少训练过程中的激活内存。
- 旋转位置嵌入（RoPE）：MLA 支持解耦的查询和键，使用 RoPE 来处理位置信息，增强模型对序列位置的敏感性。

3. MoE 架构

- 专家数量和分布：模型包含 14,906 个专家，每层有 257 个专家（1 个共享专家 + 256 个路由专家），实现了模型参数的灵活分配和计算资源的高效利用。
- 路由专家：每个 Token 选择 8 个路由专家进行处理，通过专家的专精化实现知识的深度挖掘，提升模型的推理性能。
- 共享专家：每个 Token 都会使用 1 个共享专家，提供通用知识和稳定性，确保模型在处理多样任务时的鲁棒性。

4. MTP

- 多 token 预测：MTP 机制使模型能够预测每个位置的多个后续 token，增加训练信号的密度，提高数据利用效率，加速模型的训练过程。
- 顺序预测：MTP 采用顺序预测方式，保持每个 token 预测过程中的完整因果关系链，确保生成的序列具有连贯性和逻辑性。

5. 模型参数和训练

- 总参数量：DeepSeek-V3 拥有 671B 总参数，每个 token 激活 37B 参数，庞大的参数量为模型提供了卓越的表达能力和知识储备。
- 训练策略：模型采用 AdamW 优化器，预训练阶段最大序列长度为 4K，在 14.8T token 上进行训练，确保模型能够充分学习到语言的多样性和复杂性。

6. 推理和部署

● 硬件部署：DeepSeek-V3 部署在H800 集群上，采用预填充和解码阶段分离的部署策略，以确保在线服务质量和高吞吐量。

● 负载均衡：在推理阶段，模型采用冗余专家部署策略，确保每个GPU处理相近数量的token，实现负载均衡，提升推理效率。

DeepSeek-V3 模型通过这些创新的技术和架构设计，实现了高效的长序列处理、负载均衡和多token预测，使其在性能和效率上超越了其他开源模型，并接近领先的闭源模型。

4.5.2 配置信息

（1）下面代码定义了DeepSeek-V3 模型的初始参数环境和硬件实现方式：

```
world_size = 1
rank = 0
block_size = 128
gemm_impl: Literal["bf16", "fp8"] = "bf16"
attn_impl: Literal["naive", "absorb"] = "absorb"
```

对上述代码的具体说明如下所示。

● world_size = 1：指定单个GPU的运行环境，表示程序在单一GPU上执行。

● rank = 0：定义当前的GPU或进程标识符，0 表示主节点或唯一的计算资源。

● block_size = 128：设置块尺寸为 128，用于优化计算性能，通过块级别的数据分解来加速矩阵运算，提升内存访问效率。

● gemm_impl: Literal["bf16", "fp8"] = "bf16"：配置矩阵乘法的默认实现为脑浮点 16 位，以低精度的数值格式进行数值计算。

● attn_impl: Literal["naive", "absorb"] = "absorb"：选择注意力机制的高效实现策略，absorb表示采用特定的优化方法来提升注意力计算的性能和稳定性。

（2）实现数据类ModelArgs，用于定义模型的参数和超参数。ModelArgs类包含了模型配置所需的各种属性，如最大批处理大小、最大序列长度、数据类型、词汇表大小、模型维度、层数等。这些参数用于控制模型的结构和行为，以适应不同的任务需求。

```
class ModelArgs:
    """
    定义模型参数和超参数的数据类。

    属性：
        max_batch_size (int): 最大批处理大小。
        max_seq_len (int): 最大序列长度。
        dtype (Literal["bf16", "fp8"]): 计算使用的数据类型。
        vocab_size (int): 词汇表大小。
```

```
        dim (int): 模型维度。
        inter_dim (int): MLP 层的中间维度。
        moe_inter_dim (int): MoE 层的中间维度。
        n_layers (int): Transformer 层数。
        n_dense_layers (int): 模型中的稠密层数。
        n_heads (int): 注意力头数。
        n_routed_experts (int): MoE 层的路由专家数。
        n_shared_experts (int): MoE 层的共享专家数。
        n_activated_experts (int): MoE 层中激活的专家数。
        n_expert_groups (int): 专家组数。
        n_limited_groups (int): MoE 路由的受限组数。
        score_func (Literal["softmax", "sigmoid"]): MoE 路由的评分函数。
        route_scale (float): 路由分数的缩放因子。
        q_lora_rank (int): 查询投影的 LoRA 等级。
        kv_lora_rank (int): 键值投影的 LoRA 等级。
        qk_nope_head_dim (int): 无位置嵌入的查询 - 键投影维度。
        qk_rope_head_dim (int): 旋转位置嵌入的查询 - 键投影维度。
        v_head_dim (int): 值投影的维度。
        original_seq_len (int): 原始序列长度。
        rope_theta (float): 旋转位置编码的基数。
        rope_factor (float): 扩展序列长度的缩放因子。
        beta_fast (int): 快速 beta 校正因子。
        beta_slow (int): 慢速 beta 校正因子。
        mscale (float): 扩展注意力的缩放因子。
    """
    max_batch_size: int = 8
    max_seq_len: int = 4096 * 4
    dtype: Literal["bf16", "fp8"] = "bf16"
    vocab_size: int = 102400
    dim: int = 2048
    inter_dim: int = 10944
    moe_inter_dim: int = 1408
    n_layers: int = 27
    n_dense_layers: int = 1
    n_heads: int = 16
    # moe
    n_routed_experts: int = 64
    n_shared_experts: int = 2
    n_activated_experts: int = 6
    n_expert_groups: int = 1
    n_limited_groups: int = 1
    score_func: Literal["softmax", "sigmoid"] = "softmax"
    route_scale: float = 1.
    # mla
    q_lora_rank: int = 0
    kv_lora_rank: int = 512
```

```
    qk_nope_head_dim: int = 128
    qk_rope_head_dim: int = 64
    v_head_dim: int = 128
    # yarn
    original_seq_len: int = 4096
    rope_theta: float = 10000.0
    rope_factor: float = 40.
    beta_fast: int = 32
    beta_slow: int = 1
    mscale: float = 1.
```

这些参数的设置对于模型的性能和适用性至关重要。例如，max_seq_len决定了模型可以处理的最大序列长度，vocab_size决定了模型的词汇表大小，dim和inter_dim等参数则影响模型的容量和计算需求。通过调整这些参数，可以优化模型以满足特定任务的需求。

4.5.3 并行嵌入

类ParallelEmbedding是一个并行嵌入层，用于在分布式训练中分割词汇表（vocab_size）并进行嵌入操作。它将词汇表根据世界大小（world_size）分割成多个部分，每个部分由不同的进程处理。这样可以有效减少单个进程的内存占用，并提高训练效率。通过给每个进程分配较小的权重矩阵，在训练期间，可以利用分布式训练框架（如PyTorch的DistributedDataParallel）来同步梯度和参数更新，从而实现高效的并行计算。

```
class ParallelEmbedding(nn.Module):
    """
    支持分布式进程的并行嵌入层。

    参数:
        vocab_size (int): 词汇表大小。
        dim (int): 嵌入维度。
    """
    def __init__(self, vocab_size: int, dim: int):
        super().__init__()
        self.vocab_size = vocab_size
        self.dim = dim
        assert vocab_size % world_size == 0, f"词汇表大小必须能被世界大小整除
                                              (world_size={world_size})"
        self.part_vocab_size = (vocab_size // world_size)
        self.vocab_start_idx = rank * self.part_vocab_size
        self.vocab_end_idx = self.vocab_start_idx + self.part_vocab_size
        self.weight = nn.Parameter(torch.empty(self.part_vocab_size, self.dim))

    def forward(self, x: torch.Tensor) -> torch.Tensor:
        """
```

```
并行嵌入层的前向传播。

参数:
    x (torch.Tensor): 包含标记索引的输入张量。

返回:
    torch.Tensor: 嵌入表示。

异常:
    ValueError: 如果未定义 `world_size`。
"""
if world_size > 1:
    mask = (x < self.vocab_start_idx) | (x >= self.vocab_end_idx)
    x = x - self.vocab_start_idx
    x[mask] = 0
y = F.embedding(x, self.weight)
if world_size > 1:
    y[mask] = 0
    dist.all_reduce(y)
return y
```

上述代码定义了一个用于分布式训练的嵌入层，通过将词汇表分割为多个部分，每个部分由不同的进程处理，从而提高了训练效率。对上述代码的具体说明如下所示。

- 初始化：在构造函数中，首先验证词汇表大小是否能被世界大小（world_size）整除，以确保每个进程处理的词汇量相等。然后根据当前进程的排名（rank），确定该进程负责的词汇范围，并为该范围内的词汇创建嵌入权重。
- 前向传播：在前向传播过程中，首先根据当前进程的词汇范围对输入的索引进行筛选和调整。然后使用PyTorch的嵌入函数（F.embedding）获取对应的嵌入表示。若存在多个进程，还需要对结果进行归约操作，以确保每个进程都能访问到完整的嵌入表示。

4.5.4 线性变换

定义函数linear()，用于对输入张量x执行线性变换：$y = xA^\wedge T + b$。该函数根据权重张量weight的量化状态和全局变量gemm_impl的值，选择不同的计算路径。如果weight的元素大小大于1，则直接使用PyTorch的线性函数F.linear进行计算。如果gemm_impl的值为 "bf16"，则对权重进行去量化处理，并使用F.linear进行计算。否则，首先对输入张量x进行量化处理，然后使用fp8_gemm函数进行计算。 如果提供了偏置张量bias，则将其添加到结果中。

```
def linear(x: torch.Tensor, weight: torch.Tensor, bias: Optional[torch.
        Tensor] = None) -> torch.Tensor:
    """
    对传入的数据应用线性变换: y = xA^T + b。
```

```
    该函数根据量化和张量格式的不同，支持特定的实现方式。

    参数：
        x (torch.Tensor)：输入张量。
        weight (torch.Tensor)：权重张量。可能是量化的，需要在某些情况下进行去量化处理。
        bias (Optional[torch.Tensor])：要添加的偏置张量。默认为 None。

    返回：
        torch.Tensor：线性变换的结果，可能涉及量化感知计算，具体取决于输入参数。

    注意：
        - 如果 `weight` 是量化的（例如，`element_size()` 等于 1），则使用去量化版本
          进行计算。
        - 如果 `gemm_impl` 等于 `"bf16"`,则进行去量化处理,并使用 `bf16` 的GEMM操作。
        - 对于其他情况，函数对 `x` 进行量化处理，并使用 `fp8_gemm` 进行计算。
    """
    if weight.element_size() > 1:
        return F.linear(x, weight, bias)
    elif gemm_impl == "bf16":
        weight = weight_dequant(weight, weight.scale)
        return F.linear(x, weight, bias)
    else:
        x, scale = act_quant(x, block_size)
        y = fp8_gemm(x, scale, weight, weight.scale)
        if bias is not None:
            y += bias
        return y
```

4.5.5 线性层

在DeepSeek-V3模型中，Linear是标准的线性层，用于实现线性变换。ColumnParallelLinear和RowParallelLinear是基于Linear的扩展类，分别实现了列并行和行并行的线性变换。

（1）下面的代码定义了一个自定义的线性层Linear，该层支持量化权重和可选的偏置项。在初始化方法 __init__ 中，首先定义了输入特征数in_features和输出特征数out_features。权重weight被初始化为一个形状为 (out_features, in_features) 的张量，并根据指定的数据类型dtype进行初始化。如果权重的元素大小为1（量化权重），则计算输出特征和输入特征的块大小，并初始化相应的缩放因子scale。 如果指定了偏置项，则初始化偏置bias。在前向传播方法forward中，调用之前定义的linear函数，传入输入张量x、权重weight和偏置bias，返回线性变换后的结果。

```
class Linear(nn.Module):
    """
    自定义线性层，支持量化权重和可选的偏置项。
```

```
    参数:
        in_features (int): 输入特征数。
        out_features (int): 输出特征数。
        bias (bool): 是否包含偏置项。默认为 False。
        dtype (optional): 层的数据类型。默认为 `torch.bfloat16`。
    """
    dtype = torch.bfloat16

    def __init__(self, in_features: int, out_features: int, bias: bool =
                 False, dtype = None):
        super().__init__()
        self.in_features = in_features
        self.out_features = out_features
        self.weight = nn.Parameter(torch.empty(out_features, in_features,
                                   dtype=dtype or Linear.dtype))
        if self.weight.element_size() == 1:
            scale_out_features = (out_features + block_size - 1) // block_size
            scale_in_features = (in_features + block_size - 1) // block_size
            self.weight.scale =
                self.scale =
                nn.Parameter(torch.empty(scale_out_features,
                             scale_in_features, dtype=torch.float32))
        else:
            self.register_parameter("scale", None)
        if bias:
            self.bias = nn.Parameter(torch.empty(out_features))
        else:
            self.register_parameter("bias", None)

    def forward(self, x: torch.Tensor) -> torch.Tensor:
        """
        自定义线性层的前向传播。

        参数:
            x (torch.Tensor): 输入张量。

        返回:
            torch.Tensor: 经过线性变换后的张量。
        """
        return linear(x, self.weight, self.bias)
```

（2）类ColumnParallelLinear是DeepSeek-V3 模型中的一个线性层，支持列并行，即将输出特征分割到多个分布式进程中。类ColumnParallelLinear继承自Linear类，并重写了初始化方法和前向传播方法，以确保输出特征能够在分布式环境中被均匀分割。

```
class ColumnParallelLinear(Linear):
    """
    列并行线性层，将输出特征分割到分布式进程中。

    参数:
        in_features (int): 输入特征数。
        out_features (int): 输出特征总数。
        bias (bool): 是否包含偏置项。默认为 False。
        dtype (optional): 层的数据类型。默认为 `torch.bfloat16`。
    """
    def __init__(self, in_features: int, out_features: int, bias: bool =
                 False, dtype = None):
        # 确保输出特征能够被世界大小整除
        assert out_features % world_size == 0, f"输出特征必须是世界大小的倍数
                                                (world_size={world_size})"
        self.part_out_features = out_features // world_size
        # 调用父类的初始化方法，设置部分输出特征
        super().__init__(in_features, self.part_out_features, bias, dtype)

    def forward(self, x: torch.Tensor) -> torch.Tensor:
        """
        列并行线性层的前向传播。

        参数:
            x (torch.Tensor): 输入张量。

        返回:
            torch.Tensor: 经过列并行计算后的张量。
        """
        y = linear(x, self.weight, self.bias)
        return y
```

类ColumnParallelLinear通过列并行技术，将线性层的输出特征分割到不同的进程中，从而实现分布式计算，提高训练和推理的效率。

（3）类RowParallelLinear是PyTorch中的自定义线性层，实现了行并行。它将输入特征分割到多个分布式进程中，从而允许在分布式环境下高效地计算线性变换。该类继承了Linear的属性和方法，并在初始化和前向传播过程中进行了扩展。

```
class RowParallelLinear(Linear):
    """
    具有行并行支持的线性层，将输入特征在分布式进程之间拆分。

    参数:
        in_features (int): 输入特征的总数量。
        out_features (int): 输出特征的数量。
```

```
        bias (bool): 是否包含偏置项。默认为 False。
        dtype (optional): 层的数值类型。默认为 `torch.bfloat16`。
    """
    def __init__(self, in_features: int, out_features: int, bias: bool =
                False, dtype=None):
        assert in_features % world_size == 0, f"输入特征数量必须能被世界大小整除
                                    (world_size={world_size})"
        self.part_in_features = in_features // world_size
        super().__init__(self.part_in_features, out_features, bias, dtype)

    def forward(self, x: torch.Tensor) -> torch.Tensor:
        """
        具有行并行支持的线性层的前向传播。

        参数:
            x (torch.Tensor): 输入张量。

        返回:
            torch.Tensor: 经行并行计算后的转换张量。
        """
        y = linear(x, self.weight)
        if world_size > 1:
            dist.all_reduce(y)
        if self.bias is not None:
            y += self.bias
        return y
```

4.5.6 RMSNorm（均方根层归一化）

RMSNorm是对传统LayerNorm的改进，旨在降低计算开销，同时保持模型性能。与LayerNorm需要计算均值和方差不同，RMSNorm仅需计算输入的均方根，从而降低了计算复杂度。在RMSNorm中，输入张量的每个元素除以其均方根值，并乘以可学习的缩放参数weight。这种方法保留了缩放不变性，舍弃了平移不变性。在前向传播中，RMSNorm对输入张量进行归一化处理，输出与输入形状相同的张量。这种归一化方法在训练深度神经网络时有助于提高稳定性和收敛速度。

在下面的实现代码中，类RMSNorm继承自nn.Module，包含一个可学习的参数weight，用于对归一化后的输入进行缩放。在前向传播方法中，使用F.rms_norm函数对输入张量进行归一化处理。这种实现方式使得RMSNorm可以方便地集成到PyTorch模型中。

```
class RMSNorm(nn.Module):
    """
    均方根层归一化（RMSNorm）。

    参数:
```

```
        dim (int): 输入张量的维度。
        eps (float): 用于数值稳定性的 epsilon 值。默认为 1e-6。
    """
    def __init__(self, dim: int, eps: float = 1e-6):
        super().__init__()
        self.dim = dim
        self.eps = eps
        self.weight = nn.Parameter(torch.ones(dim))

    def forward(self, x: torch.Tensor):
        """
        RMSNorm 的前向传播。

        参数:
            x (torch.Tensor): 输入张量。

        返回:
            torch.Tensor: 与输入形状相同的归一化张量。
        """
        return F.rms_norm(x, (self.dim,), self.weight, self.eps)
```

注意

（1）RMSNorm主要用于替代LayerNorm，在某些模型中可能会带来性能提升。然而，具体效果可能因任务和模型架构而异。因此，在实际应用中，建议根据具体情况进行实验和评估。

（2）RMSNorm并非DeepSeek-V3 模型的独创，它最早由Biao Zhu和Rico Sennrich在 2019 年提出，旨在简化LayerNorm的计算过程，降低计算复杂度，同时保持或提升模型性能。DeepSeek-V3 模型用RMSNorm替代传统的LayerNorm，仅使用均方根统计进行输入缩放，这种简化设计不仅减少了计算量，还提升了训练稳定性。此外，RMSNorm在其他模型（如LLaMA）中也有应用，进一步证明了其通用性和有效性。

4.5.7 RoPE 计算

在DeepSeek-V3模型中，RoPE用于将位置信息融入到输入序列的表示中。与传统的绝对位置编码不同，RoPE通过对嵌入向量应用旋转变换，使模型能够捕捉序列中元素的相对位置信息。这种方法提高了模型对相对位置的敏感性，有助于更好地理解序列数据的结构和顺序。

（1）函数precompute_freqs_cis()用于预计算RoPE所需的频率复数指数值。RoPE是一种将相对位置信息集成到自注意力机制中的位置编码方法，能够提升Transformer架构的性能。这些值在后续的注意力机制中用于对查询和键进行旋转操作，以引入位置信息。

```
def precompute_freqs_cis(args: ModelArgs) -> torch.Tensor:
    """
```

```
    预先计算旋转位置嵌入的频率相关复指数值。

    参数：
        args (ModelArgs): 包含位置嵌入参数的模型参数。

    返回：
        torch.Tensor: 预先计算的用于位置嵌入的复指数值。
    """
    dim = args.qk_rope_head_dim    # 嵌入维度
    seqlen = args.max_seq_len      # 最大序列长度
    beta_fast = args.beta_fast     # 快速旋转的 beta 参数
    beta_slow = args.beta_slow     # 慢速旋转的 beta 参数
    base = args.rope_theta         # 旋转位置编码的基数
    factor = args.rope_factor      # 旋转位置编码的因子

    def find_correction_dim(num_rotations, dim, base, max_seq_len):
        """
        计算旋转位置嵌入中给定旋转次数的校正维度。

        参数：
            num_rotations (float): 要计算校正的旋转次数。
            dim (int): 嵌入空间的维度。
            base (float): 指数计算的基数。
            max_seq_len (int): 最大序列长度。

        返回：
            float: 基于输入参数的校正维度。
        """
        return dim * math.log(max_seq_len / (num_rotations * 2 * math.pi)) / (2
                            * math.log(base))

    def find_correction_range(low_rot, high_rot, dim, base, max_seq_len):
        """
        计算旋转位置嵌入的校正维度范围。

        参数：
            low_rot (float): 旋转次数的下限。
            high_rot (float): 旋转次数的上限。
            dim (int): 嵌入空间的维度。
            base (float): 指数计算的基数。
            max_seq_len (int): 最大序列长度。

        返回：
            Tuple[int, int]: 校正维度的范围（低，高），限制在有效索引范围内。
        """
        low = math.floor(find_correction_dim(low_rot, dim, base, max_seq_len))
```

```
        high = math.ceil(find_correction_dim(high_rot, dim, base, max_seq_len))
        return max(low, 0), min(high, dim-1)

    def linear_ramp_factor(min, max, dim):
        """
        计算用于在最小值和最大值之间平滑值的线性坡道函数。

        参数：
            min (float): 坡道函数的最小值。
            max (float): 坡道函数的最大值。
            dim (int): 坡道张量的维度。

        返回：
            torch.Tensor: 形状为 (dim,) 的张量，值在 0 和 1 之间线性插值，
                并限制在 [0, 1] 范围内。
        """
        if min == max:
            max += 0.001
        linear_func = (torch.arange(dim, dtype=torch.float32) - min) / (max - min)
        ramp_func = torch.clamp(linear_func, 0, 1)
        return ramp_func

    freqs = 1.0 / (base ** (torch.arange(0, dim, 2, dtype=torch.float32) / dim))
    if seqlen > args.original_seq_len:
        low, high = find_correction_range(beta_fast, beta_slow, dim, base,
                                          args.original_seq_len)
        smooth = 1 - linear_ramp_factor(low, high, dim // 2)
        freqs = freqs / factor * (1 - smooth) + freqs * smooth

    t = torch.arange(seqlen)
    freqs = torch.outer(t, freqs)
    freqs_cis = torch.polar(torch.ones_like(freqs), freqs)
    return freqs_cis
```

函数precompute_freqs_cis()首先根据模型参数计算频率，然后生成一个时间序列，并计算频率和时间的外积。最后使用这些值计算复数形式的旋转位置嵌入。

（2）函数apply_rotary_emb()用于将RoPE应用到输入张量上。通过将输入张量与预计算的复数指数值相乘，为模型引入位置信息，提高模型对序列中位置关系的感知能力。

```
def apply_rotary_emb(x: torch.Tensor, freqs_cis: torch.Tensor) -> torch.
Tensor:
    """
    将旋转位置嵌入应用到输入张量。

    参数：
```

```
        x (torch.Tensor): 要应用位置嵌入的输入张量。
        freqs_cis (torch.Tensor): 预计算的位置嵌入复数指数值。

    返回:
        torch.Tensor: 应用了旋转嵌入的张量。
    """
    dtype = x.dtype
    x = torch.view_as_complex(x.float().view(*x.shape[:-1], -1, 2))
    freqs_cis = freqs_cis.view(1, x.size(1), 1, x.size(-1))
    y = torch.view_as_real(x * freqs_cis).flatten(3)
    return y.to(dtype)
```

函数apply_rotary_emb()首先将输入张量x转换为复数表示，然后与预先计算的旋转位置嵌入freqs_cis相乘，最后将结果转换回实数表示并返回。这种方法有效地将位置信息融入输入张量中，有助于模型更好地捕捉序列数据的位置信息。

4.5.8 多头注意力层

类MLA实现了DeepSeek-V3模型中的多头注意力层，负责处理输入序列的注意力机制。MLA层通过低秩联合压缩技术，减少了推理时的键值缓存，从而在保持性能的同时显著降低了内存占用。MLA层支持多种注意力头的并行计算，能够捕捉输入序列中的不同特征，并通过RoPE提高模型对位置信息的感知能力。

```
class MLA(nn.Module):
    """
    多头注意力层（MLA）。

    属性：
        dim (int): 输入特征的维度。
        n_heads (int): 注意力头的数量。
        n_local_heads (int): 分布式系统中本地注意力头的数量。
        q_lora_rank (int): 查询低秩投影的秩。
        kv_lora_rank (int): 键/值低秩投影的秩。
        qk_nope_head_dim (int): 非位置查询/键投影的维度。
        qk_rope_head_dim (int): 旋转位置查询/键投影的维度。
        qk_head_dim (int): 查询/键投影的总维度。
        v_head_dim (int): 值投影的维度。
        softmax_scale (float): 在注意力计算中用于 softmax 的缩放因子。
    """
    def __init__(self, args: ModelArgs):
        super().__init__()
        self.dim = args.dim
        self.n_heads = args.n_heads
        self.n_local_heads = args.n_heads // world_size
```

```
        self.q_lora_rank = args.q_lora_rank
        self.kv_lora_rank = args.kv_lora_rank
        self.qk_nope_head_dim = args.qk_nope_head_dim
        self.qk_rope_head_dim = args.qk_rope_head_dim
        self.qk_head_dim = args.qk_nope_head_dim + args.qk_rope_head_dim
        self.v_head_dim = args.v_head_dim

        if self.q_lora_rank == 0:
            self.wq = ColumnParallelLinear(self.dim, self.n_heads * self.qk_
                                           head_dim)
        else:
            self.wq_a = Linear(self.dim, self.q_lora_rank)
            self.q_norm = RMSNorm(self.q_lora_rank)
            self.wq_b = ColumnParallelLinear(self.q_lora_rank, self.n_heads
                                             * self.qk_head_dim)
        self.wkv_a = Linear(self.dim, self.kv_lora_rank + self.qk_rope_head_
                            dim)
        self.kv_norm = RMSNorm(self.kv_lora_rank)
        self.wkv_b = ColumnParallelLinear(self.kv_lora_rank, self.n_heads *
                                          (self.qk_nope_head_dim +
                                          self.v_head_dim))
        self.wo = RowParallelLinear(self.n_heads * self.v_head_dim, self.dim)
        self.softmax_scale = self.qk_head_dim ** -0.5
        if args.max_seq_len > args.original_seq_len:
            mscale = 0.1 * args.mscale * math.log(args.rope_factor) + 1.0
            self.softmax_scale = self.softmax_scale * mscale * mscale

        if attn_impl == "naive":
            self.register_buffer("k_cache", torch.zeros(args.max_batch_size,
                args.max_seq_len, self.n_local_heads, self.qk_head_dim),
                persistent=False)
            self.register_buffer("v_cache", torch.zeros(args.max_batch_size,
                args.max_seq_len, self.n_local_heads, self.v_head_dim),
                persistent=False)
        else:
            self.register_buffer("kv_cache", torch.zeros(args.max_batch_size,
                args.max_seq_len, self.kv_lora_rank), persistent=False)
            self.register_buffer("pe_cache", torch.zeros(args.max_batch_size,
                args.max_seq_len, self.qk_rope_head_dim), persistent=False)

    def forward(self, x: torch.Tensor, start_pos: int, freqs_cis: torch.
        Tensor, mask: Optional[torch.Tensor]):
        """
        多头注意力层（MLA）的前向传播。
```

```
参数：
    x (torch.Tensor): 形状为 (batch_size, seq_len, dim) 的输入张量。
    start_pos (int): 用于缓存的序列起始位置。
    freqs_cis (torch.Tensor): 预计算的旋转嵌入的复数指数值。
    mask (Optional[torch.Tensor]): 用于在注意力中排除特定位置的掩码张量。

返回：
    torch.Tensor: 与输入形状相同的输出张量。
"""
bsz, seqlen, _ = x.size()
end_pos = start_pos + seqlen
if self.q_lora_rank == 0:
    q = self.wq(x)
else:
    q = self.wq_b(self.q_norm(self.wq_a(x)))
q = q.view(bsz, seqlen, self.n_local_heads, self.qk_head dim)
q_nope, q_pe = torch.split(q, [self.qk_nope_head_dim, self.qk_rope_
                           head_dim], dim=-1)
q_pe = apply_rotary_emb(q_pe, freqs_cis)
kv = self.wkv_a(x)
kv, k_pe = torch.split(kv, [self.kv_lora_rank, self.qk_rope_head_
                        dim], dim=-1)
k_pe = apply_rotary_emb(k_pe.unsqueeze(2), freqs_cis)
if attn_impl == "naive":
    q = torch.cat([q_nope, q_pe], dim=-1)
    kv = self.wkv_b(self.kv_norm(kv))
    kv = kv.view(bsz, seqlen, self.n_local_heads,
                 self.qk_nope_head_dim + self.v_head_dim)
    k_nope, v = torch.split(kv, [self.qk_nope_head_dim, self.v_head_
                            dim], dim=-1)
    k = torch.cat([k_nope, k_pe.expand(-1, -1, self.n_local_heads,
                  -1)], dim=-1)
    self.k_cache[:bsz, start_pos:end_pos] = k
    self.v_cache[:bsz, start_pos:end_pos] = v
    scores = torch.einsum("bshd,bthd->bsht", q, self.k_cache[:bsz,
                          :end_pos]) * self.softmax_scale
else:
    wkv_b = self.wkv_b.weight if self.wkv_b.scale is None else
            weight_dequant(self.wkv_b.weight, self.wkv_b.scale,
            block_size)
    wkv_b = wkv_b.view(self.n_local_heads, -1, self.kv_lora_rank)
    q_nope = torch.einsum("bshd,hdc->bshc", q_nope, wkv_b[:, :self.
                          qk_nope_head_dim])
    self.kv_cache[:bsz, start_pos:end_pos]
::contentReference[oaicite:0]{index=0}
```

4.5.9 多层感知器

多层感知器（Multilayer Perceptron，MLP）类实现了一个前馈神经网络层，用于实现神经网络中的非线性特征变换。类MLP包含三个线性层：w1、w2 和w3，其中w1 和w3 将输入特征映射到隐藏层空间，w2 将隐藏层特征映射回原始维度。在前向传播过程中，输入首先经过w1 线性变换，然后应用SiLU激活函数，并与w3 的输出逐元素相乘，最后通过w2 线性变换得到输出。

```
class MLP(nn.Module):
    """
    多层感知器（MLP），用作前馈层。

    属性：
        w1 (nn.Module): 输入隐藏层的线性变换层。
        w2 (nn.Module): 隐藏层到输出的线性变换层。
        w3 (nn.Module): 额外的特征变换线性层。
    """
    def __init__(self, dim: int, inter_dim: int):
        """
        初始化 MLP 层。

        参数：
            dim (int): 输入和输出的维度。
            inter_dim (int): 隐藏层的维度。
        """
        super().__init__()
        self.w1 = ColumnParallelLinear(dim, inter_dim)
        self.w2 = RowParallelLinear(inter_dim, dim)
        self.w3 = ColumnParallelLinear(dim, inter_dim)

    def forward(self, x: torch.Tensor) -> torch.Tensor:
        """
        MLP 层的前向传播。

        参数：
            x (torch.Tensor): 输入张量。

        返回：
            torch.Tensor: 经过 MLP 计算后的输出张量。
        """
        return self.w2(F.silu(self.w1(x)) * self.w3(x))
```

在上述代码中，ColumnParallelLinear和RowParallelLinear是自定义的线性层，用于在分布式训练中实现列并行和行并行，以提高计算效率。在前向传播函数中，输入x首先经过w1 线性变换，然后应用SiLU激活函数（F.silu），并与w3 的输出逐元素相乘，最后通过w2 线性变换得到最终输出。

4.5.10 DeepSeek-V3 中的 MoE 架构实现

DeepSeek-V3 是一款采用MoE架构的语言模型，总参数量达到 6710 亿（671B），但每个输入仅激活约 370 亿（37B）参数进行计算。这种设计使模型在保持高性能的同时，显著降低了计算成本。MoE架构的核心思想是将模型划分为多个专注于特定任务的子网络（专家），通过门控机制为每个输入选择最相关的专家，从而提高模型的效率和专业化程度。

在DeepSeek-V3的实现中，MoE架构模块由以下主要组件组成。

（1）门控机制（Gate）。

● 作用：Gate类负责将输入样本路由到最合适的专家。它通过计算输入样本与各个专家的兼容性分数，确定输入应被路由到哪些专家，从而实现专家网络的稀疏激活。

● 原理：门控机制根据输入特征和专家权重计算路由分数，并通过评分函数（如Softmax或Sigmoid）进行归一化处理。如果有多个分组，会进一步限制输入被路由到部分分组。最终返回路由权重和选中的专家索引。

（2）专家层（Expert）。

● 作用：Expert类是MoE模型中的专家层，每个专家包含三个线性层，用于输入隐藏层的变换、隐藏层到输出层的变换及特征变换。它实现了专家层的前向传播，通过非线性变换对输入数据进行处理，提高模型的表达能力。

● 结构：专家层包含输入隐藏层的线性变换、隐藏层到输出层的线性变换及特征变换的线性层。输入张量通过这些线性层和SiLU激活函数进行处理，产生专家的输出。

（3）混合专家模型（MoE）。

● 作用：MoE类是混合专家模型的实现，负责将输入样本路由到不同的专家进行处理，并将专家的输出进行整合。通过路由机制和专家网络的组合，实现高效的特征提取和知识分割。

● 原理：MoE模型首先通过门控机制确定输入样本应路由到的专家，然后将输入样本分配到相应的专家进行处理。每个专家的输出按照路由权重进行加权累加，并与共享专家模块的输出合并，产生最终的输出。

（4）工作流程。

● 输入路由：输入样本通过门控机制被路由到最合适的专家。门控机制根据输入特征和专家权重计算路由分数，并选择得分最高的专家。

● 专家计算：被选中的专家对输入样本进行非线性变换，通过线性层和激活函数提取特征。

● 输出整合：专家的输出按照路由权重进行加权累加，并与共享专家模块的输出合并，产生最终的输出。

（5）优势。

● 稀疏激活：通过门控机制实现专家的稀疏激活，减少了计算量和内存占用。

● 并行计算：专家层可以在分布式系统中并行计算，提高了计算效率。

● 知识分割：每个专家专注于处理特定类型的输入数据，实现了知识的分割和专业化。

总之，DeepSeek-V3 的MoE模型通过以上组件和机制，实现了高效的分布式计算和大规模知识的存储，提高了模型的性能和可扩展性。接下来将详细讲解DeepSeek-V3 的MoE模型的实现过程。

1. 路由选择控制

类Gate实现了一个用于MoE模型的门控机制。在MoE模型中，输入数据通过门控网络被路由到不同的专家网络，以实现模型容量的扩展和计算效率的提升。该类根据输入特征计算得分，并选择最适合的专家进行处理，从而实现高效的路由和计算。

```
class Gate(nn.Module):
    """
    专家模型（MoE）中用于路由输入的门控机制。

    属性：
        dim (int): 输入特征的维度。
        topk (int): 每个输入激活的顶级专家数量。
        n_groups (int): 路由的组数。
        topk_groups (int): 路由输入的组数。
        score_func (str): 评分函数（'softmax' 或 'sigmoid'）。
        route_scale (float): 路由权重的缩放因子。
        weight (torch.nn.Parameter): 门控的可学习权重。
        bias (Optional[torch.nn.Parameter]): 门控的可选偏置项。
    """
    def __init__(self, args: ModelArgs):
        """
        初始化 Gate 模块。

        参数：
            args (ModelArgs): 包含门控参数的模型参数。
        """
        super().__init__()
        self.dim = args.dim
        self.topk = args.n_activated_experts
        self.n_groups = args.n_expert_groups
        self.topk_groups = args.n_limited_groups
        self.score_func = args.score_func
        self.route_scale = args.route_scale
        self.weight = nn.Parameter(torch.empty(args.n_routed_experts, args.dim))
        self.bias = nn.Parameter(torch.empty(args.n_routed_experts)) if
                              self.dim == 7168 else None

    def forward(self, x: torch.Tensor) -> Tuple[torch.Tensor, torch.Tensor]:
        """
        门控机制的前向传播。
```

```
        参数：
            x (torch.Tensor): 输入张量。

        返回：
            Tuple[torch.Tensor, torch.Tensor]: 路由权重和选择的专家索引。
        """
        scores = linear(x, self.weight)
        if self.score_func == "softmax":
            scores = scores.softmax(dim=-1, dtype=torch.float32)
        else:
            scores = scores.sigmoid()
        original_scores = scores
        if self.bias is not None:
            scores = scores + self.bias
        if self.n_groups > 1:
            scores = scores.view(x.size(0), self.n_groups, -1)
            if self.bias is None:
                group_scores = scores.amax(dim=-1)
            else:
                group_scores = scores.topk(2, dim=-1)[0].sum(dim=-1)
            indices = group_scores.topk(self.topk_groups, dim=-1)[1]
            mask = torch.zeros_like(scores[..., 0]).scatter_(1, indices, True)
            scores = (scores * mask.unsqueeze(-1)).flatten(1)
        indices = torch.topk(scores, self.topk, dim=-1)[1]
        weights = original_scores.gather(1, indices)
        if self.score_func == "sigmoid":
            weights /= weights.sum(dim=-1, keepdim=True)
        weights *= self.route_scale
        return weights.type_as(x), indices
```

对上述代码的具体说明如下所示。

- __init__ 方法：初始化门控机制，设置输入特征的维度、激活的专家数量、评分函数、路由权重的缩放因子等参数。权重和偏置项是可学习的参数，用于计算路由分数。
- forward方法：计算输入样本与专家的兼容性分数，通过评分函数进行归一化处理，然后根据分数选择得分最高的专家。如果有多个分组，会进一步限制输入被路由到部分分组。最终返回路由权重和选中的专家索引。

2. 专家层

类Expert用于实现MoE模型中的专家层。在MoE模型中，多个专家网络共同处理输入数据，每个专家专注于特定的子任务。Expert类实现了一个包含三个线性层的神经网络，用于对输入数据进行特定的变换和处理。

```
class Expert(nn.Module):
    """
    专家层，用于混合专家模型。

    属性：
        w1 (nn.Module): 输入隐藏层的线性变换层。
        w2 (nn.Module): 隐藏层到输出的线性变换层。
        w3 (nn.Module): 额外的线性变换层，用于特征变换。
    """
    def __init__(self, dim: int, inter_dim: int):
        """
        初始化专家层。

        参数：
            dim (int): 输入和输出的维度。
            inter_dim (int): 隐藏层的维度。
        """
        super().__init__()
        self.w1 = Linear(dim, inter_dim)
        self.w2 = Linear(inter_dim, dim)
        self.w3 = Linear(dim, inter_dim)

    def forward(self, x: torch.Tensor) -> torch.Tensor:
        """
        专家层的前向传播。

        参数：
            x (torch.Tensor): 输入张量。

        返回：
            torch.Tensor: 经过专家层计算后的输出张量。
        """
        return self.w2(F.silu(self.w1(x)) * self.w3(x))
```

在上述代码中，类Expert包含三个线性层：w1、w2 和w3。它们是专家层（Expert）中的三个线性变换层，也是nn.Module的子模块。这些线性层用于对输入数据进行特定的线性变换，以实现专家层的前向传播。在前向传播过程中，输入x首先经过线性变换w1，然后应用SiLU激活函数。同时，输入x也经过线性变换w3。这两个结果相乘后，再经过线性变换w2，最终得到输出。这种结构使得专家层能够对输入数据进行复杂的特征变换，从而在MoE模型中发挥特定的作用。

3. MoE 结构

类MoE实现了混合专家模型的结构。在这种模型中，输入数据通过一个门控机制被路由到多个专家网络中的一个或多个，以实现条件计算。该模块包含一个门控机制和一组专家网络，每个专家网络都可以独立处理输入数据。

```
class MoE(nn.Module):
    """
    混合专家 (MoE) 模块。

    属性:
        dim (int): 输入特征的维度。
        n_routed_experts (int): 模型中的专家总数。
        n_local_experts (int): 分布式系统中本地处理的专家数量。
        n_activated_experts (int): 每个输入激活的专家数量。
        gate (nn.Module): 用于将输入路由到专家的路由机制。
        experts (nn.ModuleList): 专家模块列表。
        shared_experts (nn.Module): 应用于所有输入的共享专家。
    """
    def __init__(self, args: ModelArgs):
        """
        初始化 MoE 模块。

        参数:
            args (ModelArgs): 包含 MoE 参数的模型参数。
        """
        super().__init__()
        self.dim = args.dim
        assert args.n_routed_experts % world_size == 0, f"专家数量必须是世界大
            小的倍数(world_size={world_size})"
        self.n_routed_experts = args.n_routed_experts
        self.n_local_experts = args.n_routed_experts // world_size
        self.n_activated_experts = args.n_activated_experts
        self.experts_start_idx = rank * self.n_local_experts
        self.experts_end_idx = self.experts_start_idx + self.n_local_experts
        self.gate = Gate(args)
        self.experts = nn.ModuleList([Expert(args.dim, args.moe_inter_dim)
                       if self.experts_start_idx <= i < self.experts_end_idx
                       else None
                       for i in range(self.n_routed_experts)])
        self.shared_experts = MLP(args.dim, args.n_shared_experts * args.
                                  moe_inter_dim)

    def forward(self, x: torch.Tensor) -> torch.Tensor:
        """
        MoE 模块的前向传播。

        参数:
            x (torch.Tensor): 输入张量。

        返回:
            torch.Tensor: 经过专家路由和计算后的输出张量。
```

```
        """
        shape = x.size()
        x = x.view(-1, self.dim)
        weights, indices = self.gate(x)
        y = torch.zeros_like(x)
        counts = torch.bincount(indices.flatten(),
                                minlength=self.n_routed_experts).tolist()
        for i in range(self.experts_start_idx, self.experts_end_idx):
            if counts[i] == 0:
                continue
            expert = self.experts[i]
            idx, top = torch.where(indices == i)
            y[idx] += expert(x[idx]) * weights[idx, top, None]
        z = self.shared_experts(x)
        if world_size > 1:
            dist.all_reduce(y)
        return (y + z).view(shape)
```

MoE是一种用于DeepSeek-V3 模型的混合专家架构，旨在通过多个专家网络的组合来提高模型的表达能力和推理性能。每个专家网络可以专注于处理输入数据的不同方面，从而实现高效的特征提取和知识分割。

4.5.11 Transformer 模型

类Transformer是一个基于Transformer架构的模型，适用于自然语言处理中的任务。类Transformer包含嵌入层、多个Transformer层和一个输出投影层，其中嵌入层用于将输入的token映射到高维向量空间。Transformer块执行自注意力机制，以便模型能够关注输入序列的不同部分。最后，输出投影层将Transformer块的输出映射回词汇表空间，生成最终的logits。

```
class Transformer(nn.Module):
    """
    带有位置嵌入、多层结构和输出投影的 Transformer 模型。

    属性：
        max_seq_len (int): Transformer 的最大序列长度。
        embed (nn.Module): 输入标记的嵌入层。
        layers (torch.nn.ModuleList): Transformer 块的列表。
        norm (nn.Module): 应用于所有块之后的层归一化。
        head (nn.Module): 映射到词汇表大小的输出投影层。
        freqs_cis (torch.Tensor): 为旋转嵌入预先计算的复指数值。
    """
    def __init__(self, args: ModelArgs):
        """
        初始化 Transformer 模型。
```

```
        参数：
            args (ModelArgs): 包含 Transformer 参数的模型参数。
        """
        global world_size, rank
        world_size = dist.get_world_size() if dist.is_initialized() else 1
        rank = dist.get_rank() if dist.is_initialized() else 0
        Linear.dtype = torch.float8_e4m3fn if args.dtype == "fp8" else
                       torch.bfloat16
        super().__init__()
        self.max_seq_len = args.max_seq_len
        self.embed = ParallelEmbedding(args.vocab_size, args.dim)
        self.layers = torch.nn.ModuleList()
        for layer_id in range(args.n_layers):
            self.layers.append(Block(layer_id, args))
        self.norm = RMSNorm(args.dim)
        self.head = ColumnParallelLinear(args.dim, args.vocab_size,
                                         dtype=torch.get_default_dtype())
        self.register_buffer("freqs_cis", precompute_freqs_cis(args),
                             persistent=False)

    @torch.inference_mode()
    def forward(self, tokens: torch.Tensor, start_pos: int = 0):
        """
        Transformer 模型的前向传播。

        参数：
            tokens (torch.Tensor): 形状为 (batch_size, seq_len) 的输入标记张量。
            start_pos (int, optional): 旋转嵌入的序列起始位置。默认为 0。

        返回：
            torch.Tensor: 形状为 (batch_size, vocab_size) 的 logits 张量。
        """
        seqlen = tokens.size(1)
        h = self.embed(tokens)
        freqs_cis = self.freqs_cis[start_pos:start_pos+seqlen]
        mask = None
        if seqlen > 1:
            mask = torch.full((seqlen, seqlen), float("-inf"), device=tokens.
                              device).triu_(1)
        for layer in self.layers:
            h = layer(h, start_pos, freqs_cis, mask)
        h = self.norm(h)[:, -1]
        logits = self.head(h)
        if world_size > 1:
```

```
            all_logits = [torch.empty_like(logits) for _ in range(world_size)]
            dist.all_gather(all_logits, logits)
            logits = torch.cat(all_logits, dim=-1)
        return logits
```

上述Transformer模型的设计结合了并行嵌入、RMS归一化、列并行线性层和Dilated RoPE，能够高效地处理大规模的输入序列，并在分布式训练中实现高效的通信和计算。

4.5.12 验证和测试

下面的代码用于测试和验证Transformer模型的功能。当直接运行model.py文件时，代码会设置默认的张量数据类型为bfloat16，默认设备为GPU，并设置随机种子以确保结果的可重复性。然后，代码创建一个包含随机整数的输入张量x，并实例化一个Transformer模型。最后，代码打印模型对输入x的输出张量的尺寸，以验证模型的输出是否符合预期。

```
if __name__ == "__main__":
    # 设置默认的张量数据类型为 bfloat16
    torch.set_default_dtype(torch.bfloat16)
    # 设置默认设备为 GPU
    torch.set_default_device("cuda")
    # 设置随机种子为 0
    torch.manual_seed(0)
    # 创建模型参数实例
    args = ModelArgs()
    # 生成形状为 (2, 128) 的随机输入张量，元素值在 0 到词汇表大小之间
    x = torch.randint(0, args.vocab_size, (2, 128))
    # 实例化 Transformer 模型
    model = Transformer(args)
    # 打印模型对输入 x 的输出张量的尺寸
    print(model(x).size())
```

注意

torch.set_default_dtype用于设置默认的浮点数数据类型，而torch.set_default_device用于设置默认的设备（如CPU或GPU），这对于在不同硬件环境下运行模型非常有用。

4.6 量化计算

在DeepSeek-V3项目中，文件kernel.py是深度学习框架中用于实现模型量化计算的核心模块，通过利用Triton和Torch的并行计算能力，实现高效的数据量化、反量化和矩阵乘法操作，从而加速模型推理并降低内存占用。

在量化计算方面，DeepSeek-V3 首次在大规模语言模型训练中系统性地部署了FP8（8 位浮点）量化技术。这种技术显著降低了训练对显卡内存的需求，加快了训练过程。此外，DeepSeek-V3 还采用了FP4 量化训练，以进一步突破计算瓶颈，实现更高效的训练和推理。通过这些创新，DeepSeek-V3 在保持高性能的同时，大幅度降低了计算成本，展现出卓越的训练和推理效率。

4.6.1 输入张量进行量化处理

下面的代码定义了一个名为act_quant_kernel()的Triton内核函数，用于对输入张量进行量化处理。具体而言，函数act_quant_kernel()将输入张量的每个块（由BLOCK_SIZE指定）进行处理，计算每个块的最大绝对值并据此确定缩放因子s，然后使用该缩放因子对块内的元素进行量化，最终将量化后的值和对应的缩放因子分别存储在输出指针y_ptr和s_ptr所指向的内存位置。这种方法有助于在保持数值精度的同时减少存储和计算的开销。

```
@triton.jit
def act_quant_kernel(x_ptr, y_ptr, s_ptr, BLOCK_SIZE: tl.constexpr):
    """
    对输入张量 `x_ptr` 进行量化，并将结果存储在 `y_ptr` 和缩放因子存储在 `s_ptr` 中。

    参数：
        x_ptr (triton.Pointer): 指向输入张量的指针。
        y_ptr (triton.Pointer): 指向存储量化值的输出张量的指针。
        s_ptr (triton.Pointer): 指向存储缩放因子的输出张量的指针。
        BLOCK_SIZE (tl.constexpr): 每个程序实例处理的块大小。

    返回：
        None
    """
    pid = tl.program_id(axis=0)
    offs = pid * BLOCK_SIZE + tl.arange(0, BLOCK_SIZE)
    x = tl.load(x_ptr + offs).to(tl.float32)
    s = tl.max(tl.abs(x)) / 448.  # 缩放因子计算
    y = x / s  # 量化操作
    y = y.to(y_ptr.dtype.element_ty)  # 转换数据类型
    tl.store(y_ptr + offs, y)  # 存储量化值
    tl.store(s_ptr + pid, s)  # 存储缩放因子
```

4.6.2 块级量化处理

下面的代码定义了函数act_quant()，用于对输入的PyTorch张量x进行块级量化。首先，函数act_quant()检查输入张量是否连续，并确保其最后一维的大小可以被指定的块大小block_size整除。然后，它初始化一个与输入张量形状相同但数据类型为torch.float8_e4m3fn的空张量y，用于存储量化后的值。同时，初始化一个用于存储缩放因子的张量s。接着，函数计算网格大小，并调用之前

定义的act_quant_kernel内核函数，对输入张量进行量化处理。最后，函数返回量化后的张量y和对应的缩放因子s。

```
def act_quant(x: torch.Tensor, block_size: int = 128) -> Tuple[torch.Tensor, 
torch.Tensor]:
    """
    使用块量化对输入张量 `x` 进行量化。

    参数:
        x (torch.Tensor): 要量化的输入张量。必须是连续的，并且末尾维度大小必须能被
                          `block_size` 整除。
        block_size (int, optional): 用于量化的块大小。默认是 128。

    返回:
        Tuple[torch.Tensor, torch.Tensor]: 包含:
            - 数据类型为 `torch.float8_e4m3fn` 的量化张量。
            - 数据类型为 `torch.float32` 的缩放因子张量。
    """
    assert x.is_contiguous(), '输入张量必须是连续的'
    assert x.size(-1) % block_size == 0, f'末尾维度大小必须能被块大小整除(block_
                                          size={block_size})'
    y = torch.empty_like(x, dtype=torch.float8_e4m3fn)      # 初始化量化张量
    s = x.new_empty(*x.size()[:-1], x.size(-1) // block_size, dtype=torch.
float32)  # 初始化缩放因子张量
    grid = lambda meta: (triton.cdiv(x.numel(), meta['BLOCK_SIZE']), )
                                                                  # 确定网格尺寸
    act_quant_kernel[grid](x, y, s, BLOCK_SIZE=block_size)  # 调用量化内核
    return y, s  # 返回量化张量和缩放因子
```

4.6.3 权重矩阵的反量化

下面的代码定义了一个名为weight_dequant_kernel()的Triton内核函数，用于对量化后的权重矩阵进行反量化。函数接受量化权重指针x_ptr、缩放因子指针s_ptr、输出缓冲区指针y_ptr，以及权重矩阵的行数M、列数N和块大小BLOCK_SIZE。在执行过程中，函数根据程序ID计算当前处理的块在矩阵中的位置，加载对应的量化权重和缩放因子，将量化值乘以缩放因子以恢复原始权重值，并将结果存储到输出缓冲区。这种方法通过块级处理提高了反量化操作的并行效率。

```
@triton.jit
def weight_dequant_kernel(x_ptr, s_ptr, y_ptr, M, N, BLOCK_SIZE: 
tl.constexpr):
    """
    使用提供的缩放因子对权重进行反量化，并存储结果。

    参数:
```

```
        x_ptr (tl.pointer): 指向量化权重的指针。
        s_ptr (tl.pointer): 指向缩放因子的指针。
        y_ptr (tl.pointer): 指向存储反量化权重的输出缓冲区的指针。
        M (int): 权重矩阵的行数。
        N (int): 权重矩阵的列数。
        BLOCK_SIZE (tl.constexpr): 用于平铺的块大小。

    返回:
        None
    """
    pid_m = tl.program_id(axis=0)
    pid_n = tl.program_id(axis=1)
    n = tl.cdiv(N, BLOCK_SIZE)
    offs_m = pid_m * BLOCK_SIZE + tl.arange(0, BLOCK_SIZE)
    offs_n = pid_n * BLOCK_SIZE + tl.arange(0, BLOCK_SIZE)
    offs = offs_m[:, None] * N + offs_n[None, :]
    mask = (offs_m[:, None] < M) & (offs_n[None, :] < N)
    x = tl.load(x_ptr + offs, mask=mask).to(tl.float32)
    s = tl.load(s_ptr + pid_m * n + pid_n)
    y = x * s
    tl.store(y_ptr + offs, y, mask=mask)
```

4.6.4 对激活值和权重的量化与反量化

定义函数weight_dequant()，使用Triton库对神经网络的激活值和权重进行量化和反量化操作，以提高模型的计算效率。具体而言，act_quant_kernel函数对输入张量进行块级量化，将结果存储在输出张量和对应的缩放因子中；act_quant函数是其对应的Python接口。类似地，weight_dequant_kernel函数使用提供的缩放因子对量化后的权重进行反量化，而weight_dequant函数提供了相应的Python接口。这些操作有助于在保持模型性能的同时降低计算资源的消耗。

```
def weight_dequant(x: torch.Tensor, s: torch.Tensor, block_size: int = 128)
                   -> torch.Tensor:
    """
    使用提供的缩放张量对给定的权重张量进行反量化。

    参数:
        x (torch.Tensor): 形状为 (M, N) 的量化权重张量。
        s (torch.Tensor): 形状为 (M, N) 的缩放张量。
        block_size (int, optional): 用于反量化的块大小。默认为 128。

    返回:
        torch.Tensor: 与 `x` 形状相同的反量化权重张量。
```

```
    引发:
        AssertionError: 如果 `x` 或 `s` 不是连续的，或者它们的维度不是 2。
    """
    assert x.is_contiguous() and s.is_contiguous(), '输入张量必须是连续的'
    assert x.dim() == 2 and s.dim() == 2, '输入张量必须有 2 个维度'
    M, N = x.size()
    y = torch.empty_like(x, dtype=torch.get_default_dtype())
    grid = lambda meta: (triton.cdiv(M, meta['BLOCK_SIZE']), triton.cdiv(N,
                         meta['BLOCK_SIZE']))
    weight_dequant_kernel[grid](x, s, y, M, N, BLOCK_SIZE=block_size)
    return y
```

4.6.5 调优参数

下面的代码定义了一个名为fp8_gemm_configs的列表，其中包含多个triton.Config对象。每个Config对象指定了用于矩阵乘法（General Matrix Multiplication，GEMM）运算的不同块大小（BLOCK_SIZE_M和BLOCK_SIZE_N）、流水线阶段数（num_stages）和使用的warps数量（num_warps）。这些配置将用于自动调优，以确定在特定硬件上执行FP8 精度矩阵乘法的最佳参数组合。

```
fp8_gemm_configs = [
    Config({'BLOCK_SIZE_M': block_m, 'BLOCK_SIZE_N': block_n, 'BLOCK_SIZE_K':
128}, num_stages=num_stages, num_warps=8)
    for block_m in [16, 32, 64] for block_n in [32, 64, 128] for num_stages
in [3, 4, 5, 6]
]
```

在高性能计算中，选择合适的块大小和其他参数对于优化矩阵乘法的性能至关重要。通过定义不同的配置，程序可以在运行时测试并选择最适合当前硬件和问题规模的参数组合，从而实现最佳性能。

例如，Triton的文档中提到，块大小的选择会显著影响矩阵乘法内核的性能。通过自动调优，可以找到最适合特定硬件架构的配置，从而提高计算效率。

此外，自动调优还可以帮助确定最佳的流水线阶段数和warps数量，以充分利用GPU资源，实现更高的吞吐量和性能。

4.6.6 FP8 矩阵乘法内核

下面的代码定义了一个使用Triton编写的FP8 矩阵乘法内核函数fp8_gemm_kernel，并通过@triton.autotune装饰器对其进行自动调优，以优化性能。通过使用 @triton.autotune装饰器，Triton会自动尝试不同的配置参数，以确定在特定硬件上执行该内核函数时的最佳性能配置。

```
@triton.autotune(configs=fp8_gemm_configs, key=['N', 'K'])
@triton.jit
```

```
def fp8_gemm_kernel(a_ptr, b_ptr, c_ptr,
                    a_s_ptr, b_s_ptr,
                    M, N: tl.constexpr, K: tl.constexpr,
                    BLOCK_SIZE_M: tl.constexpr,
                    BLOCK_SIZE_N: tl.constexpr,
                    BLOCK_SIZE_K: tl.constexpr):
    """
    在具有缩放因子的 FP8 矩阵上执行矩阵乘法运算。

    参数:
        a_ptr (tl.tensor): 指向第一个输入矩阵 A 的指针。
        b_ptr (tl.tensor): 指向第二个输入矩阵 B 的指针。
        c_ptr (tl.tensor): 指向输出矩阵 C 的指针。
        a_s_ptr (tl.tensor): 指向矩阵 A 的缩放因子的指针。
        b_s_ptr (tl.tensor): 指向矩阵 B 的缩放因子的指针。
        M (int): 矩阵 A 和 C 的行数。
        N (tl.constexpr): 矩阵 B 和 C 的列数。
        K (tl.constexpr): 矩阵 A 的列数和矩阵 B 的行数。
        BLOCK_SIZE_M (tl.constexpr): M 维度的块大小。
        BLOCK_SIZE_N (tl.constexpr): N 维度的块大小。
        BLOCK_SIZE_K (tl.constexpr): K 维度的块大小。

    返回:
        None
    """
    pid_m = tl.program_id(axis=0)
    pid_n = tl.program_id(axis=1)
    k = tl.cdiv(K, BLOCK_SIZE_K)
    offs_m = (pid_m * BLOCK_SIZE_M + tl.arange(0, BLOCK_SIZE_M)) % M
    offs_n = (pid_n * BLOCK_SIZE_N + tl.arange(0, BLOCK_SIZE_N)) % N
    offs_k = tl.arange(0, BLOCK_SIZE_K)
    a_ptrs = a_ptr + offs_m[:, None] * K + offs_k[None, :]
    b_ptrs = b_ptr + offs_n[None, :] * K + offs_k[:, None]
    a_s_ptrs = a_s_ptr + offs_m * k
    b_s_ptrs = b_s_ptr + (offs_n // BLOCK_SIZE_K) * k

    accumulator = tl.zeros((BLOCK_SIZE_M, BLOCK_SIZE_N), dtype=tl.float32)
    for i in range(k):
        a = tl.load(a_ptrs, mask=offs_k[None, :] < K - i * BLOCK_SIZE_K,
                    other=0.0)
        b = tl.load(b_ptrs, mask=offs_k[:, None] < K - i * BLOCK_SIZE_K,
                    other=0.0)
        a_s = tl.load(a_s_ptrs)
        b_s = tl.load(b_s_ptrs)
        accumulator += tl.dot(a, b) * a_s[:, None] * b_s[None, :]
        a_ptrs += BLOCK_SIZE_K
```

```
        b_ptrs += BLOCK_SIZE_K
        a_s_ptrs += 1
        b_s_ptrs += 1
    c = accumulator.to(c_ptr.dtype.element_ty)
    offs_m = pid_m * BLOCK_SIZE_M + tl.arange(0, BLOCK_SIZE_M)
    offs_n = pid_n * BLOCK_SIZE_N + tl.arange(0, BLOCK_SIZE_N)
    c_ptrs = c_ptr + offs_m[:, None] * N + offs_n[None, :]
    mask = (offs_m[:, None] < M) & (offs_n[None, :] < N)
    tl.store(c_ptrs, c, mask=mask)
```

4.6.7 FP8 矩阵乘法实现

下面的代码定义了函数fp8_gemm()，用于在FP8 精度下执行矩阵乘法。函数fp8_gemm()接受如下所示的4个输入参数。

- a：第一个输入矩阵，必须是连续的。
- a_s：第一个输入矩阵的缩放因子，必须是连续的。
- b：第二个输入矩阵，必须是连续的。
- b_s：第二个输入矩阵的缩放因子，必须是连续的。

首先，该函数验证输入张量和缩放因子张量是否连续。然后，计算矩阵乘法的维度，并创建一个新的张量c来存储结果。接下来，使用Triton的JIT编译内核fp8_gemm_kernel执行矩阵乘法操作。最后，返回计算得到的结果张量c。

```
def fp8_gemm(a: torch.Tensor, a_s: torch.Tensor, b: torch.Tensor, b_s:
torch.Tensor):
    """
    使用 FP8 精度执行矩阵乘法。

    参数:
        a (torch.Tensor): 第一个输入矩阵，必须是连续的。
        a_s (torch.Tensor): 第一个输入矩阵的缩放因子，必须是连续的。
        b (torch.Tensor): 第二个输入矩阵，必须是连续的。
        b_s (torch.Tensor): 第二个输入矩阵的缩放因子，必须是连续的。

    返回:
        torch.Tensor: 矩阵乘法的结果。
    """
    assert a.is_contiguous() and b.is_contiguous(), '输入张量必须是连续的'
    assert a_s.is_contiguous() and b_s.is_contiguous(), '缩放因子张量必须是连续
的'
    K = a.size(-1)
    M = a.numel() // K
    N = b.size(0)
    c = a.new_empty(*a.size()[:-1], N, dtype=torch.get_default_dtype())
```

```
    grid = lambda META: (triton.cdiv(M, META['BLOCK_SIZE_M']), triton.
                        cdiv(N, META['BLOCK_SIZE_N']))
    fp8_gemm_kernel[grid](a, b, c, a_s, b_s, M, N, K)
    return c
```

4.7 权重转换

DeepSeek-V3模型采用了FP8（8位浮点数）训练，以提高训练速度并降低GPU内存占用。为方便社区适配和拓展应用场景，DeepSeek-V3提供了从FP8到BF16（16位浮点数）的权重转换脚本。此外，DeepSeek-V3还开源了原生FP8权重，并提供了从FP8到BF16的转换脚本，方便社区适配和拓展应用场景。

4.7.1 权重格式转换

文件convert.py的功能是将Hugging Face格式的DeepSeek-V3 模型权重转换为特定格式，以便在不同的推理框架中使用。该脚本通过解析输入的模型权重文件，按照预定义的映射关系，将模型参数重命名并拆分，以适应模型并行和专家模型的需求。转换后的权重文件将被保存到指定的目录中，供后续推理使用。

```
import os
import shutil
from argparse import ArgumentParser
from glob import glob
from tqdm import tqdm, trange

import torch
from safetensors.torch import safe_open, save_file

# 参数映射关系
mapping = {
    "embed_tokens": ("embed", 0),
    "input_layernorm": ("attn_norm", None),
    "post_attention_layernorm": ("ffn_norm", None),
    "q_proj": ("wq", 0),
    "q_a_proj": ("wq_a", None),
    "q_a_layernorm": ("q_norm", None),
    "q_b_proj": ("wq_b", 0),
    "kv_a_proj_with_mqa": ("wkv_a", None),
    "kv_a_layernorm": ("kv_norm", None),
    "kv_b_proj": ("wkv_b", 0),
    "o_proj": ("wo", 1),
```

```
    "gate": ("gate", None),
    "gate_proj": ("w1", 0),
    "down_proj": ("w2", 1),
    "up_proj": ("w3", 0),
    "norm": ("norm", None),
    "lm_head": ("head", 0),
    "scale": ("scale", None),
}

def main(hf_ckpt_path, save_path, n_experts, mp):
    """
    将模型检查点文件转换并保存为指定格式。

    参数:
        hf_ckpt_path (str): 输入检查点文件所在的目录路径。
        save_path (str): 转换后的检查点文件保存的目录路径。
        n_experts (int): 模型中的专家数量。
        mp (int): 模型并行度。

    返回:
        None
    """
    torch.set_num_threads(8)
    n_local_experts = n_experts // mp
    state_dicts = [{} for _ in range(mp)]

    # 遍历所有的 safetensors 文件
    for file_path in tqdm(glob(os.path.join(hf_ckpt_path, "*.safetensors"))):
        with safe_open(file_path, framework="pt", device="cpu") as f:
            for name in f.keys():
                if "model.layers.61" in name:
                    continue
                param: torch.Tensor = f.get_tensor(name)
                if name.startswith("model."):
                    name = name[len("model."):]
                name = name.replace("self_attn", "attn")
                name = name.replace("mlp", "ffn")
                name = name.replace("weight_scale_inv", "scale")
                name = name.replace("e_score_correction_bias", "bias")
                key = name.split(".")[-2]
                assert key in mapping, f"Key {key} not found in mapping"
                new_key, dim = mapping[key]
                name = name.replace(key, new_key)
                for i in range(mp):
                    new_param = param
                    if "experts" in name and "shared_experts" not in name:
```

```
                    idx = int(name.split(".")[-3])
                    if idx < i * n_local_experts or idx >= (i + 1) * n_
                                                            local_experts:
                        continue
                elif dim is not None:
                    assert param.size(dim) % mp == 0, f"Dimension {dim}
                        must be divisible by {mp}"
                    shard_size = param.size(dim) // mp
                    new_param = param.narrow(dim, i * shard_size, shard_
                        size).contiguous()
                state_dicts[i][name] = new_param

    os.makedirs(save_path, exist_ok=True)

    # 保存转换后的模型权重
    for i in trange(mp):
        save_file(state_dicts[i], os.path.join(save_path, f"model{i}-mp{mp}.
             safetensors"))

    # 复制token文件
    for file_path in glob(os.path.join(hf_ckpt_path, "*token*")):
        new_file_path = os.path.join(save_path, os.path.basename(file_path))
        shutil.copyfile(file_path, new_file_path)

if __name__ == "__main__":
    parser = ArgumentParser()
    parser.add_argument("--hf-ckpt-path", type=str, required=True)
    parser.add_argument("--save-path", type=str, required=True)
    parser.add_argument("--n-experts", type=int, required=True)
    parser.add_argument("--model-parallel", type=int, required=True)
    args = parser.parse_args()
    assert args.n_experts % args.model_parallel == 0, "专家数量必须能被模型并行
度整除"
    main(args.hf_ckpt_path, args.save_path, args.n_experts, args.model_
parallel)
```

在模型权重转换过程中，用户需要下载Hugging Face上的模型权重，并将其放入指定文件夹。然后，运行转换脚本，将HuggingFace模型权重转换为DeepSeek-V3特定格式。转换命令如下：

```
python convert.py --hf-ckpt-path /path/to/DeepSeek-V3 --save-path /path/to/
DeepSeek-V4-Demo --n-experts 256 --model-parallel 16
```

对上述命令参数的具体说明如下所示。

- --hf-ckpt-path /path/to/DeepSeek-V3：指定Hugging Face格式的DeepSeek-V3 模型权重文件的路径。

- --save-path /path/to/DeepSeek-V4-Demo：指定转换后模型权重保存的目标路径。
- --n-experts 256：设置模型中每个专家的数量。
- --model-parallel 16：设置模型并行度，即模型在多个设备上的分布方式。

这些参数共同作用，确保模型权重从Hugging Face格式成功转换为DeepSeek-V3所需的格式，并根据指定的配置进行优化。

通过上述操作，用户可以将Hugging Face模型权重转换为DeepSeek-V3所需的格式，以便在DeepSeek-V3框架中进行推理和部署。

4.7.2 权重精度转换

文件fp8_cast_bf16.py的功能是将FP8权重转换为BF16格式，并保存转换后的权重。它会读取指定目录中的FP8权重，将其转换为BF16，然后保存到另一个指定目录。此外，它还会更新模型索引文件，以反映这些更改。如果缺少某个权重的scale_inv张量，脚本会引发KeyError异常。请注意，脚本假设FP8权重存储在safetensor文件中，并会缓存已加载的safetensor文件以优化内存使用。它还会更新模型索引文件，移除对scale_inv张量的引用。

```
import os
import json
from glob import glob
from tqdm import tqdm
from argparse import ArgumentParser
import torch
from safetensors.torch import load_file, save_file

def main(fp8_path, bf16_path):
    """
    将 FP8 权重转换为 BF16 并保存转换后的权重。

    此函数从指定目录读取 FP8 权重，将其转换为 BF16，并将转换后的权重保存到另一个指定的目录。
它还会更新模型索引文件，以反映这些更改。

    参数：
    fp8_path (str): 包含 FP8 权重和模型索引文件的目录路径。
    bf16_path (str): 转换后的 BF16 权重将保存的目录路径。

    异常：
    KeyError: 如果某个权重缺少所需的 scale_inv 张量。

    注意：
    - 该函数假设 FP8 权重存储在 safetensor 文件中。
    - 该函数会缓存已加载的 safetensor 文件，以优化内存使用。
    - 该函数会更新模型索引文件，删除指向 scale_inv 张量的引用。
```

```
"""
torch.set_default_dtype(torch.bfloat16)    # 设置默认数据类型为 BF16
os.makedirs(bf16_path, exist_ok=True)     # 创建保存 BF16 权重的目录
model_index_file = os.path.join(fp8_path, "model.safetensors.index.
                                json")    # 获取模型索引文件路径
with open(model_index_file, "r") as f:
    model_index = json.load(f)            # 读取模型索引文件
weight_map = model_index["weight_map"]

# 加载 safetensor 文件的缓存
loaded_files = {}
fp8_weight_names = []

# 辅助函数：获取正确文件中的张量
def get_tensor(tensor_name):
    """
    从缓存的 safetensor 文件中检索张量，如果没有缓存，则从磁盘加载。

    参数：
        tensor_name (str): 要检索的张量名称。

    返回：
        torch.Tensor: 检索到的张量。

    异常：
        KeyError: 如果张量在 safetensor 文件中不存在。
    """
    file_name = weight_map[tensor_name]
    if file_name not in loaded_files:
        file_path = os.path.join(fp8_path, file_name)
        loaded_files[file_name] = load_file(file_path, device="cuda")
                                           # 从磁盘加载 safetensor 文件
    return loaded_files[file_name][tensor_name]

safetensor_files = list(glob(os.path.join(fp8_path, "*.safetensors")))
safetensor_files.sort()
for safetensor_file in tqdm(safetensor_files):      # 遍历所有 safetensor 文
件
    file_name = os.path.basename(safetensor_file)
    current_state_dict = load_file(safetensor_file, device="cuda")
                                           # 加载当前 safetensor 文件
    loaded_files[file_name] = current_state_dict

    new_state_dict = {}
    for weight_name, weight in current_state_dict.items():
        if weight_name.endswith("_scale_inv"):  # 跳过 scale_inv 权重
```

```
                continue
            elif weight.element_size() == 1:          # FP8 权重
                scale_inv_name = f"{weight_name}_scale_inv"
                try:
                    # 获取 scale_inv 权重
                    scale_inv = get_tensor(scale_inv_name)
                    fp8_weight_names.append(weight_name)
                    new_state_dict[weight_name] = weight_dequant(weight,
                        scale_inv)   # 转换 FP8 为 BF16
                except KeyError:
                    print(f"警告：找不到 {weight_name} 的 scale_inv 张量，
                          跳过转换")
                    new_state_dict[weight_name] = weight # 如果缺少 scale_inv，
                                                           则保留原权重
            else:
                new_state_dict[weight_name] = weight  # 非 FP8 权重直接保存

        new_safetensor_file = os.path.join(bf16_path, file_name)
        save_file(new_state_dict, new_safetensor_file)  # 保存转换后的权重

        # 内存管理：仅保留最近使用的两个文件
        if len(loaded_files) > 2:
            oldest_file = next(iter(loaded_files))
            del loaded_files[oldest_file]
            torch.cuda.empty_cache()

    # 更新模型索引文件
    new_model_index_file = os.path.join(bf16_path, "model.safetensors.index.
                                         json")
    for weight_name in fp8_weight_names:
        scale_inv_name = f"{weight_name}_scale_inv"
        if scale_inv_name in weight_map:
            weight_map.pop(scale_inv_name)  # 移除 scale_inv 权重
    with open(new_model_index_file, "w") as f:
        json.dump({"metadata": {}, "weight_map": weight_map}, f, indent=2)
# 更新索引文件

if __name__ == "__main__":
    parser = ArgumentParser()
    parser.add_argument("--input-fp8-hf-path", type=str, required=True)
    parser.add_argument("--output-bf16-hf-path", type=str, required=True)
    args = parser.parse_args()
    main(args.input_fp8_hf_path, args.output_bf16_hf_path)
```

通过运行如下命令，可以将DeepSeek-V3 模型的FP8 权重转换为BF16 格式，以便在支持BF16

的硬件上进行推理。

```
cd inference
python fp8_cast_bf16.py --input-fp8-hf-path /path/to/fp8_weights --output-
bf16-hf-path /path/to/bf16_weights
```

具体来说，cd inference命令将当前工作目录切换到inference目录，然后执行python fp8_cast_bf16.py脚本，并通过 --input-fp8-hf-path和 --output-bf16-hf-path参数指定输入和输出路径。以下是各个参数的含义。

- --input-fp8-hf-path /path/to/fp8_weights：指定包含FP8 权重文件的目录路径。
- --output-bf16-hf-path /path/to/bf16_weights：指定转换后BF16 权重文件保存的目录路径。

在执行此命令之前，请确保已将FP8 权重文件放置在 /path/to/fp8_weights目录中，并准备好用于保存BF16 权重文件的 /path/to/bf16_weights目录。执行该命令后，脚本会读取FP8 权重文件，进行转换，并将结果保存到指定的输出目录。

4.7.3 不同硬件平台的转换

DeepSeek-V3 模型的权重转换是模型部署和推理过程中的关键步骤，主要涉及将模型的权重从FP8 格式转换为BF16 格式，以适应不同的硬件平台和推理需求。

1. GPU 权重转换

对于GPU用户，可以通过以下步骤进行权重转换：

（1）克隆仓库，命令如下：

```
git clone https://github.com/deepseek-ai/DeepSeek-V3.git
cd DeepSeek-V3/inference
```

（2）执行权重转换脚本，命令如下：

```
python fp8_cast_bf16.py --input-fp8-hf-path /path/to/DeepSeek-V3 --output-
bf16-hf-path /path/to/deepseek-v4-bf16
```

2. NPU 权重转换

对于NPU用户，可以通过以下步骤进行权重转换。

（1）克隆仓库，命令如下：

```
git clone https://gitee.com/ascend/ModelZoo-PyTorch.git
cd ModelZoo-PyTorch/MindIE/LLM/DeepSeek/DeepSeek-V2/NPU_inference
```

（2）执行权重转换脚本，命令如下：

```
python fp8_cast_bf16.py --input-fp8-hf-path /path/to/DeepSeek-V3 --output-
bf16-hf-path /path/to/deepseek-v4-bf16
```

"/path/to/DeepSeek-V3"表示DeepSeek-V3原始权重路径，"/path/to/deepseek-v4-bf16"表示权重转换后的新权重路径。

注意

由于模型权重较大，需确保磁盘有足够的空间。例如，DeepSeek-V3 **在转换前权重约为** 640G，**转换后权重约为** 1.3T。

3. 权重转换后的使用

在完成权重转换后，用户需要下载适配DeepSeek-V3的镜像包，并使用docker命令进行加载。

（1）下载镜像包，命令如下：

```
docker load -i mindie:1.0.T71-800I-A2-py311-ubuntu22.04-arm64
```

（2）确认镜像名称与标签，命令如下：

```
docker images
```

（3）启动容器：将转换后的权重放置在模型代码的主目录下，修改模型文件夹的属组和执行权限。然后使用预置了DeepSeek-V3推理脚本的MindIE镜像启动容器。

4. 服务化测试

配置服务化环境变量，启用内存池扩展段功能，当屏幕上出现"Daemon start success!"提示时，表示服务已成功启动。

通过以上步骤，用户可以在昇腾社区高效获取并使用DeepSeek-V3模型，实现一键部署和快速推理。

4.8 测试模型

DeepSeek-V3是一款基于PyTorch的深度学习模型，主要用于文本生成任务。在加载模型时，首先需要从指定路径加载模型的配置文件和预训练权重。加载完成后，模型被设置为评估模式，并移动到GPU上以加速推理过程。在生成文本时，用户可以输入提示文本，模型会根据这些提示生成相应的文本输出。生成过程支持交互式输入和批量处理两种模式，用户可以根据需要选择适合的方式进行文本生成。此外，模型的生成过程还支持设置温度参数，以控制生成文本的多样性和创造性。通过这些功能，DeepSeek-V3能够高效地加载模型并生成符合用户需求的文本内容。

4.8.1 模型加载与文本生成

在DeepSeek-V3项目中，脚本文件generate.py用于加载模型并执行文本生成任务。该脚本提供了交互式和批量处理两种模式，允许用户根据输入提示生成相应的文本。在交互式模式下，用户可以输入提示，模型将生成相应的文本；在批量处理模式下，脚本从指定的文件中读取多个提示，并生成对应的文本。此外，脚本还支持设置生成文本的最大长度和温度参数，以控制生成文本的多样性和随机性。

```
from model import Transformer, ModelArgs

def sample(logits, temperature: float = 1.0):
    """
    使用温度缩放从 logits 中采样一个 token。

    参数:
        logits (torch.Tensor): 用于 token 预测的 logits 张量。
        temperature (float, 可选): 用于缩放 logits 的温度。默认为 1.0。

    返回:
        torch.Tensor: 采样得到的 token。
    """
    logits = logits / max(temperature, 1e-5)
    probs = torch.softmax(logits, dim=-1)
    return probs.div_(torch.empty_like(probs).exponential_(1)).argmax(dim=-1)


@torch.inference_mode()
def generate(
    model: Transformer,
    prompt_tokens: List[List[int]],
    max_new_tokens: int,
    eos_id: int,
    temperature: float = 1.0
) -> List[List[int]]:
    """
    基于给定的提示 tokens 使用指定的模型生成新的 tokens。

    参数:
        model (Transformer): 用于 token 生成的 transformer 模型。
        prompt_tokens (List[List[int]]): 包含每个序列提示 tokens 的列表。
        max_new_tokens (int): 最大生成的新 tokens 数量。
        eos_id (int): 结束序列的 token ID。
        temperature (float, 可选): 采样的温度值。默认为 1.0。

    返回:
        List[List[int]]: 包含每个序列生成的 tokens 的列表。
    """
    prompt_lens = [len(t) for t in prompt_tokens]
    assert max(prompt_lens) <= model.max_seq_len, f"提示长度超过模型最大序列长度
                                  (max_seq_len={model.max_seq_len})"
```

```
    total_len = min(model.max_seq_len, max_new_tokens + max(prompt_lens))
    tokens = torch.full((len(prompt_tokens), total_len), -1, dtype=torch.
                        long, device="cuda")
    for i, t in enumerate(prompt_tokens):
        tokens[i, :len(t)] = torch.tensor(t, dtype=torch.long, device="cuda")
    prev_pos = 0
    finished = torch.tensor([False] * len(prompt_tokens), device="cuda")
    prompt_mask = tokens != -1
    for cur_pos in range(min(prompt_lens), total_len):
        logits = model.forward(tokens[:, prev_pos:cur_pos], prev_pos)
        if temperature > 0:
            next_token = sample(logits, temperature)
        else:
            next_token = logits.argmax(dim=-1)
        next_token = torch.where(prompt_mask[:, cur_pos], tokens[:, cur_
                                 pos], next_token)
        tokens[:, cur_pos] = next_token
        finished |= torch.logical_and(~prompt_mask[:, cur_pos], next_token
                                      == eos_id)
        prev_pos = cur_pos
        if finished.all():
            break
    completion_tokens = []
    for i, toks in enumerate(tokens.tolist()):
        toks = toks[prompt_lens[i]:prompt_lens[i]+max_new_tokens]
        if eos_id in toks:
            toks = toks[:toks.index(eos_id)]
        completion_tokens.append(toks)
    return completion_tokens


def main(
    ckpt_path: str,
    config: str,
    input_file: str = "",
    interactive: bool = True,
    max_new_tokens: int = 100,
    temperature: float = 1.0,
) -> None:
    """
    加载模型并执行交互式或批量文本生成。

    参数:
        ckpt_path (str): 模型检查点目录的路径。
        config (str): 模型配置文件的路径。
        input_file (str, 可选): 包含输入提示的文件路径。默认为 ""。
        interactive (bool, 可选): 是否以交互模式运行。默认为 True。
```

```
        max_new_tokens (int, 可选): 最大生成的新 tokens 数量。默认为 100。
        temperature (float, 可选): 采样的温度。默认为 1.0。
    """
    world_size = int(os.getenv("WORLD_SIZE", "1"))
    rank = int(os.getenv("RANK", "0"))
    local_rank = int(os.getenv("LOCAL_RANK", "0"))
    if world_size > 1:
        dist.init_process_group("nccl")
    global print
    if rank != 0:
        print = lambda *_, **__: None
    torch.cuda.set_device(local_rank)
    torch.set_default_dtype(torch.bfloat16)
    torch.set_num_threads(8)
    torch.manual_seed(965)
    with open(config) as f:
        args = ModelArgs(**json.load(f))
    print(args)
    with torch.device("cuda"):
        model = Transformer(args)
    tokenizer = AutoTokenizer.from_pretrained(ckpt_path)
    tokenizer.decode(generate(model, [tokenizer.encode("DeepSeek")], 2, -1,
1.)[0])
    load_model(model, os.path.join(ckpt_path, f"model{rank}-mp{world_size}.
            safetensors"))

    if interactive:
        messages = []
        while True:
            if world_size == 1:
                prompt = input(">>> ")
            elif rank == 0:
                prompt = input(">>> ")
                objects = [prompt]
                dist.broadcast_object_list(objects, 0)
            else:
                objects = [None]
                dist.broadcast_object_list(objects, 0)
                prompt = objects[0]
            if prompt == "/exit":
                break
            elif prompt == "/clear":
                messages.clear()
                continue
            messages.append({"role": "user", "content": prompt})
            prompt_tokens = tokenizer.apply_chat_template(messages, add_
                        generation_prompt=True)
```

```
            completion_tokens = generate(model, [prompt_tokens],
                                         max_new_tokens,
                                         tokenizer.eos_token_id,
                                         temperature)
            completion = tokenizer.decode(completion_tokens[0],
                                          skip_special_tokens=True)
            print(completion)
            messages.append({"role": "assistant", "content": completion})
    else:
        with open(input_file) as f:
            prompts = [line.strip() for line in f.readlines()]
        assert len(prompts) <= args.max_batch_size, f"提示数量超过最大批次大小
                              ({args.max_batch_size})"
        prompt_tokens = [tokenizer.apply_chat_template([{"role": "user",
                         "content": prompt}], add_generation_prompt=True)
                         for prompt in
::contentReference[oaicite:0]{index=0}
```

文件generate.py实现了基于给定提示词生成新文本的功能，主要实现流程如下所示。

（1）加载模型和配置：从指定路径加载模型检查点和配置文件。

（2）初始化模型：使用加载的配置初始化Transformer模型，并设置默认数据类型为bfloat16。

（3）加载分词器：从模型检查点加载分词器，用于处理输入和输出的文本。

（4）加载模型权重：加载模型的权重文件。

（5）生成文本：根据输入的提示词，使用模型生成指定数量的新词。

（6）输出结果：将生成的文本输出到控制台。

4.8.2 测试功能

DeepSeek-V3 模型提供了如下两种对话模式。

● 交互式模式（Interactive Mode）：使用 --interactive参数启动。

● 批量推理模式（Batch Inference Mode）：使用 --input-file参数指定输入文件路径。此外，通过开启联网搜索功能，模型能够实时获取最新信息，为用户提供更准确及时的答案。

1. 交互式对话模式

通过下面的命令启动DeepSeek-V3 模型的文本生成任务，实现交互式对话模式功能。在这种模式下，用户可以输入提示，模型将生成相应的文本回复。

```
torchrun --nnodes 2 --nproc-per-node 8 --node-rank $RANK --master-addr
$ADDR generate.py --ckpt-path /path/to/DeepSeek-V4-Demo --config configs/
config_671B.json --interactive --temperature 0.7 --max-new-tokens 200
```

对上述命令的具体说明如下所示。

● torchrun：用于启动分布式训练或推理的命令行工具。

- --nnodes 2：指定总共有 2 个节点参与分布式任务。
- --nproc-per-node 8：在每个节点上启动 8 个进程。
- --node-rank $RANK：当前节点的排名，$RANK 是一个环境变量，表示当前节点在集群中的编号。
- --master-addr $ADDR：主节点的地址，$ADDR 是主节点的 IP 地址或主机名。
- generate.py：用于生成文本的脚本。
- --ckpt-path /path/to/DeepSeek-V4-Demo：指定模型检查点的路径。
- --config configs/config_671B.json：指定模型配置文件的路径。
- --interactive：启用交互式模式，允许用户输入提示并获取模型的响应。
- --temperature 0.7：设置采样的温度参数，控制生成文本的随机性。
- --max-new-tokens 200：设置每次生成的最大新词数。

如果在命令行中运行了文件generate.py并启用了交互模式，会显示提示符 >>> 等待用户输入。用户可以输入提示，脚本会生成响应。下面是一个交互的示例：

```
>>> 用户：你好，我先想问一下，DeepSeek-V3 模型的主要特点是什么？
** pharmacies##########**
DeepSeek-V3 模型的主要特点是它采用了混合专家模型（MoE）、多头注意力机制和量化的权重，以实现高效的大规模语言建模。它的 MoE 架构允许模型有效地处理大规模数据集，同时保持较高的精度。多头注意力机制则使得模型能够关注输入序列的不同部分，从而更好地理解上下文。量化权重有助于减少模型的内存占用，同时保持良好的性能。

此外，DeepSeek-V3 模型还通过独特的多 token 预测目标（MTP）扩展了多 token 预测的能力，显著提高了训练数据的效率。预训练和后训练阶段的创新策略进一步提升了模型的性能。

>>> 用户：请解释一下 DeepSeek-V3 模型的多头注意力机制是如何工作的？
** pharmacies##########**
多头注意力机制是一种用于处理序列数据的技术，它允许模型同时关注输入序列的不同部分。在 DeepSeek-V3 模型中，每个头都负责处理输入序列的一个特定方面。例如，一个头可能专注于理解句子的语法结构，而另一个头可能专注于理解句子的语义含义。

这种多头注意力机制通过将输入序列分解为多个头，并分别计算每个头的注意力权重，然后将结果合并来生成输出。这样，模型可以捕捉到输入序列中的复杂模式和关系，从而提高其性能。同时，多头注意力机制还具有并行计算的优势，可以显著提高模型的训练和推理速度。

>>> 用户：/exit
```

2. 批量推理模式

通过下面的命令启动DeepSeek-V3 模型的批量推理模式，模型将读取指定文件中的每一行作为输入提示，生成相应的文本输出。

```
torchrun --nnodes 2 --nproc-per-node 8 --node-rank $RANK --master-addr
```

```
$ADDR generate.py --ckpt-path /path/to/DeepSeek-V4-Demo --config configs/
config_671B.json --input-file $FILE
```

对上述命令的具体说明如下所示。

- torchrun：用于启动分布式训练或推理的命令行工具。
- --nnodes 2：指定总共有 2 个节点参与分布式任务。
- --nproc-per-node 8：在每个节点上启动 8 个进程。
- --node-rank $RANK：当前节点的排名，$RANK是一个环境变量，表示当前节点在集群中的编号。
- --master-addr $ADDR：主节点的地址，$ADDR是主节点的IP地址或主机名。
- generate.py：用于生成文本的脚本。
- --ckpt-path /path/to/DeepSeek-V4-Demo：指定模型检查点的路径。
- --config configs/config_671B.json：指定模型配置文件的路径。
- --input-file $FILE：指定包含输入提示的文件路径，模型将对文件中的每一行进行推理。

如果在命令行中运行了文件generate.py并启用了参数input_file，会从文件中读取输入提示并生成响应。假设input_file的内容如下：

```
请解释一下 DeepSeek-V3 模型的多头注意力机制是如何工作的?
```

运行后可能得到下面的交互信息：

```
Prompt: 请解释一下 DeepSeek-V3 模型的多头注意力机制是如何工作的?
Completion: 多头注意力机制是一种用于处理序列数据的技术，它允许模型同时关注输入序列的不同部分。在 DeepSeek-V3 模型中，每个头都负责处理输入序列的一个特定方面。例如，一个头可能专注于理解句子的语法结构，而另一个头可能专注于理解句子的语义含义。

这种多头注意力机制通过将输入序列分解为多个头，并分别计算每个头的注意力权重，然后将结果合并来生成输出。这样，模型可以捕捉到输入序列中的复杂模式和关系，从而提高其性能。同时，多头注意力机制还具有并行计算的优势，可以显著提高模型的训练和推理速度。
```

4.9 DeepSeek-V3 模型总结

DeepSeek-V3 是一款强大的开源语言模型，其设计和实现涵盖多个关键方面，包括模型架构、训练、预训练、知识蒸馏和高性能推理等。下面是对DeepSeek-V3的详细总结。

1. 模型架构

- Transformer基础：DeepSeek-V3 基于Transformer架构，包含 61 层Transformer，每层包含自注意力机制和前馈神经网络。
- 多头注意力机制（MLA）：采用多头注意力机制，通过低秩联合压缩技术减少推理中的内存占用，同时保持性能。

- 混合专家模型（MoE）：引入混合专家模型，通过专家路由和负载均衡策略提高模型的表达能力和推理效率。
- 多token预测目标（MTP）：扩展了模型在每个位置预测多个未来token的能力范围，提高了训练数据的效率。

2. 训练

- 优化器：使用AdamW优化器进行训练，预训练阶段的学习率为0.0001，后训练阶段的学习率为0.00005。
- 训练策略：采用DualPipe算法，通过前向和反向计算的重叠显著减少通信开销，提高训练效率。
- 训练稳定性：训练过程非常稳定，没有出现任何不可恢复的损失峰值，也未进行过回滚操作。

3. 预训练

- 预训练数据：使用14.8万亿个token的多样化和高质量数据进行预训练。
- 预训练阶段：预训练阶段的超参数设置为学习率0.0001，batch size1024，训练200个epoch。
- 后训练阶段：后训练阶段的超参数设置为学习率0.00005，batch size1024，训练100个epoch。

4. 知识蒸馏

- 知识蒸馏策略：从DeepSeek-R1模型中提炼推理能力，通过特定的蒸馏策略将长链推理能力传递给标准语言模型。
- 蒸馏效果：知识蒸馏显著提高了模型的推理性能，同时保持了输出风格和长度的控制。

5. 高性能推理

- 量化计算：支持FP8混合精度训练和推理，通过量化权重和激活值减少内存占用，提高计算效率。
- 推理优化：通过Triton和PyTorch的结合，实现了高效的FP8矩阵乘法，显著提高了推理速度。
- 推理加速：在推理阶段，模型的端到端生成速度比上一代提升了两倍以上。

6. 模型性能

- 数学任务：在AIME 2024、MATH-500等数学任务中，DeepSeek-V3能达到OpenAI-o1-1217的性能水平。
- 编程任务：在编程任务（如Codeforces和LiveCodeBench）上，DeepSeek-V3表现优于大多数对比模型。
- 知识任务：在MMLU和GPQA Diamond等多学科基准测试中，DeepSeek-V3展现了卓越的知识推理能力。
- 生成任务：在AlpacaEval和ArenaHard等开放式生成任务中，DeepSeek-V3的胜率分别达到

87.6% 和 92.3%，展现了强大的文本生成能力。

7. 部署和优化

- 长上下文扩展：支持最大上下文长度从 32K 扩展至 128K，使模型更适用于长文档处理。
- 推理加速：通过量化和优化技术，模型的端到端生成速度比上一代提升了两倍以上。
- 权重转换：提供convert.py和fp8_cast_bf16.py脚本，用于将模型权重从FP8 格式转换为BF16 格式，以适应不同的硬件平台和推理需求。

8. 项目开源代码解析

- 模型架构：通过model.py文件定义了模型的结构和前向传播逻辑，包括嵌入层、Transformer 块、归一化层和输出投影层。
- 训练和推理：通过generate.py文件实现了模型的训练和推理功能，支持交互式和批量文本生成。
- 权重转换：通过convert.py和fp8_cast_bf16.py文件实现了模型权重的转换和优化，支持FP8 和BF16 格式的转换。

总之，DeepSeek-V3 通过创新的模型架构、高效的训练策略、知识蒸馏技术和高性能推理优化，实现了在多个基准测试中超越其他开源模型的性能，达到了与领先闭源模型相当的水平。

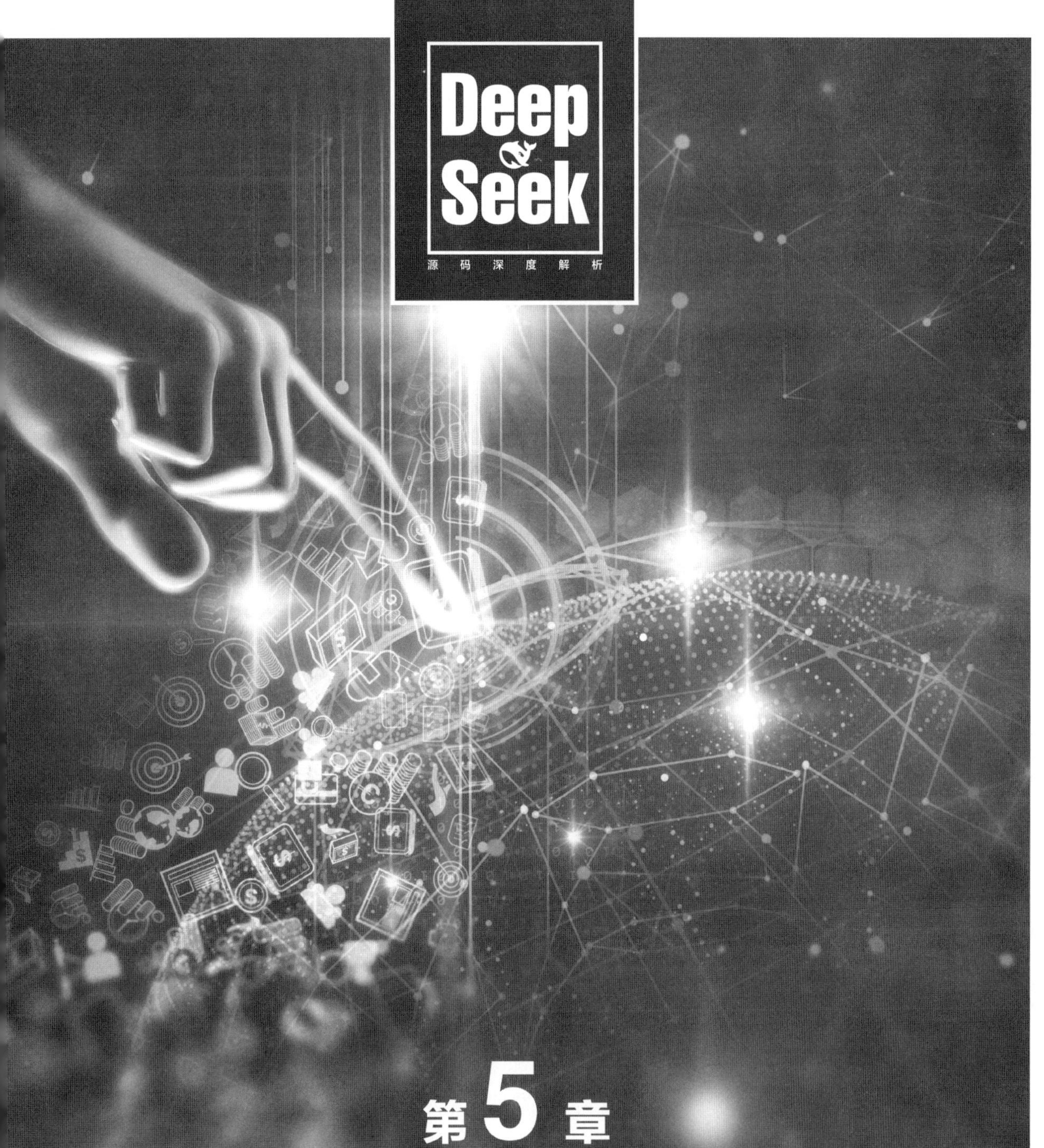

第5章

统一多模态大模型

Janus项目由DeepSeek推出，是一套统一多模态理解与生成模型，包括Janus、Janus-Pro和JanusFlow三个版本。该项目旨在通过统一的自回归Transformer架构，将文本与图像信息高效融合，同时采用“解耦视觉编码”设计，使模型既能精准理解多模态数据，又能生成高质量图像，满足从文本指令到图像生成等多种应用需求。

5.1 项目介绍

DeepSeek的Janus项目是一个开源的多模态理解与生成框架，致力于构建统一且高效的视觉—语言模型。整个项目主要由以下三个版本组成。

- Janus：这是最初的版本，提出了“视觉编码解耦”的思想，通过分别设计用于图像理解和图像生成的编码器，并利用统一的自回归Transformer架构，将文本和图像信息融合，从而在多模态任务中实现协同处理。
- Janus-Pro：在Janus的基础上，Janus-Pro通过优化训练策略、扩充训练数据和扩大模型规模，进一步提升了多模态理解和文本到图像生成的能力。改进后的版本在指令跟随、视觉生成稳定性及生成图像的细节质量上均有明显进步，部分指标甚至超越了专用生成模型。
- JanusFlow：JanusFlow则是对多模态生成的进一步探索，通过将自回归语言模型与Rectified Flow生成模型结合，采用基于常微分方程（Ordinary Differential Equations，ODE）的流匹配方法进行图像生成，显著提升了生成质量和采样效率。同时，JanusFlow依然采用了任务解耦的视觉编码器设计，以避免不同任务间的干扰，从而在视觉理解和生成任务中都展现出优异性能。

总之，整个项目的代码结构较为丰富，包含演示文件、应用程序入口以及多个模型组件（如Patchify、Downsample2D、Upsample2D、ConvNeXtBlock、UVitBlock、ShallowUViTEncoder等），这些模块协同构成了一个完整的视觉—语言处理管道（pipeline）。Janus项目不仅在技术上提出了新颖的架构设计，同时也为多模态AI领域提供了一个可扩展、灵活且高效的开源平台，适用于图像生成、图像理解、图像编辑等众多任务。

5.2 架构原理与核心技术

Janus项目基于统一的自回归Transformer架构，通过解耦视觉编码器将图像理解和生成任务分离处理，从而有效解决了不同任务间的信息冲突。Janus-Pro在此基础上采用了优化的训练策略、数据扩展和模型规模放大技术，显著提升了多模态理解和文本到图像生成的性能。JanusFlow则进一步引入了基于常微分方程的Rectified Flow生成方法，与自回归语言模型相结合，实现了高效、稳定且高质量的图像生成。

5.2.1 Janus架构

Janus是一个用于多模态理解和生成的自回归框架，其核心技术原理包括视觉编码解耦和统一变换器架构。

1. 视觉编码解耦（Decoupled Visual Encoding）

传统的多模态模型（如Chameleon）通常使用单一视觉编码器来同时处理多模态理解（Multimodal Understanding）和视觉生成（Visual Generation），这会导致信息粒度的冲突。

- 多模态理解需要高层语义信息（如图像中的对象类别、属性等）。
- 视觉生成需要保留图像局部细节和全局的一致性，编码方式更偏向于低维表示。

Janus的创新在于分别为理解和生成任务设计独立的视觉编码路径，具体说明如下：

- 理解编码器（Understanding Encoder）：采用SigLIP提取高维语义特征，以提高图像理解能力。
- 生成编码器（Generation Encoder）：采用VQ（Vector Quantization）Tokenization方法，将图像离散化为token，以提高图像生成能力。
- 统一变换器（Unified Transformer）：作为核心处理架构，统一处理两种任务的信息。

2. 统一变换器架构（Unified Transformer Architecture）

Janus采用自回归变换器（Autoregressive Transformer），在多模态理解和生成任务之间共享相同的架构。

- 语言任务（文本理解）：使用LLM内置Tokenizer处理文本。
- 多模态理解任务：将图像编码为高维特征，再通过适配器（Adaptor）映射到LLM输入空间。
- 视觉生成任务：将图像离散化为token，经过适配器转换后输入LLM，并使用专门的预测头进行图像生成。

这种设计不仅避免了不同任务的相互干扰，而且保持了架构的简单性和灵活性。

3. 三阶段训练策略（Three-Stage Training Procedure）

Janus的训练过程分为三个阶段，具体说明如下。

（1）适配器和图像预测头训练（Training Adaptors and Image Head）：冻结LLM和视觉编码器，仅训练理解适配器、生成适配器和图像预测头，建立视觉—语言之间的联系。

（3）统一预训练（Unified Pretraining）：采用多模态数据（文本、图像理解、图像生成数据）进行联合训练，使模型具备强大的多模态能力。

（3）监督微调（Supervised Fine-tuning）：通过指令微调（Instruction Tuning）提高模型的指令跟随和对话能力。

4. 推理与扩展性

- 推理采用自回归方式（Autoregressive Inference），对于文本任务，按token逐步解码；对于图像生成任务，使用条件扩散模型（Classifier-Free Guidance，CFG）提升生成质量。
- 扩展性强，可以引入新的编码器来处理 3D点云、EEG信号、音频数据等，实现更通用的多模态建模。

5. 实验结果

Janus在多模态理解和生成任务上均超越了现有的统一模型，并在某些基准测试中超越了任务专用模型。

- 在多模态理解任务（如POPE、MMBench、SEED-Bench）上，Janus（1.3B）优于LLaVA-v1.5（7B）。
- 在视觉生成任务（如GenEval、COCO-30K）上，Janus生成的图像质量接近甚至超越了DALL·E 2、SDXL等生成专用模型。

总之，Janus通过解耦视觉编码，成功解决了多模态理解和生成任务之间的冲突，提供了统一、多功能且易扩展的多模态架构，为下一代多模态模型的发展开辟了创新路径。

5.2.2 Janus-Pro 架构

Janus-Pro的核心创新在于将视觉编码过程分为两个独立的路径，从而解决传统统一编码中“理解”和“生成”任务之间的冲突。

1. 解耦视觉编码

- 理解编码器：利用SigLIP视觉编码器提取图像的高维语义特征。这些特征经过二维网格采样后被展平为一维序列。通过专门设计的理解适配器，将图像特征映射到LLM的输入空间，从而使模型能更好地理解图像内容。
- 生成编码器：使用VQ Tokenizer将图像转化为离散的ID序列。同样地，通过生成适配器将每个离散码对应的嵌入映射到LLM的输入空间。最终，将文本和图像两部分的特征序列拼接后，统一输入自回归Transformer中进行处理。

这种解耦方法使得模型能够在多模态理解（如图像分类、问答等）和文本到图像生成任务之间达到更好的平衡，同时利用单一自回归框架完成统一处理。

2. 优化训练策略

DeepSeek的报告指出，原始Janus采用了三阶段训练流程，存在训练效率和数据利用率的问题。针对这一点，Janus-Pro对训练策略做了如下改进。

阶段I：适配器与图像预测头训练。

增加了在ImageNet数据上的训练步数，确保模型即使在冻结LLM参数的情况下，也能充分学习图像的像素依赖关系，从而生成较为合理的图像。

阶段II：统一预训练。

取消了原先依赖ImageNet分类提示的训练部分，直接使用常规文本—图像数据进行训练，重点学习如何根据密集的文本描述生成图像。

这种方式不仅提高了训练效率，也使得模型在处理复杂描述时表现得更稳定。

阶段 Ⅲ：监督微调。

调整了训练数据的比例，将原来的多模态理解数据、纯文本数据和文本—图像数据的比例 7:3:10，调整为 5:1:4。

这一调整既保证了模型保持较强的图像生成能力，又使多模态理解性能得到提升。

3. 数据扩展策略

为进一步提升模型性能，Janus-Pro 在数据方面也做出了大幅扩展。

- 多模态理解数据扩展：在预训练阶段，新增约 9000 万个样本，包括图像字幕数据（如 YFCC 数据集）和针对表格、图表、文档理解的专用数据集（如 Docmatix），使模型能应对更丰富多样的视觉任务。
- 视觉生成数据扩展：为了解决真实数据中存在的噪声问题，模型在训练时引入了大约 7200 万条合成美学数据，并将真实数据与合成数据的比例设定为 1:1。

此举不仅加快了模型的收敛速度，还使得生成的图像在稳定性和美观度上有显著提升。

4. 模型规模扩展

除了架构和训练策略的改进，Janus-Pro 还通过扩展模型规模来进一步提升性能。

（1）多种模型规模。

- Janus-Pro 提供了 1B 和 7B 两个版本。
- 7B 版本在嵌入维度、注意力头数量及层数上均有所提升（例如，词汇量保持 100K，但 1B 版本的嵌入维度为 2048，而 7B 版本则提升至 4096；注意力头数由 16 增至 32，层数也相应从 24 增加到 30），使得大模型在多模态理解和文本到图像生成上的表现更加出色。

（2）收敛速度与性能提升。

实验结果表明，较大规模的模型不仅在收敛速度上有明显加快，而且在多个评测基准上均能取得领先成绩，这验证了模型扩展策略的有效性。

总之，Janus-Pro 通过以下几大关键技术实现了多模态理解与生成任务性能的全面提升。

- 解耦视觉编码：分别设计理解和生成编码路径，避免了传统统一模型中两任务之间的互相干扰。
- 优化训练策略：采用延长阶段 I 训练、聚焦阶段 II 的文本—图像数据训练及微调阶段的数据比例调整，显著提高了训练效率和模型性能。
- 数据扩展：大规模引入多模态理解及合成美学数据，使模型在处理复杂场景时更具鲁棒性和美观性。
- 模型规模扩展：通过扩展参数规模（从 1B 到 7B），进一步提升了模型的理解和生成能力，带来更快的收敛速度和更好的实验结果。

这些技术共同构成了 Janus-Pro 的优势，使其在多模态任务上不仅能精准理解视觉和文本信息，还能根据文本指令生成高质量的图像，代表了统一多模态模型发展的一大进步。

5.2.3 JanusFlow 架构

JanusFlow是一种融合自回归（Autoregression）语言模型和Rectified Flow生成模型的统一多模态理解与生成框架。该框架旨在实现高效的图像理解（Vision Understanding）和文本到图像生成（Text-to-Image Generation），同时避免以往统一模型的架构复杂性和任务冲突问题。

1. 统一的大语言模型（LLM）

JanusFlow将DeepSeek-LLM (1.3B参数量) 作为核心，自回归地处理文本输入，并与图像编码特征相结合，实现多模态理解和文本到图像的生成。

2. 结合自回归语言模型与 Rectified Flow

● 自回归建模（Autoregressive Modeling）：传统LLM模型主要用于多模态理解，通过逐步预测下一个token来处理文本及图像描述任务。

● Rectified Flow生成（Rectified Flow Generation）：一种基于流匹配（Flow Matching）的生成方法，在大语言模型框架内训练，无需额外的复杂网络结构。其核心原理是利用常微分方程进行图像的递归生成，提升采样效率和生成质量。

3. 采用任务解耦的编码器（Decoupled Encoder Design）

JanusFlow采用了独立的视觉编码器来分别处理理解 (Understanding) 和生成 (Generation) 任务，避免任务之间的干扰。

● 理解任务（Vision Understanding Encoder，f_{enc}）：采用SigLIP-Large-Patch/16 预训练模型，专门用于语义特征提取。

● 生成任务（Generation Encoder，g_{enc}）：采用ConvNeXt结构，独立负责生成任务。

● 解码（Generation Decoder，g_{dec}）：负责将潜变量映射回图像空间，并生成最终输出。

4. 三阶段训练策略

JanusFlow采用三阶段训练策略，同时优化自回归模型和Rectified Flow生成。

阶段 1：适配。

训练新增模块（生成编码器、解码器、线性变换层）以适配预训练的LLM和SigLIP编码器。

阶段 2：统一预训练 。

在大规模数据上进行端到端训练，训练数据包含多模态理解数据（Vision-Language Data）、图像生成数据（Text-to-Image Pairs）、纯文本数据（Text-Only Data）。

阶段 3：监督微调。

使用高质量指令微调数据（Instruction Tuning Data），提升模型在对话、问答、文本到图像生成等任务上的能力。

5. 实验结果

（1）在多模态理解任务上的表现。

JanusFlow在多个视觉理解基准测试（如POPE、VQAv2、GQA、MMBench等）上取得了超越现有方法的成绩，尤其是在文本—图像对齐和复杂视觉推理任务上表现优异。

（2）在文本到图像生成任务上的表现。

- JanusFlow在GenEval和DPG-Bench任务中的表现优于多个专用生成模型，包括SDXL和DALL·E 2。
- 在MJHQ FID-30k评测中，JanusFlow以1.3B参数规模取得了9.51FID分数，超越其他同等规模的模型，如Janus（10.10FID）和Show-o（15.18FID）。
- 在语义一致性（Semantic Consistency）方面，JanusFlow在多个任务（如颜色匹配、对象关系等）上取得了较高分数。

总之，JanusFlow是一种创新性的多模态大模型，通过将自回归LLM与Rectified Flow生成模型相结合，在多模态理解和生成任务上达到了领先水平。它的任务解耦设计、三阶段训练方案及表示对齐正则化，使其在现有方法中脱颖而出，为未来的高效统一多模态AI研究提供了新的思路。

5.2.4 核心技术对比

Janus、Janus-Pro和JanusFlow的核心技术对比如表5-1所示。

表5-1 Janus、Janus-Pro和JanusFlow的核心技术对比

技术点	Janus	Janus-Pro	JanusFlow
核心架构	统一自回归变换器	优化的自回归变换器，增强视觉编码解耦	自回归LLM + Rectified Flow生成
视觉编码方式	解耦视觉编码	增强版SigLIP + 高质量VQ Tokenizer	独立视觉编码器（SigLIP + ConvNeXt）
图像生成方法	VQ Tokenizer进行离散化	优化VQ Tokenizer，提高稳定性和细节质量	Rectified Flow生成（基于常微分方程的连续生成）
训练策略	三阶段训练	优化训练流程，提升数据利用率	三阶段训练，结合自回归损失与流匹配损失
优化目标	兼顾理解与生成	改进指令跟随，提升文本—图像生成稳定性	通过流匹配优化常微分方程轨迹，提升图像质量
文本—图像生成能力	基于VQ Tokenizer，质量接近DALL·E 2	优化VQ Tokenizer，超越DALL·E 3	基于Rectified Flow生成，超越Stable Diffusion 3 Medium

经过测试证明，Janus-Pro和JanusFlow在多个任务上超越了Janus，并在某些任务上比DALL·E 3、Stable Diffusion 3 Medium表现更优。

（1）多模态理解任务的评测结果如表5-2所示。

表5-2 多模态理解任务的评测结果

任务	Janus	Janus-Pro-7B	JanusFlow
MMBench	69.4	79.2	78.5

续表

任务	Janus	Janus-Pro-7B	JanusFlow
GQA	59.1	62.0	61.5
SEED-Bench	63.7	72.1	73.0

（2）文本—图像生成任务的评测结果如表5-3所示。

表5-3 文本—图像生成任务的评测结果

任务	Janus	Janus-Pro-7B	JanusFlow
GenEval	0.61	0.80	0.82
DPG-Bench	79.7	84.2	85.5
MJHQ FID-30k	10.10	9.51	9.12

上述评测结果说明JanusFlow在文本到图像任务上的表现比Janus-Pro更优，这主要得益于Rectified Flow生成方法，能够更精准地生成符合文本描述的高质量图像。

根据评测结果可以得出如下结论。

（1）Janus-Pro的优势。

- 在多模态理解和生成任务上均超越Janus，提升了指令跟随能力。
- 改进VQ Tokenizer，提高图像生成稳定性和细节质量，在GenEval任务上超越DALL·E 3。
- 优化训练策略和数据扩展，使其在7B规模下达到最佳性能。

（2）JanusFlow的突破。

- 采用Rectified Flow生成，超越VQ Tokenizer离散化方式，提升生成质量和采样速度。
- 引入新型任务解耦编码器，提高文本—图像对齐能力。
- 在GenEval和MJHQ FID-30k评测中的表现优于Janus-Pro和DALL·E 3。

总之，Janus-Pro更适用于提升现有的多模态理解与文本到图像生成能力。JanusFlow则通过Rectified Flow生成，使图像质量超越传统方法。

5.3 开源信息介绍

目前（作者写本书时），DeepSeek AI的Janus项目开源信息如下。

1. 开源部分

- 各版本的主程序入口文件，如app.py、app_janusflow.py、app_januspro.py和基于FastAPI的fastapi_app.py、fastapi_client.py，用于演示和部署模型。
- 实现模型架构的核心模块，分布在janus、janusflow以及多个models目录中，涵盖了视觉编码、上/下采样、ConvNeXtBlock、UVitBlock、Patchify、ShallowUViTEncoder等各个组件。

● 辅助工具和对话管理模块（如utils/conversation.py和其他工具函数），以及配置文件、依赖管理文件（如pyproject.toml、requirements.txt）和构建脚本（如Makefile）。

2. 未开源部分

● 大型预训练数据集和训练用数据没有开源，只在技术文档中描述了数据来源和数据扩充策略。

● 实际的模型训练代码并没有开源，在训练过程中使用的训练策略和超参数配置等内容在技术报告和文档中有所描述，但训练脚本和流程并未公开，模型权重则通过外部链接提供下载。

目前Janus提供了如下4个不同规模的模型变体，在Hugging Face平台公开发布。

（1）Janus-1.3B（4096 词序列长度）。

（2）JanusFlow-1.3B（4096 词序列长度）。

（3）Janus-Pro-1B（4096 词序列长度）。

（4）Janus-Pro-7B（4096 词序列长度）。

Janus-1.3B（4096 词序列长度）模型的展示如图5-1所示，大家可以从Hugging Face平台获取并部署。

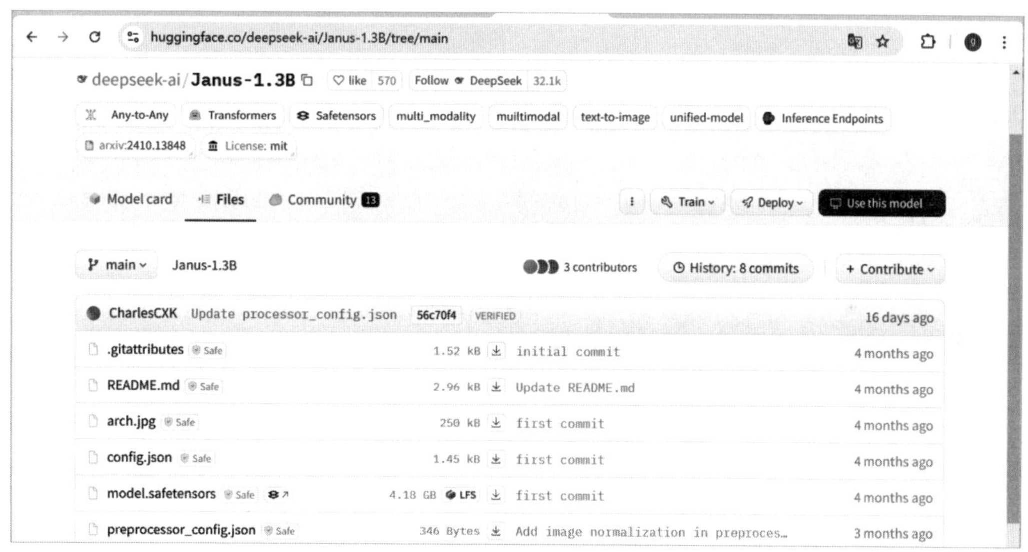

图 5-1　Hugging Face 中的 Janus-1.3B（4096 词序列长度）模型

我们可以直接从Hugging Face上获取这些训练好的模型权重，而无须重新训练模型。

5.4 工具模块

在本项目中，utils目录作为工具模块，主要包含对话管理（conversation.py）和数据处理（io.py）两部分功能，为多模态大模型提供支持。

5.4.1 对话管理

文件conversation.py是Janus项目中用于管理对话模板和历史记录的模块，定义了Conversation类，该类提供了管理系统消息、角色、消息历史及生成用于模型推理的提示等功能。此外，SeparatorStyle枚举类定义了不同的分隔符样式，以支持多种对话格式。该模块还包含一个全局字典conv_templates，用于注册和检索预定义的对话模板。

文件conversation.py的具体实现流程如下所示。

（1）下面的代码定义了SeparatorStyle枚举类，用于表示不同的分隔符风格类型。这些风格用于在对话系统中区分不同角色的消息格式，例如，在消息前后添加冒号、换行符或其他特定标记。枚举类中的每个成员（如ADD_COLON_SINGLE、LLAMA2、DeepSeek等）代表一种特定的分隔符风格，这些风格可以被对话管理系统用来生成符合特定格式要求的对话提示（prompt），从而适应不同的语言模型或应用场景的需求。

```
class SeparatorStyle(IntEnum):
    """ 分隔符风格类型 """
    ADD_COLON_SINGLE = auto()
    ADD_COLON_TWO = auto()
    ADD_COLON_SPACE_SINGLE = auto()
    NO_COLON_SINGLE = auto()
    NO_COLON_TWO = auto()
    ADD_NEW_LINE_SINGLE = auto()
    LLAMA2 = auto()
    CHATGLM = auto()
    CHATML = auto()
    CHATINTERN = auto()
    DOLLY = auto()
    RWKV = auto()
    PHOENIX = auto()
    ROBIN = auto()
    DeepSeek = auto()
    PLAIN = auto()
    ALIGNMENT = auto()
```

（2）下面的代码定义了一个名为Conversation的数据类，用于管理提示模板并保存所有对话历史。它提供了多种方法来设置系统消息、追加新消息、重置消息、更新最后一条消息，以及将对话转换为不同格式。此外，还可以根据不同的分隔符风格和配置，生成用于模型生成的提示。

```
@dataclasses.dataclass
class Conversation:
    """ 一个用于管理提示模板并保存所有对话历史的类 """

    # 模板的名称
    name: str
```

```
# 系统提示的模板
system_template: str = "{system_message}"
# 系统消息
system_message: str = ""
# 两个角色的名称
roles: List[str] = (("USER", "ASSISTANT"),)
# 所有消息。每个元素是 （角色， 消息）
messages: List[List[str]] = ()
# 少样本示例的数量
offset: int = 0
# 分隔符风格和配置
sep_style: SeparatorStyle = SeparatorStyle.ADD_COLON_SINGLE
sep: str = "\n"
sep2: str = None
# 停止条件（默认为 EOS 令牌）
stop_str: str = None
# 如果遇到此列表中的任何令牌，则停止生成
stop_token_ids: List[int] = None

def get_prompt(self) -> str:
    """ 获取用于生成的提示 """
    system_prompt = self.system_template.format(system_message=self.
                                                system_message)

    if self.sep_style == SeparatorStyle.DeepSeek:
        seps = [self.sep, self.sep2]
        if system_prompt == "" or system_prompt is None:
            ret = ""
        else:
            ret = system_prompt + seps[0]
        for i, (role, message) in enumerate(self.messages):
            if message:
                ret += role + ": " + message + seps[i % 2]
            else:
                ret += role + ":"
        return ret
    elif self.sep_style == SeparatorStyle.LLAMA2:
        seps = [self.sep, self.sep2]
        if self.system_message:
            ret = system_prompt
        else:
            ret = "[INST] "
        for i, (role, message) in enumerate(self.messages):
            tag = self.roles[i % 2]
            if message:
                if type(message) is tuple:  # 多模态消息
```

```
                message, _ = message
            if i == 0:
                ret += message + " "
            else:
                ret += tag + " " + message + seps[i % 2]
        else:
            ret += tag
    return ret
elif self.sep_style == SeparatorStyle.PLAIN:
    seps = [self.sep, self.sep2]
    ret = ""
    for i, (role, message) in enumerate(self.messages):
        if message:
            if type(message) is tuple:
                message, _, _ = message
            if i % 2 == 0:
                ret += message + seps[i % 2]
            else:
                ret += message + seps[i % 2]
        else:
            ret += ""
    return ret
elif self.sep_style == SeparatorStyle.ALIGNMENT:
    seps = [self.sep, self.sep2]
    ret = ""
    for i, (role, message) in enumerate(self.messages):
        if message:
            if type(message) is tuple:
                message, _, _ = message
            if i % 2 == 0:
                ret += "<image>\n" + seps[i % 2]
            else:
                ret += message + seps[i % 2]
        else:
            ret += ""
    return ret
else:
    raise ValueError(f"无效的风格: {self.sep_style}")

def get_prompt_for_current_round(self, content=None):
    """获取当前轮次格式化的问答提示（用于 SFT 训练）"""
    if self.sep_style == SeparatorStyle.PLAIN:
        formatted_question = "<image>\n"
    elif self.sep_style == SeparatorStyle.DeepSeek:
        formatted_question = (
            f"{self.roles[0]}: " + content.strip() + self.sep + f"{self.
```

```
                roles[1]}:"
        )
    else:
        raise ValueError(f"不支持的 sep_style: {self.sep_style}")
    return formatted_question

def set_system_message(self, system_message: str):
    """设置系统消息"""
    self.system_message = system_message

def append_message(self, role: str, message: str):
    """追加一条新消息"""
    self.messages.append([role, message])

def reset_message(self):
    """重置消息"""
    self.messages = []

def update_last_message(self, message: str):
    """更新最后一条消息

    构建提示时，最后一条消息通常被设置为 None，
    因此在从模型获取响应后，需要原地更新它。
    """
    self.messages[-1][1] = message

def to_gradio_chatbot(self):
    """将对话转换为 Gradio 聊天机器人的格式"""
    ret = []
    for i, (role, msg) in enumerate(self.messages[self.offset :]):
        if i % 2 == 0:
            ret.append([msg, None])
        else:
            ret[-1][-1] = msg
    return ret

def to_openai_api_messages(self):
    """将对话转换为 OpenAI 聊天补全的格式"""
    system_prompt = self.system_template.format(system_message=self.
                                                system_message)
    ret = [{"role": "system", "content": system_prompt}]

    for i, (_, msg) in enumerate(self.messages[self.offset :]):
        if i % 2 == 0:
            ret.append({"role": "user", "content": msg})
        else:
```

```
                if msg is not None:
                    ret.append({"role": "assistant", "content": msg})
        return ret

    def copy(self):
        return Conversation(
            name=self.name,
            system_template=self.system_template,
            system_message=self.system_message,
            roles=self.roles,
            messages=[[x, y] for x, y in self.messages],
            offset=self.offset,
            sep_style=self.sep_style,
            sep=self.sep,
            sep2=self.sep2,
            stop_str=self.stop_str,
            stop_token_ids=self.stop_token_ids,
        )

    def dict(self):
        return {
            "template_name": self.name,
            "system_message": self.system_message,
            "roles": self.roles,
            "messages": self.messages,
            "offset": self.offset,
        }
```

类Conversation中各个成员方法的具体说明如下所示。

- get_prompt(self) -> str：生成用于模型生成的提示字符串。根据sep_style的不同，采用不同的格式将系统消息和对话历史组合成一个字符串。
- get_prompt_for_current_round(self, content=None)：获取当前轮次格式化的问答提示（用于SFT训练）。根据sep_style的不同，生成特定格式的字符串，通常用于监督微调训练。
- set_system_message(self, system_message: str)：设置系统消息。将传入的system_message字符串赋值给实例的system_message属性。
- append_message(self, role: str, message: str)：追加一条新消息。将角色和消息组成的列表添加到messages属性中。
- reset_message(self)：重置消息。将messages属性重置为空列表，清除所有对话历史。
- update_last_message(self, message: str)：更新最后一条消息。通常在从模型获取响应后，更新最后一条消息的内容。
- to_gradio_chatbot(self)：将对话转换为Gradio聊天机器人的格式。将对话历史转换为适合

Gradio显示的列表格式。

- to_openai_api_messages(self)：将对话转换为OpenAI聊天补全的格式。生成一个包含系统消息和对话历史的字典列表，适用于OpenAI的聊天模型API。
- copy(self)：创建当前对话的副本。返回一个新的Conversation实例，包含当前实例的所有属性值。
- dict(self)：将对话转换为字典格式。返回一个字典，包含模板名称、系统消息、角色、消息和偏移量等信息。

（3）下面的代码定义了一个全局注册表conv_templates，用于存储所有的对话模板。通过register_conv_template函数，可以在该注册表中添加新的对话模板。在添加时，如果未设置override参数或将其设为False，则函数会检查模板名称是否已存在，防止重复注册。

```
# 全局注册表，用于保存所有对话模板
conv_templates: Dict[str, Conversation] = {}

def register_conv_template(template: Conversation, override: bool = False):
    """ 注册一个新的对话模板 """
    if not override:
        assert (
            template.name not in conv_templates
        ), f"{template.name} 已经被注册。"

    conv_templates[template.name] = template
```

（4）函数get_conv_template()的功能是从全局注册表conv_templates中获取指定名称的对话模板，并返回其副本。通过返回副本，确保对返回的模板进行的修改不会影响全局注册表中的原始模板。

```
def get_conv_template(name: str) -> Conversation:
    """ 获取一个对话模板 """
    return conv_templates[name].copy()
```

（5）下面的代码定义了一个全局注册表conv_templates，用于存储所有的对话模板。通过函数register_conv_template()，可以在该注册表中添加新的对话模板。在添加模板时，如果未设置override参数或将其设置为False，则函数register_conv_template()会检查模板名称是否已存在于注册表中，若存在则抛出异常。否则，函数register_conv_template()会将模板添加到注册表中。

```
# 注册 llava_llama2 模板
register_conv_template(
    Conversation(
        name="llava_llama2",
        system_message="You are a helpful language and vision assistant. "
        "You are able to understand the visual content that the user provides, "
        "and assist the user with a variety of tasks using natural language.",
```

```
        system_template="[INST] <<SYS>>\n{system_message}\n<</SYS>>\n\n",
        roles=("[INST]", "[/INST]"),
        messages=(),
        offset=0,
        sep_style=SeparatorStyle.LLAMA2,
        sep=" ",
        sep2=" </s><s>",
        stop_token_ids=[2],
    )
)
```

（6）下面代码的功能是注册一个名为llama-2的对话模板，该模板用于管理对话中的系统信息、角色及消息格式。该模板为基于llama-2结构的对话提供了格式化支持，并能够按预定格式生成对话消息。

```
# 注册 llama2 模板
register_conv_template(
    Conversation(
        name="llama-2",
        system_template="[INST] <<SYS>>\n{system_message}\n<</SYS>>\n\n",
        roles=("[INST]", "[/INST]"),
        messages=(),
        offset=0,
        sep_style=SeparatorStyle.LLAMA2,
        sep=" ",
        sep2=" </s><s>",
        stop_token_ids=[2],
    )
)
```

（7）下面代码的功能是注册一个名为 "deepseek_old" 的对话模板，模板包括系统消息的格式化（通过system_message变量进行配置）、对话角色（设置为User和Assistant）、消息分隔符的样式（使用SeparatorStyle.DeepSeek），以及特定的停止标识符（stop_token_ids和stop_str）。在此模板下，对话的每轮消息之间会通过\n\n分隔，并且每个对话结束时会插入一个标记 < | end＿of＿sentence | >。此模板用于处理用户和助手之间的对话，且通过该模板创建的对话会遵循特定的结构和停止条件。

```
# 注册 deepseek 模板
register_conv_template(
    Conversation(
        name="deepseek_old",
        system_template="{system_message}",
        system_message="",
        roles=("User", "Assistant"),
```

```
        messages=(),
        offset=0,
        sep_style=SeparatorStyle.DeepSeek,
        sep="\n\n",
        sep2="< | end_of_sentence | >",
        stop_token_ids=[100001],
        stop_str=["User:", "< | end_of_sentence | >"],
    )
)
```

(8)下面的代码注册了一个名为"deepseek"的对话模板，配置了系统消息格式(通过system_message)，定义了对话角色为"<|User|>"和"<|Assistant|>"，并设置了消息分隔符为"\n\n"及对话结束标记为"< | end_of_sentence | >"。该模板还使用了SeparatorStyle.DeepSeek样式，确保每轮对话中的消息都按此格式分隔。

```
register_conv_template(
    Conversation(
        name="deepseek",
        system_template="{system_message}",
        system_message="",
        roles=("<|User|>", "<|Assistant|>"),
        messages=(),
        offset=0,
        sep_style=SeparatorStyle.DeepSeek,
        sep="\n\n",
        sep2="< | end_of_sentence | >",
        stop_token_ids=[100001],
        stop_str=["<|User|>", "< | end_of_sentence | >"]
    )
)
```

(9)下面的代码注册了一个名为"plain"的对话模板，适用于plain(简单)风格的对话。

```
register_conv_template(
    Conversation(
        name="plain",
        system_template="",
        system_message="",
        roles=("", ""),
        messages=(),
        offset=0,
        sep_style=SeparatorStyle.PLAIN,
        sep="",
        sep2="",
        stop_token_ids=[2],
```

```
        stop_str=["</s>"],
    )
)
```

（10）下面的代码注册了一个名为 "alignment" 的对话模板，适用于ALIGNMENT（对齐）风格的对话。

```
register_conv_template(
    Conversation(
        name="alignment",
        system_template="",
        system_message="",
        roles=("", ""),
        messages=(),
        offset=0,
        sep_style=SeparatorStyle.ALIGNMENT,
        sep="",
        sep2="",
        stop_token_ids=[2],
        stop_str=["</s>"],
    )
)
```

（11）下面代码的功能是测试 "deepseek" 对话模板的行为，并打印出最终生成的对话提示。首先，通过get_conv_template("deepseek") 获取DeepSeek风格的对话模板，然后，使用append_message(role, message) 依次在对话中添加用户（<|User|>）和助手（<|Assistant|>）的消息：

- "Hello!" → "Hi! This is Tony."。
- "Who are you?" → "I am a helpful assistant."。
- "How are you?" → 助手的回复设为None（表示等待生成的部分）。

最后，通过conv.get_prompt() 生成并打印最终的对话提示，用于输入大模型中进行推理。

```
if __name__ == "__main__":
    print("deepseek template:")
    conv = get_conv_template("deepseek")
    conv.append_message(conv.roles[0], "Hello!")
    conv.append_message(conv.roles[1], "Hi! This is Tony.")
    conv.append_message(conv.roles[0], "Who are you?")
    conv.append_message(conv.roles[1], "I am a helpful assistant.")
    conv.append_message(conv.roles[0], "How are you?")
    conv.append_message(conv.roles[1], None)
    print(conv.get_prompt())
```

5.4.2 数据加载

文件io.py的主要功能是加载预训练的多模态大模型，并支持图像和JSON数据的处理，适用于视觉—语言任务，包括以下功能。

- 加载预训练模型：使用VLChatProcessor和MultiModalityCausalLM读取预训练的视觉—语言模型，并将其移动到GPU上运行。
- 加载并解析图像：支持从文件路径或Base64编码读取图像，并将其转换为PIL.Image对象，方便后续处理。
- 读取JSON数据：提供从文件中加载JSON数据的函数。

```
def load_pretrained_model(model_path: str):
    """
    加载预训练的多模态（视觉—语言）模型，并移动到 GPU。

    Args:
        model_path (str): 预训练模型的路径。

    Returns:
        tokenizer: 加载的分词器（tokenizer）。
        vl_chat_processor: 视觉—语言处理器（VLChatProcessor）。
        vl_gpt: 预训练的大模型（MultiModalityCausalLM）。
    """
    vl_chat_processor: VLChatProcessor = VLChatProcessor.from_
                                         pretrained(model_path)
    tokenizer = vl_chat_processor.tokenizer

    vl_gpt: MultiModalityCausalLM = AutoModelForCausalLM.from_pretrained(
        model_path, trust_remote_code=True
    )
    vl_gpt = vl_gpt.to(torch.bfloat16).cuda().eval()
                                          # 使用bfloat16 精度并移动到 GPU

    return tokenizer, vl_chat_processor, vl_gpt

def load_pil_images(conversations: List[Dict[str, str]]) -> List[PIL.Image.
                                                                   Image]:
    """
    解析会话中的图像数据，支持文件路径或 Base64 格式的图片。

    Args:
        conversations (List[Dict[str, str]]): 包含对话信息的列表，每个元素格式如下：
            [
                {
                    "role": "User",
```

```
                    "content": "<image_placeholder>\n 提取此图像中的所有信息，并
                                将其转换为 Markdown 格式。",
                    "images": ["./examples/table_datasets.png"]
                },
                {"role": "Assistant", "content": ""}
            ]

    Returns:
        pil_images (List[PIL.Image.Image]): 解析得到的 PIL 图像对象列表。
    """
    pil_images = []

    for message in conversations:
        if "images" not in message:
            continue  # 如果当前消息不包含 "images" 字段，则跳过

        for image_data in message["images"]:
            if image_data.startswith("data:image"):
                # 如果图像数据是 Base64 编码格式
                _, image_data = image_data.split(",", 1)
                image_bytes = base64.b64decode(image_data)
                pil_img = PIL.Image.open(io.BytesIO(image_bytes))
            else:
                # 如果图像数据是文件路径
                pil_img = PIL.Image.open(image_data)

            pil_img = pil_img.convert("RGB")  # 转换为 RGB 格式
            pil_images.append(pil_img)

    return pil_images

def load_json(filepath: str):
    """
    加载 JSON 文件并返回解析后的数据。

    Args:
        filepath (str): JSON 文件的路径。

    Returns:
        data (dict): 解析后的 JSON 数据。
    """
    with open(filepath, "r") as f:
        data = json.load(f)
        return data
```

5.5 构建多模态模型

在本项目中，“janus/models”目录包含Janus项目中多模态模型的核心实现，涵盖视觉和语言模态的编码器、解码器、特征提取器、投影器及多模态融合模块。这些组件共同构成了支持多模态理解和生成任务的架构基础。此外，目录还提供数据预处理工具，用于将输入图像和文本转换为模型可处理的格式。然而，该目录并未包含完整的训练流程，训练相关的代码可能位于项目的其他部分。

5.5.1 向量量化模型

文件vq_model.py实现了一个基于向量量化（Vector Quantization）的模型，用于图像的编码和解码。该模型包括编码器、解码器和向量量化器，支持图像的压缩和重建。这是多模态模型中图像处理的核心组件，为图像的嵌入和生成奠定了基础。

文件的核心思想是借助向量量化变分自编码器（Vector Quantized Variational Auto Encoder，VQ-VAE）对输入数据进行编码、离散化，并使用向量量化方法来学习更好的表示。具体实现流程如下所示。

（1）ModelArgs类主要用于定义VQ-VAE模型的关键超参数，包括码本（Codebook）相关参数（如大小、嵌入维度、是否归一化等）、损失函数超参数（如提交损失权重、熵损失比例等），以及编码器与解码器结构（如通道倍增因子、中间特征通道数等）。这些参数共同决定了模型的编码能力、离散化策略及重建质量，并可通过调整超参数来优化模型性能。

```
@dataclass
class ModelArgs:
    """ 模型参数 """
    codebook_size: int = 16384                          # 码本大小
    codebook_embed_dim: int = 8                         # 码本嵌入维度
    codebook_l2_norm: bool = True                       # 是否对码本进行 L2 归一化
    codebook_show_usage: bool = True                    # 是否显示码本的使用情况
    commit_loss_beta: float = 0.25                      # 码本提交损失的权重
    entropy_loss_ratio: float = 0.0                     # 熵损失的比例

    encoder_ch_mult: List[int] = field(default_factory=lambda: [1, 1, 2, 2, 
4])
                                                        # 编码器通道倍增因子
    decoder_ch_mult: List[int] = field(default_factory=lambda: [1, 1, 2, 2, 
4])
                                                        # 解码器通道倍增因子
    z_channels: int = 256                               # 中间特征的通道数
    dropout_p: float = 0.0                              # Dropout 概率
```

（2）Encoder类是VQ-VAE（或类似自编码器）中的编码器，负责将输入图像转换为紧凑的特征表示（潜变量）。整个过程包括特征提取、分辨率逐步降低（下采样）、残差学习、注意力机制等，以提取关键信息并减少冗余，为后续的离散化和解码提供高质量的潜在特征。

Encoder类中的成员如下。

- 方法__init__()：初始化编码器的所有模块，包括输入卷积层、多个下采样层、残差块和注意力块。它根据输入通道数、基础通道数、通道倍增因子等参数动态创建网络结构。该方法还定义了是否使用卷积进行上/下采样、Dropout概率、归一化类型及中间特征的通道数等，以确保网络在不同分辨率下有效提取信息并进行特征压缩。
- 方法forward()：前向传播方法，用于处理输入图像的前向传播过程。首先，输入通过conv_in卷积层进行初步处理。然后，图像依次通过每个分辨率级别的残差块和注意力块，在逐步降低分辨率的同时提取更深层次的特征。接下来，图像通过中间层进一步处理，最后，经过输出层的归一化和卷积，生成潜在特征图输出。

Encoder的结构如下。

- conv_in层：编码器的第一层卷积层，它接收输入图像并通过3×3的卷积核进行特征提取，将输入转换为指定通道数ch的特征图，作为后续处理的起点。
- conv_blocks层：包含多个下采样模块，每个模块由ResnetBlock和AttnBlock组成，用于逐层提取高级特征并降低图像分辨率。每个下采样模块的输出由不同数量的残差块和可能的注意力机制块组成，确保模型在各个分辨率下有效地提取信息。
- mid层：它是中间特征处理模块，包含若干ResnetBlock和AttnBlock，用于在最小分辨率级别后进一步处理和优化特征表示。这些模块帮助模型捕获更深层次的语义信息，并增强特征的表现力。
- conv_out层：编码器的最后一层卷积层，它将经过多次处理后的特征图进行归一化、非线性激活，并通过3×3的卷积层生成z_channels维度的输出，作为最终的潜在特征图供后续解码使用。

```
class Encoder(nn.Module):
    """ 编码器 """
    def __init__(
        self,
        in_channels=3,                              # 输入通道数
        ch=128,                                     # 基础通道数
        ch_mult=(1, 1, 2, 2, 4),                    # 通道倍增因子
        num_res_blocks=2,                           # 每个分辨率的残差块数量
        norm_type="group",                          # 归一化类型
        dropout=0.0,                                # Dropout 概率
        resamp_with_conv=True,                      # 是否使用卷积进行上 / 下采样
        z_channels=256,                             # 中间特征的通道数
    ):
        super().__init__()
        self.num_resolutions = len(ch_mult)     # 分辨率级别数量
        self.num_res_blocks = num_res_blocks    # 每个分辨率的残差块数量
        self.conv_in = nn.Conv2d(in_channels, ch, kernel_size=3, stride=1,
                                padding=1)     # 输入卷积
```

```
        # 下采样
        in_ch_mult = (1,) + tuple(ch_mult)
        self.conv_blocks = nn.ModuleList()
        for i_level in range(self.num_resolutions):
            conv_block = nn.Module()
            # 残差块和注意力块
            res_block = nn.ModuleList()
            attn_block = nn.ModuleList()
            block_in = ch * in_ch_mult[i_level]
            block_out = ch * ch_mult[i_level]
            for _ in range(self.num_res_blocks):
                res_block.append(
                    ResnetBlock(
                        block_in, block_out, dropout=dropout,
                        norm_type=norm_type
                    )
                )
                block_in = block_out
                if i_level == self.num_resolutions - 1:
                    attn_block.append(AttnBlock(block_in, norm_type))
            conv_block.res = res_block
            conv_block.attn = attn_block
            # 下采样
            if i_level != self.num_resolutions - 1:
                conv_block.downsample = Downsample(block_in, resamp_with_conv)
            self.conv_blocks.append(conv_block)

        # 中间层
        self.mid = nn.ModuleList()
        self.mid.append(
            ResnetBlock(block_in, block_in, dropout=dropout,
                        norm_type=norm_type)
        )
        self.mid.append(AttnBlock(block_in, norm_type=norm_type))
        self.mid.append(
            ResnetBlock(block_in, block_in, dropout=dropout, norm_type=norm_type)
        )

        # 输出层
        self.norm_out = Normalize(block_in, norm_type)
        self.conv_out = nn.Conv2d(
            block_in, z_channels, kernel_size=3, stride=1, padding=1
        )

    def forward(self, x):
```

```
        h = self.conv_in(x)
        # 下采样
        for i_level, block in enumerate(self.conv_blocks):
            for i_block in range(self.num_res_blocks):
                h = block.res[i_block](h)
                if len(block.attn) > 0:
                    h = block.attn[i_block](h)
            if i_level != self.num_resolutions - 1:
                h = block.downsample(h)

        # 中间层
        for mid_block in self.mid:
            h = mid_block(h)

        # 输出层
        h = self.norm_out(h)
        h = nonlinearity(h)
        h = self.conv_out(h)
        return h
```

（3）类Decoder是一种神经网络解码器结构，主要用于将编码器产生的潜在空间特征图z转换为最终的输出图像。它通过若干残差块和注意力机制模块逐层恢复图像的分辨率，同时保持输入特征图的语义信息。解码器的核心目标是利用输入的潜在特征图z生成和输入图像维度相同的输出，通常用于图像生成任务。

类Decoder中的成员如下。

- __init__ ()方法：负责初始化解码器的各个模块，包括输入卷积层、多个上采样模块、残差块和注意力块。它通过设定参数，如中间特征通道数、基础通道数、通道倍增因子等来定义解码器的结构。通过这一方法，解码器能够根据输入的潜在特征生成最终输出图像。
- last_layer()方法（最后一层的权重）：是一个属性方法，它返回解码器最后一层卷积层 (conv_out) 的权重。这可以用于检查或操作解码器的最后一层，通常在模型分析或特定的优化任务中使用。
- forward()方法（前向传播）：forward方法定义了解码器的前向传播流程。首先，潜在特征图z通过输入卷积层 (conv_in) 转换为初始特征图。接下来，特征图经过中间层的处理，这些层包括残差块和注意力机制模块。然后，特征图通过一系列上采样模块逐渐恢复分辨率。最后，经过输出层生成最终的输出图像。该方法是解码器的核心操作，确保特征图从潜在空间恢复到图像空间。

```
class Decoder(nn.Module):
    """ 解码器 """
    def __init__(
        self,
        z_channels=256,              # 中间特征的通道数
        ch=128,                      # 基础通道数
        ch_mult=(1, 1, 2, 2, 4),     # 通道倍增因子
```

```
    num_res_blocks=2,          # 每个分辨率的残差块数量
    norm_type="group",         # 归一化类型
    dropout=0.0,               # Dropout 概率
    resamp_with_conv=True,     # 是否使用卷积进行上 / 下采样
    out_channels=3,            # 输出通道数
):
    super().__init__()
    self.num_resolutions = len(ch_mult)    # 分辨率级别数量
    self.num_res_blocks = num_res_blocks   # 每个分辨率的残差块数量

    block_in = ch * ch_mult[self.num_resolutions - 1]
    # z 到 block_in 的转换
    self.conv_in = nn.Conv2d(
        z_channels, block_in, kernel_size=3, stride=1, padding=1
    )

    # 中间层
    self.mid = nn.ModuleList()
    self.mid.append(
        ResnetBlock(block_in, block_in, dropout=dropout,
                    norm_type=norm_type)
    )
    self.mid.append(AttnBlock(block_in, norm_type=norm_type))
    self.mid.append(
        ResnetBlock(block_in, block_in,
                    dropout=dropout, norm_type=norm_type)
    )

    # 上采样
    self.conv_blocks = nn.ModuleList()
    for i_level in reversed(range(self.num_resolutions)):
        conv_block = nn.Module()
        # 残差块和注意力块
        res_block = nn.ModuleList()
        attn_block = nn.ModuleList()
        block_out = ch * ch_mult[i_level]
        for _ in range(self.num_res_blocks + 1):
            res_block.append(
                ResnetBlock(
                    block_in, block_out, dropout=dropout,
                    norm_type=norm_type
                )
            )
            block_in = block_out
            if i_level == self.num_resolutions - 1:
                attn_block.append(AttnBlock(block_in, norm_type))
```

```
            conv_block.res = res_block
            conv_block.attn = attn_block
            # 上采样
            if i_level != 0:
                conv_block.upsample = Upsample(block_in, resamp_with_conv)
            self.conv_blocks.append(conv_block)

        # 输出层
        self.norm_out = Normalize(block_in, norm_type)
        self.conv_out = nn.Conv2d(
            block_in, out_channels, kernel_size=3, stride=1, padding=1
        )

    @property
    def last_layer(self):
        return self.conv_out.weight

    def forward(self, z):
        # z到block_in的转换
        h = self.conv_in(z)

        # 中间层
        for mid_block in self.mid:
            h = mid_block(h)

        # 上采样
        for i_level, block in enumerate(self.conv_blocks):
            for i_block in range(self.num_res_blocks + 1):
                h = block.res[i_block](h)
                if len(block.attn) > 0:
                    h = block.attn[i_block](h)
            if i_level != self.num_resolutions - 1:
                h = block.upsample(h)

        # 输出层
        h = self.norm_out(h)
        h = nonlinearity(h)
        h = self.conv_out(h)
        return h
```

（4）VectorQuantizer是一种用于量化的神经网络模块，主要用于将连续的向量表示映射到一个离散的码本空间。这种技术通常用于变分自编码器（VAE）和生成模型中，通过向量量化来提高模型的表达能力。该模块使用嵌入式码本来实现向量量化，同时支持可调的损失函数和不同的正则化策略，帮助模型在训练过程中有效地逼近目标分布。

VectorQuantizer中的成员如下所示。

● __init__()方法（构造函数）：用于初始化VectorQuantizer类的各个参数和组件。它定义了码本的大小（n_e）、嵌入维度（e_dim）、码本损失权重（beta）及是否进行L2归一化等选项，通过nn.Embedding层初始化码本，并对其进行必要的初始化操作。如果启用了L2归一化，则嵌入的权重会进行归一化处理。此外，show_usage参数用于决定是否跟踪码本的使用情况，若启用，则会额外注册一个缓冲区。

● forward()方法（前向传播）：定义了向量量化的核心操作。首先，输入张量z被重塑为适合量化的形状。然后，计算z与码本嵌入之间的距离（通过欧几里得距离的方式），并找到最近的码本向量。量化后的向量z_q被用于计算量化损失（vq_loss）、提交损失（commit_loss）及熵损失（entropy_loss）。这些损失会在训练时被用来优化模型，同时z_q通过加入detach操作保持梯度流动。最后，返回量化后的结果和损失值。

● get_codebook_entry()方法（获取码本项）：用于根据给定的索引从码本中提取对应的向量。它可以接收索引indices和目标形状shape，并根据channel_first参数来决定如何重塑输出的形状。如果启用了L2归一化，则嵌入向量会进行归一化处理。这一方法通常用于检索特定的码本向量，在推理和生成过程中可能被调用以获取相应的离散表示。

```
class VectorQuantizer(nn.Module):
    """ 向量量化器 """
    def __init__(self, n_e, e_dim, beta, entropy_loss_ratio, l2_norm, show_
            usage):
        super().__init__()
        self.n_e = n_e                      # 码本大小
        self.e_dim = e_dim                  # 码本嵌入维度
        self.beta = beta                    # 码本提交损失的权重
        self.entropy_loss_ratio = entropy_loss_ratio        # 熵损失的比例
        self.l2_norm = l2_norm              # 是否对码本进行 L2 归一化
        self.show_usage = show_usage        # 是否显示码本的使用情况

        self.embedding = nn.Embedding(self.n_e, self.e_dim)  # 码本嵌入
        self.embedding.weight.data.uniform_(-1.0 / self.n_e, 1.0 / self.n_e)
                                                            # 初始化码本
        if self.l2_norm:
            self.embedding.weight.data = F.normalize(
                self.embedding.weight.data, p=2, dim=-1
            )  # 对码本进行 L2 归一化
        if self.show_usage:
            self.register_buffer("codebook_used",
                                 nn.Parameter(torch.zeros(65536)))

    def forward(self, z):
        # 将 z 重塑为 (batch, height, width, channel) 并展平
        z = torch.einsum("b c h w -> b h w c", z).contiguous()
        z_flattened = z.view(-1, self.e_dim)
```

```
        # 计算 z 和码本嵌入之间的距离 (z - e)^2 = z^2 + e^2 - 2 e * z

        if self.l2_norm:
            z = F.normalize(z, p=2, dim=-1)  # 对 z 进行 L2 归一化
            z_flattened = F.normalize(z_flattened, p=2, dim=-1)
                                                    # 对 z_flattened 进行 L2 归一化
            embedding = F.normalize(self.embedding.weight, p=2, dim=-1)
                                                          # 对码本嵌入进行 L2 归一化
        else:
            embedding = self.embedding.weight

        d = (
            torch.sum(z_flattened**2, dim=1, keepdim=True)
            + torch.sum(embedding**2, dim=1)
            - 2
            * torch.einsum(
                "bd,dn->bn", z_flattened, torch.einsum("n d -> d n", embedding)
            )
        )

        min_encoding_indices = torch.argmin(d, dim=1)  # 找到最近的码本索引
        z_q = embedding[min_encoding_indices].view(z.shape)  # 获取量化后的向量
        perplexity = None       # 困惑度
        min_encodings = None    # 最小编码
        vq_loss = None          # 码本损失
        commit_loss = None      # 提交损失
        entropy_loss = None     # 熵损失

        # 计算码本的损失
        if self.training:
            vq_loss = torch.mean((z_q - z.detach()) ** 2)  # 码本损失
            commit_loss = self.beta * torch.mean((z_q.detach() - z) ** 2)   #
提交损失
            entropy_loss = self.entropy_loss_ratio * compute_entropy_
                           loss(-d)  # 熵损失

        # 保留梯度
        z_q = z + (z_q - z).detach()

        # 重塑回原始输入的形状
        z_q = torch.einsum("b h w c -> b c h w", z_q)

        return (
            z_q,
            (vq_loss, commit_loss, entropy_loss),
            (perplexity, min_encodings, min_encoding_indices),
```

```
        )

    def get_codebook_entry(self, indices, shape=None, channel_first=True):
        # shape = (batch, channel, height, width) if channel_first else (batch,
                    height, width, channel)
        if self.l2_norm:
            embedding = F.normalize(self.embedding.weight, p=2, dim=-1)
        else:
            embedding = self.embedding.weight
        z_q = embedding[indices]  # (b*h*w, c)

        if shape is not None:
            if channel_first:
                z_q = z_q.reshape(shape[0], shape[2], shape[3], shape[1])
                # 重塑回原始输入的形状
                z_q = z_q.permute(0, 3, 1, 2).contiguous()
            else:
                z_q = z_q.view(shape)
        return z_q
```

（5）ResnetBlock类是一种常见的神经网络模块，常用于构建残差网络（ResNet）。它通过引入快捷连接（Skip Connection）来解决深度网络中的梯度消失问题，从而加速训练。该模块由两个卷积层、两个归一化层、一个激活函数和一个可选的快捷连接组成。通过添加快捷连接，网络的每一层都可以直接接收前一层的输出，这有助于信息的流动并增强网络的表现力。

ResnetBlock类中的成员如下所示。

- __init__ 方法()：用于初始化ResnetBlock类的各个组件。它接受输入通道数（in_channels）、输出通道数（out_channels）、是否使用卷积快捷连接（conv_shortcut）、dropout概率（dropout）及归一化类型（norm_type）等参数。此方法初始化了两个卷积层及归一化层，在第二个卷积层后还会添加dropout操作。此外，如果输入和输出通道不匹配，则根据conv_shortcut参数，构造卷积方式的快捷连接或 1×1 卷积的快捷连接。
- forward()方法：定义了ResnetBlock的前向传播过程。首先，对输入进行归一化处理，然后通过非线性激活函数激活，并传递至第一个卷积层。接着对输出进行归一化、激活、dropout并通过第二个卷积层得到结果。如果输入和输出通道数不一致，则使用快捷连接对输入进行处理，最后将输入和卷积后的结果相加，这就是残差连接的核心。最终输出是两者的和，即 x + h。这种结构有助于解决梯度消失问题，并加速训练过程。

```
class ResnetBlock(nn.Module):
    """ 残差块 """
    def __init__(
        self,
        in_channels,
        out_channels=None,
```

```
        conv_shortcut=False,
        dropout=0.0,
        norm_type="group",
    ):
        super().__init__()
        self.in_channels = in_channels
        out_channels = in_channels if out_channels is None else out_channels
        self.out_channels = out_channels
        self.use_conv_shortcut = conv_shortcut

        self.norm1 = Normalize(in_channels, norm_type)
        self.conv1 = nn.Conv2d(
            in_channels, out_channels, kernel_size=3, stride=1, padding=1
        )
        self.norm2 = Normalize(out_channels, norm_type)
        self.dropout = nn.Dropout(dropout)
        self.conv2 = nn.Conv2d(
            out_channels, out_channels, kernel_size=3, stride=1, padding=1
        )

        if self.in_channels != self.out_channels:
            if self.use_conv_shortcut:
                self.conv_shortcut = nn.Conv2d(
                    in_channels, out_channels, kernel_size=3, stride=1, padding=1
                )
            else:
                self.nin_shortcut = nn.Conv2d(
                    in_channels, out_channels, kernel_size=1, stride=1, padding=0
                )

    def forward(self, x):
        h = x
        h = self.norm1(h)
        h = nonlinearity(h)
        h = self.conv1(h)
        h = self.norm2(h)
        h = nonlinearity(h)
        h = self.dropout(h)
        h = self.conv2(h)

        if self.in_channels != self.out_channels:
            if self.use_conv_shortcut:
                x = self.conv_shortcut(x)
            else:
                x = self.nin_shortcut(x)
        return x + h
```

（6）AttnBlock是一个实现自注意力机制的神经网络模块，该模块通过使用三个卷积层（分别用于查询、键、值）来计算输入特征图的注意力权重。首先，输入特征图经过归一化处理，生成查询（Q）、键（K）和值（V）。然后，计算查询和键之间的相似度，生成注意力权重，并使用softmax函数进行归一化。接着，将这些权重与值进行加权平均，得到加权后的特征图，并通过一个卷积层进行最终的输出投影。最后，将输出结果与原始输入特征图相加，形成残差连接。该模块的主要目的是使网络能够捕捉不同空间位置之间的依赖关系，从而提高特征的表达能力。

```
class AttnBlock(nn.Module):
    """ 注意力块 """
    def __init__(self, in_channels, norm_type="group"):
        super().__init__()
        self.norm = Normalize(in_channels, norm_type)
        self.q = nn.Conv2d(in_channels, in_channels, kernel_size=1, stride=1,
                           padding=0)
        self.k = nn.Conv2d(in_channels, in_channels, kernel_size=1, stride=1,
                           padding=0)
        self.v = nn.Conv2d(in_channels, in_channels, kernel_size=1, stride=1,
                           padding=0)
        self.proj_out = nn.Conv2d(
            in_channels, in_channels, kernel_size=1, stride=1, padding=0
        )

    def forward(self, x):
        h_ = x
        h_ = self.norm(h_)
        q = self.q(h_)
        k = self.k(h_)
        v = self.v(h_)

        # 计算注意力
        b, c, h, w = q.shape
        q = q.reshape(b, c, h * w)
        q = q.permute(0, 2, 1)  # b,hw,c
        k = k.reshape(b, c, h * w)  # b,c,hw
        w_ = torch.bmm(q, k)  # b,hw,hw    w[b,i,j]=sum_c q[b,i,c]k[b,c,j]
        w_ = w_ * (int(c) ** (-0.5))
        w_ = F.softmax(w_, dim=2)

        # 应用注意力
        v = v.reshape(b, c, h * w)
        w_ = w_.permute(0, 2, 1)  # b,hw,hw (first hw of k, second of q)
        h_ = torch.bmm(v, w_)     # b, c,hw (hw of q) h_[b,c,j] = sum_i
                                    v[b,c,i] w_[b,i,j]
        h_ = h_.reshape(b, c, h, w)
```

```
        h_ = self.proj_out(h_)

        return x + h_
```

（7）Nonlinearity()是一个自定义的激活函数，该函数通过将输入值与Sigmoid函数的输出相乘来提供非线性变换，通常可以改善梯度流动并在某些情况下提升模型性能。

```
def nonlinearity(x):
    # Swish 激活函数
    return x * torch.sigmoid(x)
```

（8）Normalize()是一个归一化层的构造函数，根据指定的norm_type类型选择不同的归一化方法。它支持两种归一化方式。

- GroupNorm：使用分组归一化（Group Normalization），适用于小批量数据，且不依赖批量大小。它将输入通道划分为多个组进行归一化。
- BatchNorm：使用批量归一化（Batch Normalization），通过计算批量数据的均值和方差来进行归一化，通常用于训练深度神经网络。

```
def Normalize(in_channels, norm_type="group"):
    """ 归一化层 """
    assert norm_type in ["group", "batch"]
    if norm_type == "group":
        return nn.GroupNorm(
            num_groups=32, num_channels=in_channels, eps=1e-6, affine=True
        )
    elif norm_type == "batch":
        return nn.SyncBatchNorm(in_channels)
```

（9）类Upsample是一个上采样层，用于对输入的张量进行上采样（增加空间分辨率）。类Upsample支持两种方式。

- 不带卷积：使用最近邻插值法（nearest）将输入的张量按比例（scale_factor=2.0）扩大两倍。
- 带卷积：除了使用最近邻插值法上采样，还会通过一个卷积层（3×3卷积）进一步处理上采样后的特征图。

此外，类Upsample会根据输入的数据类型进行转换，确保在采样过程中数据格式的兼容性。

```
class Upsample(nn.Module):
    """ 上采样层 """
    def __init__(self, in_channels, with_conv):
        super().__init__()
        self.with_conv = with_conv
        if self.with_conv:
            self.conv = nn.Conv2d(
```

```
                in_channels, in_channels, kernel_size=3, stride=1, padding=1
            )

    def forward(self, x):
        if x.dtype != torch.float32:
            x = F.interpolate(x.to(torch.float), scale_factor=2.0,
                              mode="nearest").to(
                torch.bfloat16
            )
        else:
            x = F.interpolate(x, scale_factor=2.0, mode="nearest")

        if self.with_conv:
            x = self.conv(x)
        return x
```

（10）类Downsample是一个下采样层，用于降低输入张量的空间分辨率。

```
class Downsample(nn.Module):
    """ 下采样层 """
    def __init__(self, in_channels, with_conv):
        super().__init__()
        self.with_conv = with_conv
        if self.with_conv:
            # torch.conv 不支持非对称填充，必须手动实现
            self.conv = nn.Conv2d(
                in_channels, in_channels, kernel_size=3, stride=2, padding=0
            )

    def forward(self, x):
        if self.with_conv:
            pad = (0, 1, 0, 1)
            x = F.pad(x, pad, mode="constant", value=0)
            x = self.conv(x)
        else:
            x = F.avg_pool2d(x, kernel_size=2, stride=2)
        return x
```

（11）函数compute_entropy_loss()用于计算熵损失，主要用于量化学习中的熵正则化。熵损失有助于控制模型的输出分布，使其更具不确定性。

```
def compute_entropy_loss(affinity, loss_type="softmax", temperature=0.01):
    """ 计算熵损失 """
    flat_affinity = affinity.reshape(-1, affinity.shape[-1])
    flat_affinity /= temperature
    probs = F.softmax(flat_affinity, dim=-1)
```

```
    log_probs = F.log_softmax(flat_affinity + 1e-5, dim=-1)
    if loss_type == "softmax":
        target_probs = probs
    else:
        raise ValueError("不支持的熵损失类型: {}".format(loss_type))
    avg_probs = torch.mean(target_probs, dim=0)
    avg_entropy = -torch.sum(avg_probs * torch.log(avg_probs + 1e-5))
    sample_entropy = -torch.mean(torch.sum(target_probs * log_probs, dim=-1))
    loss = sample_entropy - avg_entropy
    return loss
```

（12）类VQModel是一个向量量化模型，主要用于图像或其他数据的编码和解码过程。该模型包括编码器、解码器及向量量化模块，具体说明如下。

- 初始化：模型通过配置参数初始化编码器（encoder）、解码器（decoder）及向量量化器（VectorQuantizer）。此外，还包括了用于量化的卷积层（quant_conv）和解码后的卷积层（post_quant_conv）。
- 编码（encode）：输入数据首先通过编码器处理，然后经过卷积层，最后进行向量量化得到离散化的代码，并返回量化结果、嵌入损失及信息。
- 解码（decode）：量化后的编码通过解码器进行还原，生成重建数据。
- 解码代码（decode_code）：根据给定的编码（code_b），将其转换为具体的量化值并经过解码，生成解码后的数据。
- 前向传播（forward）：在前向传播中，输入数据经过编码和解码过程，返回解码后的输出及嵌入损失（用于正则化）。

VQModel模型通过向量量化将连续的输入映射到离散的编码空间，在图像生成和重建任务中起重要作用。

```
class VQModel(nn.Module):
    """ 向量量化模型 """
    def __init__(self, config: ModelArgs):
        super().__init__()
        self.config = config
        self.encoder = Encoder(
            ch_mult=config.encoder_ch_mult,
            z_channels=config.z_channels,
            dropout=config.dropout_p,
        )
        self.decoder = Decoder(
            ch_mult=config.decoder_ch_mult,
            z_channels=config.z_channels,
            dropout=config.dropout_p,
        )
```

```
        self.quantize = VectorQuantizer(
            config.codebook_size,
            config.codebook_embed_dim,
            config.commit_loss_beta,
            config.entropy_loss_ratio,
            config.codebook_l2_norm,
            config.codebook_show_usage,
        )
        self.quant_conv = nn.Conv2d(config.z_channels,
                                        config.codebook_embed_dim, 1)
        self.post_quant_conv = nn.Conv2d(
            config.codebook_embed_dim, config.z_channels, 1
        )

    def encode(self, x):
        h = self.encoder(x)
        h = self.quant_conv(h)
        quant, emb_loss, info = self.quantize(h)
        return quant, emb_loss, info

    def decode(self, quant):
        quant = self.post_quant_conv(quant)
        dec = self.decoder(quant)
        return dec

    def decode_code(self, code_b, shape=None, channel_first=True):
        quant_b = self.quantize.get_codebook_entry(code_b, shape,
                                                        channel_first)
        dec = self.decode(quant_b)
        return dec

    def forward(self, input):
        quant, diff, _ = self.encode(input)
        dec = self.decode(quant)
        return dec, diff
```

（13）下面的代码定义了一个名为VQ_16()的模型配置函数，用于创建一个带有特定参数的VQModel实例。配置包括编码器和解码器的通道倍增参数，均是 [1, 1, 2, 2, 4]。此外，还允许通过关键字参数 (**kwargs) 来传入其他配置。

```
def VQ_16(**kwargs):
    return VQModel(
        ModelArgs(
            encoder_ch_mult=[1, 1, 2, 2, 4], decoder_ch_mult
                            =[1, 1, 2, 2, 4], **kwargs
```

```
        )
    )

VQ_models = {"VQ-16": VQ_16}
```

VQ_models = {"VQ-16": VQ_16} 这行代码的意思是将一个名为 "VQ-16" 的字符串作为键，将函数VQ_16 作为值，存储在字典VQ_models中。这样做的好处是，可以通过名字来动态选择不同的模型配置，而无须直接调用VQ_16() 函数。

5.5.2 CLIP 视觉编码器

文件clip_encoder.py定义了一个名为CLIPVisionTower的类，用于构建并使用不同类型的视觉模型（如siglip、sam或HuggingFace的CLIP）。这个类包括图像预处理（如像素均值和标准差归一化）、选择不同层输出的功能，以及根据不同模型配置生成相应的视觉模型。具体来说，CLIPVisionTower将图像通过视觉塔（vision tower）进行特征提取，提取指定层的特征，并根据select_feature参数选择合适的特征（如“patch”或“cls_patch”）。它还支持自定义图像归一化。

```
class CLIPVisionTower(nn.Module):
    def __init__(
        self,
        model_name: str = "siglip_large_patch16_384",
        image_size: Union[Tuple[int, int], int] = 336,
        select_feature: str = "patch",
        select_layer: int = -2,
        select_layers: list = None,
        ckpt_path: str = "",
        pixel_mean: Optional[List[float]] = None,
        pixel_std: Optional[List[float]] = None,
        **kwargs,
    ):
        super().__init__()

        self.model_name = model_name
        self.select_feature = select_feature
        self.select_layer = select_layer
        self.select_layers = select_layers

        vision_tower_params = {
            "model_name": model_name,
            "image_size": image_size,
            "ckpt_path": ckpt_path,
            "select_layer": select_layer,
        }
        vision_tower_params.update(kwargs)
```

```
        self.vision_tower, self.forward_kwargs = self.build_vision_tower(
            vision_tower_params
        )

        if pixel_mean is not None and pixel_std is not None:
            image_norm = torchvision.transforms.Normalize(
                mean=pixel_mean, std=pixel_std
            )
        else:
            image_norm = None

        self.image_norm = image_norm

    def build_vision_tower(self, vision_tower_params):
        if self.model_name.startswith("siglip"):
            self.select_feature = "same"
            vision_tower = create_siglip_vit(**vision_tower_params)
            forward_kwargs = dict()

        elif self.model_name.startswith("sam"):
            vision_tower = create_sam_vit(**vision_tower_params)
            forward_kwargs = dict()

        else:  # huggingface
            from transformers import CLIPVisionModel

            vision_tower = CLIPVisionModel.from_pretrained(**vision_tower_params)
            forward_kwargs = dict(output_hidden_states=True)

        return vision_tower, forward_kwargs

    def feature_select(self, image_forward_outs):
        if isinstance(image_forward_outs, torch.Tensor):
            # 输出已经是 self.select_layer 对应的特征
            image_features = image_forward_outs
        else:
            image_features = image_forward_outs.hidden_states[self.select_layer]

        if self.select_feature == "patch":
            # 如果输出中包含 cls_token
            image_features = image_features[:, 1:]
        elif self.select_feature == "cls_patch":
            image_features = image_features
        elif self.select_feature == "same":
            image_features = image_features
```

```
        else:
            raise ValueError(f"Unexpected select feature: {self.select_
feature}")
        return image_features

    def forward(self, images):
        """
        参数:
            images (torch.Tensor): [b, 3, H, W]

        返回:
            image_features (torch.Tensor): [b, n_patch, d]
        """

        if self.image_norm is not None:
            images = self.image_norm(images)

        image_forward_outs = self.vision_tower(images, **self.forward_kwargs)
        image_features = self.feature_select(image_forward_outs)
        return image_features
```

总之，类CLIPVisionTower为多模态模型提供了图像特征提取功能，这些特征可以用于和文本特征的对齐与融合。

5.5.3 投影器

文件projector.py定义了一个名为MlpProjector的类，其作为一个多层感知机（MLP）投影器，用于将输入数据映射到指定的嵌入空间。在多模态模型中，投影器是连接不同模态特征的关键组件。MlpProjector投影器根据配置（cfg）的不同，支持多种类型的投影方式：identity、linear、mlp_gelu、和low_high_hybrid_split_mlp_gelu。如果配置为low_high_hybrid_split_mlp_gelu，则会将输入分为两部分（高分辨率和低分辨率），分别进行投影后合并。前向传播方法接收一个输入（可以是元组，也可以是单一张量），并通过配置的投影器进行转换，返回投影后的结果。

```
class MlpProjector(nn.Module):
    def __init__(self, cfg):
        super().__init__()

        self.cfg = cfg

        if cfg.projector_type == "identity":
            modules = nn.Identity()

        elif cfg.projector_type == "linear":
            modules = nn.Linear(cfg.input_dim, cfg.n_embed)
```

```
        elif cfg.projector_type == "mlp_gelu":
            mlp_depth = cfg.get("depth", 1)
            modules = [nn.Linear(cfg.input_dim, cfg.n_embed)]
            for _ in range(1, mlp_depth):
                modules.append(nn.GELU())
                modules.append(nn.Linear(cfg.n_embed, cfg.n_embed))
            modules = nn.Sequential(*modules)

        elif cfg.projector_type == "low_high_hybrid_split_mlp_gelu":
            mlp_depth = cfg.get("depth", 1)
            self.high_up_proj = nn.Linear(cfg.input_dim, cfg.n_embed // 2)
            self.low_up_proj = nn.Linear(cfg.input_dim, cfg.n_embed // 2)

            modules = []
            for _ in range(1, mlp_depth):
                modules.append(nn.GELU())
                modules.append(nn.Linear(cfg.n_embed, cfg.n_embed))
            modules = nn.Sequential(*modules)

        else:
            raise ValueError(f"Unknown projector type: {cfg.projector_type}")

        self.layers = modules

    def forward(
        self, x_or_tuple: Union[Tuple[torch.Tensor, torch.Tensor], torch.Tensor]
    ):
        """
        参数:
            x_or_tuple (Union[Tuple[torch.Tensor, torch.Tensor],
                        torch.Tensor]): 如果是一个元组，则来自混合视觉编码器，其中 x
                        = high_res_x, low_res_x;
                否则，它是来自单一视觉编码器的特征。

        返回:
            x (torch.Tensor): [b, s, c]
        """

        if isinstance(x_or_tuple, tuple):
            # self.cfg.projector_type == "low_high_hybrid_split_mlp_gelu":
            high_x, low_x = x_or_tuple
            high_x = self.high_up_proj(high_x)
            low_x = self.low_up_proj(low_x)
            x = torch.concat([high_x, low_x], dim=-1)
        else:
```

```
            x = x_or_tuple

        return self.layers(x)


if __name__ == "__main__":
    cfg = AttrDict(
        input_dim=1024,
        n_embed=2048,
        depth=2,
        projector_type="low_high_hybrid_split_mlp_gelu",
    )
    inputs = (torch.rand(4, 576, 1024), torch.rand(4, 576, 1024))

    m = MlpProjector(cfg)
    out = m(inputs)
    print(out.shape)
```

5.5.4 Vision Transformer 视觉模型

文件siglip_vit.py实现了一个基于Vision Transformer（ViT）的视觉模型，支持多种配置和预训练权重加载。该模型用于图像特征提取和分类任务，支持动态图像大小调整、位置嵌入调整等功能。此外，代码还提供了权重初始化、层归一化和注意力机制的实现，以及模型的前向传播逻辑。

文件siglip_vit.py的主要目标是利用Transformer结构进行图像识别任务。该实现参考了论文*An Image is Worth 16×16 Words: Transformers for Image Recognition at Scale*，并在标准ViT基础上进行了改进，例如：

- 支持动态图像尺寸，以适配不同输入大小。
- 可变的全局池化策略（如token、avg、map）。
- 可选的前归一化 (Pre-Norm) 结构。
- 增强的MLP结构。
- 可调的Dropout机制 以提高泛化能力。
- 可选的PatchDropout用于数据增强。
- 支持Attention Pooling以替代传统的全局池化。

文件siglip_vit.py的具体实现流程如下所示。

（1）下面的代码实现了梯度截断正态分布初始化操作，用于在区间[a,b]内按照给定均值mean和标准差std生成符合截断正态分布的随机数，并填充到tensor中。核心逻辑包括：检查均值是否超出截断范围的两倍标准差（防止分布偏移），使用逆误差函数erfinv_()将均匀分布转换为截断正态分布，并最终调整均值、标准差和范围约束clamp_()。该方法可用于深度学习模型参数的合理初始化，以避免极端值影响训练的稳定性。

```
def _no_grad_trunc_normal_(tensor, mean, std, a, b):
    if (mean < a - 2 * std) or (mean > b + 2 * std):
        warnings.warn(
            "均值超出 [a, b] 范围的两倍标准差。"
            "值的分布可能不正确。",
            stacklevel=2,
        )

    with torch.no_grad():
        # 使用截断均匀分布生成值，然后使用正态分布的逆累积分布函数。
        # 获取累积分布函数的上下值
        l = norm_cdf((a - mean) / std)  # noqa: E741
        u = norm_cdf((b - mean) / std)

        # 使用均匀分布填充张量，范围为 [l, u]，然后转换为 [2l-1, 2u-1]。
        tensor.uniform_(2 * l - 1, 2 * u - 1)

        # 使用逆累积分布函数转换为截断标准正态分布。
        tensor.erfinv_()

        # 转换为正确的均值和标准差。
        tensor.mul_(std * math.sqrt(2.0))
        tensor.add_(mean)

        # 限制范围以确保值在正确范围内。
        tensor.clamp_(min=a, max=b)
        return tensor
```

（2）下面的代码实现了截断正态分布初始化操作，用于在区间 [a, b] 内按照均值mean和标准差std随机填充tensor，并确保数值不会超出指定范围。其核心逻辑是，先将tensor转换为float32进行计算（以兼容bfloat16），然后调用_no_grad_trunc_normal_()生成符合截断正态分布的值，最后转换回原始数据类型并复制回tensor。

```
def trunc_normal_(tensor, mean=0.0, std=1.0, a=-2.0, b=2.0):
    # type: (torch.Tensor, float, float, float, float) -> torch.Tensor
    r"""原始的 timm.models.layers.weight_init.trunc_normal_ 无法处理 bfloat16,
    因此这里先将张量转换为 float32,
    应用 trunc_normal_() 后再转换回原始数据类型。
    使用截断正态分布填充输入张量。值实际上是从正态分布 :math:`\mathcal{N}(\text{mean},
    \text{std}^2)` 中抽取的，超出 :math:`[a, b]` 的值将重新抽取，直到它们在范围内。
    使用生成随机值的方法在 :math:`a \leq \text{mean} \leq b` 时效果最佳。
    参数:
        tensor: 一个 n 维的 `torch.Tensor`
        mean: 正态分布的均值
        std: 正态分布的标准差
```

```
        a：最小截断值
        b：最大截断值
    示例：
        >>> w = torch.empty(3, 5)
        >>> nn.init.trunc_normal_(w)
    """

    with torch.no_grad():
        dtype = tensor.dtype
        tensor_fp32 = tensor.float()
        tensor_fp32 = _no_grad_trunc_normal_(tensor_fp32, mean, std, a, b)
        tensor_dtype = tensor_fp32.to(dtype=dtype)
        tensor.copy_(tensor_dtype)
```

（3）函数init_weights ()用于初始化模型的权重，具体来说，对位置编码 (pos_embed) 和潜在表示 (latent) 进行截断正态分布初始化。如果pos_embed存在，则按照标准差std = 1 / sqrt(embedding_dim) 进行初始化，而latent的标准差为1 / sqrt(latent_dim)，这种初始化方式有助于控制参数的尺度，防止梯度消失或爆炸，提高训练的稳定性。

```
def init_weights(self):
    if self.pos_embed is not None:
        trunc_normal_(self.pos_embed, std=self.pos_embed.shape[1] ** -0.5)
    trunc_normal_(self.latent, std=self.latent_dim**-0.5)
```

（4）函数init_weights_vit_timm()用于初始化Vision Transformer的权重，并遵循timm库的初始化方式，以保证可复现性。具体而言，它对nn.Linear层的weight采用标准差为0.02的截断正态分布进行初始化，并将bias置为零。此外，如果module具有init_weights方法，则调用该方法执行自定义初始化，以确保不同模块的权重初始化符合预期。

```
def init_weights_vit_timm(module: nn.Module, name: str = "") -> None:
    """ViT 权重初始化，原始 timm 实现（为了可复现性）"""
    if isinstance(module, nn.Linear):
        trunc_normal_(module.weight, std=0.02)
        if module.bias is not None:
            nn.init.zeros_(module.bias)
    elif hasattr(module, "init_weights"):
        module.init_weights()
```

（5）类Attention实现了多头自注意力机制（Multi-Head Self-Attention，MHSA），用于处理输入特征并计算注意力加权的输出。类Attention的核心功能包括：

- 线性变换：使用qkv层将输入x投影为查询（Q）、键（K）、值（V）三个张量。
- 归一化：可选地对Q和K进行归一化，以增强训练的稳定性。
- 注意力计算：若self.fused_attn为True，则使用torch.F.scaled_dot_product_attention计算注意

力（高效优化）。否则，手动计算缩放点积注意力，即softmax(q @ k^T / sqrt(d))，然后与v相乘。

● 输出映射：将注意力加权后的x重新排列，并通过proj层投影回原始维度，最后经过Dropout处理，以防止过拟合。

该Attention机制用于Transformer结构，帮助模型在不同位置间建立长距离依赖，从而提高对全局信息的捕捉能力。

```
class Attention(nn.Module):
    fused_attn: Final[bool]

    def __init__(
        self,
        dim: int,
        num_heads: int = 8,
        qkv_bias: bool = False,
        qk_norm: bool = False,
        attn_drop: float = 0.0,
        proj_drop: float = 0.0,
        norm_layer: nn.Module = nn.LayerNorm,
    ) -> None:
        super().__init__()
        assert dim % num_heads == 0, "dim应该能被num_heads整除"
        self.num_heads = num_heads
        self.head_dim = dim // num_heads
        self.scale = self.head_dim**-0.5
        # self.fused_attn = use_fused_attn()
        self.fused_attn = True

        self.qkv = nn.Linear(dim, dim * 3, bias=qkv_bias)
        self.q_norm = norm_layer(self.head_dim) if qk_norm else nn.Identity()
        self.k_norm = norm_layer(self.head_dim) if qk_norm else nn.Identity()
        self.attn_drop = nn.Dropout(attn_drop)
        self.proj = nn.Linear(dim, dim)
        self.proj_drop = nn.Dropout(proj_drop) if proj_drop > 0.0 else nn.Identity()

    def forward(self, x: torch.Tensor) -> torch.Tensor:
        B, N, C = x.shape
        qkv = (
            self.qkv(x)
            .reshape(B, N, 3, self.num_heads, self.head_dim)
            .permute(2, 0, 3, 1, 4)
        )
        q, k, v = qkv.unbind(0)
        q, k = self.q_norm(q), self.k_norm(k)

        if self.fused_attn:
```

```
            x = F.scaled_dot_product_attention(
                q,
                k,
                v,
                dropout_p=self.attn_drop.p if self.training else 0.0,
            )
        else:
            q = q * self.scale
            attn = q @ k.transpose(-2, -1)
            attn = attn.softmax(dim=-1)
            attn = self.attn_drop(attn)
            x = attn @ v

        x = x.transpose(1, 2).reshape(B, N, C)
        x = self.proj(x)
        x = self.proj_drop(x)
        return x
```

（6）类LayerScale实现了一种层级缩放（Layer Scaling）技术，用于调整输入张量的尺度，通过一个可学习的参数gamma。类LayerScale的核心功能如下：

- 在初始化时，gamma是一个大小为dim的可学习参数，初始值为init_values（默认为1e-5）。
- 在forward方法中，输入张量x被乘以gamma，可选择是否使用就地（inplace）操作来进行计算。如果inplace为True，则直接修改x的值；否则，返回缩放后的新张量。

该技术可以调整激活值的尺度，在训练深度神经网络时帮助稳定训练过程。

```
class LayerScale(nn.Module):
    def __init__(
        self,
        dim: int,
        init_values: float = 1e-5,
        inplace: bool = False,
    ) -> None:
        super().__init__()
        self.inplace = inplace
        self.gamma = nn.Parameter(init_values * torch.ones(dim))

    def forward(self, x: torch.Tensor) -> torch.Tensor:
        return x.mul_(self.gamma) if self.inplace else x * self.gamma
```

（7）类Block实现了一个Transformer模块的基础构建块，包含注意力层（Attention）和前馈神经网络层（MLP）。类Block的核心功能如下。

- 归一化与注意力机制：首先对输入x进行归一化 (norm1)，然后通过注意力层进行处理，接着应用层级缩放（LayerScale）和残差连接。如果启用了DropPath，则会在此阶段引入随机的路径丢弃。

● 前馈神经网络：输入通过一个MLP层处理，经过归一化 (norm2)，并同样应用层级缩放和DropPath。

● 通过残差连接将经过两次处理后的输出加回原输入，形成最终的输出。

类Block提供了Transformer网络中的多头自注意力和前馈网络功能，并结合了层级缩放和路径丢弃技术来优化训练过程和防止过拟合。

```
class Block(nn.Module):
    def __init__(
        self,
        dim: int,
        num_heads: int,
        mlp_ratio: float = 4.0,
        qkv_bias: bool = False,
        qk_norm: bool = False,
        proj_drop: float = 0.0,
        attn_drop: float = 0.0,
        init_values: Optional[float] = None,
        drop_path: float = 0.0,
        act_layer: nn.Module = nn.GELU,
        norm_layer: nn.Module = nn.LayerNorm,
        mlp_layer: nn.Module = Mlp,
    ) -> None:
        super().__init__()
        self.norm1 = norm_layer(dim)
        self.attn = Attention(
            dim,
            num_heads=num_heads,
            qkv_bias=qkv_bias,
            qk_norm=qk_norm,
            attn_drop=attn_drop,
            proj_drop=proj_drop,
            norm_layer=norm_layer,
        )
        self.ls1 = (
            LayerScale(dim, init_values=init_values) if init_values else
                nn.Identity()
        )
        self.drop_path1 = DropPath(drop_path) if drop_path > 0.0 else
                          nn.Identity()

        self.norm2 = norm_layer(dim)
        self.mlp = mlp_layer(
            in_features=dim,
            hidden_features=int(dim * mlp_ratio),
            act_layer=act_layer,
```

```
            drop=proj_drop,
        )
        self.ls2 = (
            LayerScale(dim, init_values=init_values) if init_values else
                nn.Identity()
        )
        self.drop_path2 = DropPath(drop_path) if drop_path > 0.0 else
            nn.Identity()

    def forward(self, x: torch.Tensor) -> torch.Tensor:
        x = x + self.drop_path1(self.ls1(self.attn(self.norm1(x))))
        x = x + self.drop_path2(self.ls2(self.mlp(self.norm2(x))))
        return x
```

（8）类VisionTransformer是一个基于Transformer的视觉模型，专门用于图像分类任务。其核心思想是将输入图像划分为固定大小的块（patches），并将这些块嵌入向量，作为Transformer模型的输入。该模型支持动态调整输入图像大小、位置嵌入的处理、块的丢弃等多种配置。其结构包括图像块嵌入、多个Transformer层以及一个分类头。VisionTransformer通过学习图像块之间的关系，利用自注意力机制进行特征提取，最后输出类别预测结果。模型的不同配置选项（如注意力池化、位置嵌入等）允许在不同应用场景中灵活调整性能与效率。

● 方法 __init__()：功能是初始化VisionTransformer类的各个成员变量，包括图像尺寸、嵌入维度、Transformer层的深度、注意力头的数量等，构建模型所需的各个组件，并设定模型的基本结构。

```
class VisionTransformer(nn.Module):
    """Vision Transformer

    PyTorch实现：`An Image is Worth 16x16 Words: Transformers for Image
                                                 Recognition at Scale`
        - <url id="cum1qp64bbjjq7ihej10" type="url" status="parsed" title=""
              wc="67570">https://arxiv.org/abs/2010.11929</url>
    """

    dynamic_img_size: Final[bool]

    def __init__(
        self,
        img_size: Union[int, Tuple[int, int]] = 224,
        patch_size: Union[int, Tuple[int, int]] = 16,
        in_chans: int = 3,
        num_classes: int = 1000,
        global_pool: Literal["", "avg", "token", "map"] = "token",
        embed_dim: int = 768,
        depth: int = 12,
```

```
    num_heads: int = 12,
    mlp_ratio: float = 4.0,
    qkv_bias: bool = True,
    qk_norm: bool = False,
    init_values: Optional[float] = None,
    class_token: bool = True,
    no_embed_class: bool = False,
    reg_tokens: int = 0,
    pre_norm: bool = False,
    fc_norm: Optional[bool] = None,
    dynamic_img_size: bool = False,
    dynamic_img_pad: bool = False,
    drop_rate: float = 0.0,
    pos_drop_rate: float = 0.0,
    patch_drop_rate: float = 0.0,
    proj_drop_rate: float = 0.0,
    attn_drop_rate: float = 0.0,
    drop_path_rate: float = 0.0,
    weight_init: Literal["skip", "jax", "jax_nlhb", "moco", ""] = "",
    embed_layer: Callable = PatchEmbed,
    norm_layer: Optional[LayerType] = None,
    act_layer: Optional[LayerType] = None,
    block_fn: Type[nn.Module] = Block,
    mlp_layer: Type[nn.Module] = Mlp,
    ignore_head: bool = False,
) -> None:
    """
    参数:
        img_size: 输入图像大小。
        patch_size: 图像块大小。
        in_chans: 输入图像通道数。
        num_classes: 分类头的类别数。
        global_pool: 最终序列的全局池化类型（默认：'token'）。
        embed_dim: Transformer 嵌入维度。
        depth: Transformer 的深度。
        num_heads: 注意力头的数量。
        mlp_ratio: MLP 隐藏层维度与嵌入维度的比例。
        qkv_bias: 如果为 True，则为 qkv 投影启用偏置。
        init_values: 层归一化的初始化值（如果为 None，则不启用层归一化）。
        class_token: 是否使用类别令牌。
        no_embed_class: 不为类别（或注册）令牌包含位置嵌入。
        reg_tokens: 注册令牌的数量。
        pre_norm: 是否在 Transformer 前使用归一化。
        fc_norm: 在池化后（而不是之前）使用头部归一化，如果为 None，则在 global_
                 pool == 'avg' 时启用。
        drop_rate: 分类头的 Dropout 比率。
```

```
        pos_drop_rate: 位置嵌入的 Dropout 比率。
        attn_drop_rate: 注意力的 Dropout 比率。
        drop_path_rate: 随机深度比率。
        weight_init: 权重初始化方案。
        embed_layer: 图像块嵌入层。
        norm_layer: 归一化层。
        act_layer: MLP 激活层。
        block_fn: Transformer 块层。
    """
    super().__init__()
    assert global_pool in ("", "avg", "token", "map")
    assert class_token or global_pool != "token"
    use_fc_norm = global_pool == "avg" if fc_norm is None else fc_norm
    # norm_layer = get_norm_layer(norm_layer) or partial(nn.LayerNorm, eps=1e-6)
    # act_layer = get_act_layer(act_layer) or nn.GELU
    norm_layer = partial(nn.LayerNorm, eps=1e-6)
    act_layer = nn.GELU

    self.num_classes = num_classes
    self.global_pool = global_pool
    self.num_features = self.embed_dim = (
        embed_dim  # num_features for consistency with other models
    )
    self.num_prefix_tokens = 1 if class_token else 0
    self.num_prefix_tokens += reg_tokens
    self.num_reg_tokens = reg_tokens
    self.has_class_token = class_token
    self.no_embed_class = (
        no_embed_class  # don't embed prefix positions (includes reg)
    )
    self.dynamic_img_size = dynamic_img_size
    self.grad_checkpointing = False
    self.ignore_head = ignore_head

    embed_args = {}
    if dynamic_img_size:
        # flatten deferred until after pos embed
        embed_args.update(dict(strict_img_size=False,
                               output_fmt="NHWC"))
    self.patch_embed = embed_layer(
        img_size=img_size,
        patch_size=patch_size,
        in_chans=in_chans,
        embed_dim=embed_dim,
        bias=not pre_norm,  # disable bias if pre-norm is used (e.g. CLIP)
        dynamic_img_pad=dynamic_img_pad,
```

```
        **embed_args,
    )
    num_patches = self.patch_embed.num_patches

    self.cls_token = (
        nn.Parameter(torch.zeros(1, 1, embed_dim)) if class_token else None
    )
    self.reg_token = (
        nn.Parameter(torch.zeros(1, reg_tokens, embed_dim)) if reg_
                    tokens else None
    )
    embed_len = (
        num_patches if no_embed_class else num_patches + self.num_
            prefix_tokens
    )
    self.pos_embed = nn.Parameter(torch.randn(1, embed_len, embed_dim) *
                                  0.02)
    self.pos_drop = nn.Dropout(p=pos_drop_rate)
    if patch_drop_rate > 0:
        self.patch_drop = PatchDropout(
            patch_drop_rate,
            num_prefix_tokens=self.num_prefix_tokens,
        )
    else:
        self.patch_drop = nn.Identity()
    self.norm_pre = norm_layer(embed_dim) if pre_norm else nn.Identity()

    dpr = [
        x.item() for x in torch.linspace(0, drop_path_rate, depth)
    ]  # stochastic depth decay rule
    self.blocks = nn.Sequential(
        *[
            block_fn(
                dim=embed_dim,
                num_heads=num_heads,
                mlp_ratio=mlp_ratio,
                qkv_bias=qkv_bias,
                qk_norm=qk_norm,
                init_values=init_values,
                proj_drop=proj_drop_rate,
                attn_drop=attn_drop_rate,
                drop_path=dpr[i],
                norm_layer=norm_layer,
                act_layer=act_layer,
                mlp_layer=mlp_layer,
            )
```

```
                for i in range(depth)
            ]
        )
        self.norm = norm_layer(embed_dim) if not use_fc_norm else nn.Identity()

        # Classifier Head
        if global_pool == "map":
            AttentionPoolLatent.init_weights = init_weights
            self.attn_pool = AttentionPoolLatent(
                self.embed_dim,
                num_heads=num_heads,
                mlp_ratio=mlp_ratio,
                norm_layer=norm_layer,
            )
        else:
            self.attn_pool = None
        self.fc_norm = norm_layer(embed_dim) if use_fc_norm else nn.Identity()
        self.head_drop = nn.Dropout(drop_rate)
        self.head = (
            nn.Linear(self.embed_dim, num_classes) if num_classes > 0 else
                nn.Identity()
        )

        if weight_init != "skip":
            self.init_weights(weight_init)
```

- 方法 init_weights()：功能是初始化模型的权重，根据不同的初始化方案（如 "jax" "moco" 等）进行权重初始化，并对位置嵌入、类别令牌等进行初始化。

```
    def init_weights(self, mode: Literal["jax", "jax_nlhb", "moco", ""] =
        "") -> None:
        assert mode in ("jax", "jax_nlhb", "moco", "")
        # head_bias = -math.log(self.num_classes) if "nlhb" in mode else 0.0
        trunc_normal_(self.pos_embed, std=0.02)
        if self.cls_token is not None:
            nn.init.normal_(self.cls_token, std=1e-6)
        named_apply(init_weights_vit_timm, self)
```

- 方法 no_weight_decay()：功能是返回一个不参与权重衰减的参数集合，通常用于控制某些参数在优化过程中不进行权重衰减。

```
    @torch.jit.ignore
    def no_weight_decay(self) -> Set:
        return {"pos_embed", "cls_token", "dist_token"}
```

- 方法 group_matcher()：功能是返回一个字典，用于匹配不同组的参数，主要用于模型的优化器设置。

```
@torch.jit.ignore
def group_matcher(self, coarse: bool = False) -> Dict:
    return dict(
        stem=r"^cls_token|pos_embed|patch_embed",  # stem and embed
        blocks=[(r"^blocks\.(\d+)", None), (r"^norm", (99999,))],
    )
```

- 方法 set_grad_checkpointing()：功能是启用或禁用梯度检查点，它允许在训练过程中保存中间计算的梯度，从而减少显存的占用。

```
@torch.jit.ignore
def set_grad_checkpointing(self, enable: bool = True) -> None:
    self.grad_checkpointing = enable
```

- 方法 get_classifier()：功能是返回分类器头部模块，通常用于模型的推理或训练过程中的分类任务。

```
@torch.jit.ignore
def get_classifier(self) -> nn.Module:
    return self.head
```

- 方法 reset_classifier()：功能是重置模型的分类头部，将其更新为指定的类别数，并可以选择添加或删除全局池化层。

```
def reset_classifier(self, num_classes: int, global_pool=None) -> None:
    self.num_classes = num_classes
    if global_pool is not None:
        assert global_pool in ("", "avg", "token", "map")
        if global_pool == "map" and self.attn_pool is None:
            assert (
                False
            ), "Cannot currently add attention pooling in reset_
                classifier()."
        elif global_pool != "map " and self.attn_pool is not None:
            self.attn_pool = None  # remove attention pooling
        self.global_pool = global_pool
    self.head = (
        nn.Linear(self.embed_dim, num_classes) if num_classes > 0 else
            nn.Identity()
    )
```

- 方法 _pos_embed()：功能是对输入进行位置嵌入操作，并根据动态图像大小设置对位置嵌入的采样方式，最终将其添加到输入特征中。

```
def _pos_embed(self, x: torch.Tensor) -> torch.Tensor:
    if self.dynamic_img_size:
        B, H, W, C = x.shape
        pos_embed = resample_abs_pos_embed(
            self.pos_embed,
            (H, W),
            num_prefix_tokens=0 if self.no_embed_class else self.num_
                                  prefix_tokens,
        )
        x = x.view(B, -1, C)
    else:
        pos_embed = self.pos_embed

    to_cat = []
    if self.cls_token is not None:
        to_cat.append(self.cls_token.expand(x.shape[0], -1, -1))
    if self.reg_token is not None:
        to_cat.append(self.reg_token.expand(x.shape[0], -1, -1))

    if self.no_embed_class:
        # deit-3, updated JAX (big vision)
        # position embedding does not overlap with class token,
        # add then concat
        x = x + pos_embed
        if to_cat:
            x = torch.cat(to_cat + [x], dim=1)
    else:
        # original timm, JAX, and deit vit impl
        # pos_embed has entry for class token, concat then add
        if to_cat:
            x = torch.cat(to_cat + [x], dim=1)
        x = x + pos_embed

    return self.pos_drop(x)
```

- 方法 _intermediate_layers()：功能是获取模型中的中间层输出，允许用户选择返回指定层的输出，支持多层的中间结果提取。

```
def _intermediate_layers(
    self,
    x: torch.Tensor,
    n: Union[int, Sequence] = 1,
) -> List[torch.Tensor]:
    outputs, num_blocks = [], len(self.blocks)
    take_indices = set(
        range(num_blocks - n, num_blocks) if isinstance(n, int) else n
```

```
    )

    # forward pass
    x = self.patch_embed(x)
    x = self._pos_embed(x)
    x = self.patch_drop(x)
    x = self.norm_pre(x)
    for i, blk in enumerate(self.blocks):
        x = blk(x)
        if i in take_indices:
            outputs.append(x)

    return outputs
```

● 方法get_intermediate_layers()：功能是获取中间层输出的接口，提供更多选项来提取指定的层输出，包括是否规范化、是否返回类别令牌等。

```
def get_intermediate_layers(
    self,
    x: torch.Tensor,
    n: Union[int, Sequence] = 1,
    reshape: bool = False,
    return_prefix_tokens: bool = False,
    norm: bool = False,
) -> Tuple[Union[torch.Tensor, Tuple[torch.Tensor]]]:
    """Intermediate layer accessor (NOTE: This is a WIP experiment).
    Inspired by DINO / DINOv2 interface
    """
    # take last n blocks if n is an int, if in is a sequence, select by
      matching indices
    outputs = self._intermediate_layers(x, n)
    if norm:
        outputs = [self.norm(out) for out in outputs]
    prefix_tokens = [out[:, 0 : self.num_prefix_tokens] for out in outputs]
    outputs = [out[:, self.num_prefix_tokens :] for out in outputs]

    if reshape:
        grid_size = self.patch_embed.grid_size
        outputs = [
            out.reshape(x.shape[0], grid_size[0], grid_size[1], -1)
            .permute(0, 3, 1, 2)
            .contiguous()
            for out in outputs
        ]

    if return_prefix_tokens:
```

```
            return tuple(zip(outputs, prefix_tokens))
        return tuple(outputs)
```

● 方法 forward_features()：功能是执行模型的前向传播，计算特征表示。在这一过程中，图像输入经过嵌入、位置嵌入、注意力层等，输出最终的特征表示。

```
    def forward_features(self, x: torch.Tensor) -> torch.Tensor:
        x = self.patch_embed(x)
        x = self._pos_embed(x)
        x = self.patch_drop(x)
        x = self.norm_pre(x)
        if self.grad_checkpointing and not torch.jit.is_scripting():
            x = checkpoint_seq(self.blocks, x)
        else:
            x = self.blocks(x)
        x = self.norm(x)
        return x
```

● 方法 forward_head()：功能是执行分类头部的前向传播，包括池化操作（如全局平均池化）和分类输出。如果设置了分类头，它就会输出类别预测结果。

```
    def forward_head(self, x: torch.Tensor, pre_logits: bool = False) ->
                    torch.Tensor:
        if self.attn_pool is not None:
            x = self.attn_pool(x)
        elif self.global_pool == "avg":
            x = x[:, self.num_prefix_tokens :].mean(dim=1)
        elif self.global_pool:
            x = x[:, 0]  # class token
        x = self.fc_norm(x)
        x = self.head_drop(x)
        return x if pre_logits else self.head(x)
```

● 方法 forward()：功能是执行模型的前向传播，包括特征提取和分类头部的计算，最终输出类别预测结果。

```
    def forward(self, x: torch.Tensor) -> torch.Tensor:
        x = self.forward_features(x)
        if not self.ignore_head:
            x = self.forward_head(x)
        return x
```

（9）下面的代码定义了一个名为SigLIPVisionCfg的数据类，用于配置视觉模型的各种超参数。该类包含了多个参数：模型宽度、层数、头部数量、图像分块大小、输入图像尺寸、全局池化方式、MLP 比例、是否使用类别令牌、类别数量和是否使用检查点等。这些参数用于控制视觉模型的结构和训练过程，便于在不同配置下灵活调整模型的性能与特性。

```
@dataclass
class SigLIPVisionCfg:
    width: int = 1152
    layers: Union[Tuple[int, int, int, int], int] = 27
    heads: int = 16
    patch_size: int = 14
    image_size: Union[Tuple[int, int], int] = 336
    global_pool: str = "map"
    mlp_ratio: float = 3.7362
    class_token: bool = False
    num_classes: int = 0
    use_checkpoint: bool = False
```

（10）下面的代码定义了SigLIP模型的配置字典，其中包含多个SigLIP模型的参数设置。这些配置用于初始化不同版本的SigLIP模型，以适应不同的输入图像尺寸和计算需求。

```
SigLIP_MODEL_CONFIG = {
    "siglip_so400m_patch14_384": {
        "image_size": 336,
        "patch_size": 14,
        "width": 1152,
        "layers": 27,
        "heads": 16,
        "mlp_ratio": 3.7362,
        "global_pool": "map",
        "use_checkpoint": False,
    },
    "siglip_so400m_patch14_224": {
        "image_size": 224,
        "patch_size": 14,
        "width": 1152,
        "layers": 27,
        "heads": 16,
        "mlp_ratio": 3.7362,
        "global_pool": "map",
        "use_checkpoint": False,
    },
    "siglip_large_patch16_384": {
        "image_size": 384,
        "patch_size": 16,
        "width": 1024,
        "layers": 24,
        "heads": 16,
        "mlp_ratio": 4,
```

```
        "global_pool": "map",
        "use_checkpoint": False,
    },
}
```

（11）下面代码定义了一个名为SigLIP_MODEL_CONFIG的字典，存储了不同视觉模型配置的超参数设置。每个配置以模型名称为键，包含图像尺寸、图像分块大小、宽度、层数、头部数量、MLP比例、全局池化方式及是否使用检查点等参数。此字典提供了多种模型配置的选项，便于根据不同需求选择合适的视觉模型，并进行灵活调整。

```
def create_siglip_vit(
    model_name: str = "siglip_so400m_patch14_384",
    image_size: int = 384,
    select_layer: int = -1,
    ckpt_path: str = "",
    **kwargs,
):
    assert (
        model_name in SigLIP_MODEL_CONFIG.keys()
    ), f"model name should be in {SigLIP_MODEL_CONFIG.keys()}"

    vision_cfg = SigLIPVisionCfg(**SigLIP_MODEL_CONFIG[model_name])

    if select_layer <= 0:
        layers = min(vision_cfg.layers, vision_cfg.layers + select_layer + 1)
    else:
        layers = min(vision_cfg.layers, select_layer)

    model = VisionTransformer(
        img_size=image_size,
        patch_size=vision_cfg.patch_size,
        embed_dim=vision_cfg.width,
        depth=layers,
        num_heads=vision_cfg.heads,
        mlp_ratio=vision_cfg.mlp_ratio,
        class_token=vision_cfg.class_token,
        global_pool=vision_cfg.global_pool,
        ignore_head=kwargs.get("ignore_head", True),
        weight_init=kwargs.get("weight_init", "skip"),
        num_classes=0,
    )

    if ckpt_path:
        state_dict = torch.load(ckpt_path, map_location="cpu")
```

```
        incompatible_keys = model.load_state_dict(state_dict, strict=False)
        print(
            f"SigLIP-ViT restores from {ckpt_path},\n"
            f"\tincompatible_keys:', {incompatible_keys}."
        )

    return model
```

5.5.5 图像处理器

文件image_processing_vlm.py定义了一个图像处理器（VLMImageProcessor），用于对输入图像进行预处理，包括调整大小、归一化等操作。此文件能够确保输入图像符合模型的要求，为图像特征提取提供了标准化的输入。

```
logger = logging.get_logger(__name__)

ImageType = Union[np.ndarray, torch.Tensor, Image.Image]
IMAGENET_MEAN = (0.48145466, 0.4578275, 0.40821073)
IMAGENET_STD = (0.26862954, 0.26130258, 0.27577711)
IMAGENET_INCEPTION_MEAN = (0.5, 0.5, 0.5)
IMAGENET_INCEPTION_STD = (0.5, 0.5, 0.5)


def expand2square(pil_img, background_color):
    width, height = pil_img.size
    if width == height:
        return pil_img
    elif width > height:
        result = Image.new(pil_img.mode, (width, width), background_color)
        result.paste(pil_img, (0, (width - height) // 2))
        return result
    else:
        result = Image.new(pil_img.mode, (height, height), background_color)
        result.paste(pil_img, ((height - width) // 2, 0))
        return result


class VLMImageProcessorConfig(PretrainedConfig):
    model_type = "deepseek_vlm"
    image_size: int
    min_size: int
    image_mean: Union[Tuple[float, float, float], List[float]]
    image_std: Union[Tuple[float, float, float], List[float]]
    rescale_factor: float
```

```
    do_normalize: bool

    def __init__(
        self,
        image_size: int,
        min_size: int = 14,
        image_mean: Union[Tuple[float, float, float], List[float]] = (
            0.48145466,
            0.4578275,
            0.40821073,
        ),
        image_std: Union[Tuple[float, float, float], List[float]] = (
            0.26862954,
            0.26130258,
            0.27577711,
        ),
        rescale_factor: float = 1.0 / 255.0,
        do_normalize: bool = True,
        **kwargs,
    ):
        self.image_size = image_size
        self.min_size = min_size
        self.image_mean = image_mean
        self.image_std = image_std
        self.rescale_factor = rescale_factor
        self.do_normalize = do_normalize

        super().__init__(**kwargs)


class VLMImageProcessor(BaseImageProcessor):
    model_input_names = ["pixel_values"]

    def __init__(
        self,
        image_size: int,
        min_size: int = 14,
        image_mean: Union[Tuple[float, float, float], List[float]] = (
            0.48145466,
            0.4578275,
            0.40821073,
        ),
        image_std: Union[Tuple[float, float, float], List[float]] = (
            0.26862954,
            0.26130258,
            0.27577711,
```

```
        ),
        rescale_factor: float = 1.0 / 255.0,
        do_normalize: bool = True,
        **kwargs,
    ):
        super().__init__(**kwargs)

        self.image_size = image_size
        self.rescale_factor = rescale_factor
        self.image_mean = image_mean
        self.image_std = image_std
        self.min_size = min_size
        self.do_normalize = do_normalize

        if image_mean is None:
            self.background_color = (127, 127, 127)
        else:
            self.background_color = tuple([int(x * 255) for x in image_mean])

    def resize(self, pil_img: Image) -> np.ndarray:
        """
        调整图像大小，使其适应预设尺寸

        参数：
            pil_img (PIL.Image)：输入的 PIL 图像，格式为 [H, W, 3]，RGB

        返回：
            x (np.ndarray)：调整后的图像，格式为 [3, self.image_size, self.
                        image_size]
        """
        width, height = pil_img.size
        max_size = max(width, height)

        size = [
            max(int(height / max_size * self.image_size), self.min_size),
            max(int(width / max_size * self.image_size), self.min_size),
        ]

        if width <= 0 or height <= 0 or size[0] <= 0 or size[1] <= 0:
            print(f"orig size = {pil_img.size}, new size = {size}")
            raise ValueError("Invalid size!")

        pil_img = torchvision.transforms.functional.resize(
            pil_img,
            size,
            interpolation=torchvision.transforms.functional.
```

```
                InterpolationMode.BICUBIC,
            antialias=True,
        )

        pil_img = expand2square(pil_img, self.background_color)
        x = to_numpy_array(pil_img)

        # [H, W, 3] -> [3, H, W]
        x = np.transpose(x, (2, 0, 1))

        return x

    def preprocess(self, images, return_tensors: str = "pt", **kwargs) ->
        BatchFeature:
        # resize and pad to [self.image_size, self.image_size]
        # 然后将图像从 [H, W, 3] 转换为 [3, H, W]
        images: List[np.ndarray] = [self.resize(image) for image in images]

        # 将像素值从 [0, 255] 范围映射到 [0, 1]
        images = [
            self.rescale(
                image=image,
                scale=self.rescale_factor,
                input_data_format="channels_first",
            )
            for image in images
        ]

        # 归一化
        if self.do_normalize:
            images = [
                self.normalize(
                    image=image,
                    mean=self.image_mean,
                    std=self.image_std,
                    input_data_format="channels_first",
                )
                for image in images
            ]

        data = {"pixel_values": images}
        return BatchFeature(data=data, tensor_type=return_tensors)

    @property
    def default_shape(self):
        return [3, self.image_size, self.image_size]
```

```
AutoImageProcessor.register(VLMImageProcessorConfig, VLMImageProcessor)


if __name__ == "__main__":
    image_processor = VLMImageProcessor(
        image_size=1024,
        image_mean=IMAGENET_INCEPTION_MEAN,
        image_std=IMAGENET_INCEPTION_STD,
        do_normalize=True,
    )
```

在上述代码中，VLMImageProcessor 继承自 BaseImageProcessor，用于对输入图像进行标准化处理，包括调整大小、填充和归一化，以确保符合模型输入要求。同时，VLMImageProcessorConfig 负责配置相关超参数，如图像尺寸、最小尺寸、均值和标准差等。

5.5.6 多模态因果语言模型

文件modeling_vlm.py定义了多模态因果语言模型，整合了视觉和语言模态，支持多模态理解和生成任务。此文件将图像和文本信息融合在一起，通过各个模块（视觉头、对齐器、语言模型等）进行处理，最终生成图像与文本结合的嵌入，并支持推理任务。

文件modeling_vlm.py的具体实现流程如下所示。

（1）类vision_head是一个神经网络模块，用于处理视觉信息。它包含三个主要组件：一个线性层（output_mlp_projector）将输入嵌入（embedding）转换为图像令牌嵌入，接着通过GELU激活函数（vision_activation）先进行非线性转换，最后通过另一个线性层（vision_head）将嵌入转换为最终的图像特征输出。该模块的目的是将输入的特征映射到一个特定尺寸的图像表示。

```
class vision_head(torch.nn.Module):
    def __init__(self, params):
        super().__init__()
        self.output_mlp_projector = torch.nn.Linear(
            params.n_embed, params.image_token_embed
        )
        self.vision_activation = torch.nn.GELU()
        self.vision_head = torch.nn.Linear(
            params.image_token_embed, params.image_token_size
        )

    def forward(self, x):
        x = self.output_mlp_projector(x)
        x = self.vision_activation(x)
        x = self.vision_head(x)
```

```
        return x
```

（2）函数model_name_to_cls()根据传入的类名（cls_name）返回对应的模型类。如果类名包含"MlpProjector" "CLIPVisionTower" "VQ" 或 "vision_head"，则分别返回相应的类。如果类名不匹配这些条件，则会抛出一个ValueError错误。

```
def model_name_to_cls(cls_name):
    if "MlpProjector" in cls_name:
        cls = MlpProjector

    elif "CLIPVisionTower" in cls_name:
        cls = CLIPVisionTower

    elif "VQ" in cls_name:
        from janus.models.vq_model import VQ_models

        cls = VQ_models[cls_name]
    elif "vision_head" in cls_name:
        cls = vision_head
    else:
        raise ValueError(f"class_name {cls_name} is invalid.")

    return cls
```

（3）类VisionConfig是一个配置类，继承自PretrainedConfig，用于定义和初始化与视觉模型相关的配置信息。它包含以下两个主要属性。

- cls：一个字符串，用于指定视觉模型的类名。如果传入的cls不是字符串类型，则将其转换为类名字符串。
- params：一个AttrDict类型的字典，用于存储与视觉模型相关的参数。

在 __init__()方法中，cls和params会根据传入的关键字参数（kwargs）进行初始化，并处理cls为字符串类型的情况。

```
class VisionConfig(PretrainedConfig):
    model_type = "vision"
    cls: str = ""
    params: AttrDict = {}

    def __init__(self, **kwargs):
        super().__init__(**kwargs)

        self.cls = kwargs.get("cls", "")
        if not isinstance(self.cls, str):
            self.cls = self.cls.__name__
```

```
        self.params = AttrDict(kwargs.get("params", {}))
```

（4）类AlignerConfig是一个配置类，继承自PretrainedConfig，用于定义和初始化与对齐模型（Aligner）相关的配置信息。在 __init__()方法中，cls和params会根据传入的关键字参数（kwargs）进行初始化，并处理cls为字符串类型的情况。

```
class AlignerConfig(PretrainedConfig):
    model_type = "aligner"
    cls: str = ""
    params: AttrDict = {}

    def __init__(self, **kwargs):
        super().__init__(**kwargs)

        self.cls = kwargs.get("cls", "")
        if not isinstance(self.cls, str):
            self.cls = self.cls.__name__

        self.params = AttrDict(kwargs.get("params", {}))
```

（5）类GenVisionConfig是一个配置类，继承自PretrainedConfig，用于定义和初始化生成视觉模型（GenVision）相关的配置信息。

```
class GenVisionConfig(PretrainedConfig):
    model_type = "gen_vision"
    cls: str = ""
    params: AttrDict = {}

    def __init__(self, **kwargs):
        super().__init__(**kwargs)

        self.cls = kwargs.get("cls", "")
        if not isinstance(self.cls, str):
            self.cls = self.cls.__name__

        self.params = AttrDict(kwargs.get("params", {}))
```

（6）类GenAlignerConfig是一个配置类，继承自PretrainedConfig，用于定义和初始化生成对齐模型（GenAligner）相关的配置信息。

```
class GenAlignerConfig(PretrainedConfig):
    model_type = "gen_aligner"
    cls: str = ""
    params: AttrDict = {}

    def __init__(self, **kwargs):
```

```
        super().__init__(**kwargs)

        self.cls = kwargs.get("cls", "")
        if not isinstance(self.cls, str):
            self.cls = self.cls.__name__

        self.params = AttrDict(kwargs.get("params", {}))
```

（7）类GenHeadConfig是一个配置类，继承自PretrainedConfig，用于定义和初始化生成头模型（GenHead）相关的配置信息。

```
class GenHeadConfig(PretrainedConfig):
    model_type = "gen_head"
    cls: str = ""
    params: AttrDict = {}

    def __init__(self, **kwargs):
        super().__init__(**kwargs)

        self.cls = kwargs.get("cls", "")
        if not isinstance(self.cls, str):
            self.cls = self.cls.__name__

        self.params = AttrDict(kwargs.get("params", {}))
```

（8）类MultiModalityConfig是一个配置类，继承自PretrainedConfig，用于实现多模态模型的配置信息。类MultiModalityConfig整合了多个子模块的配置，包括视觉模型（vision_config）、对齐模型（aligner_config）、生成视觉模型（gen_vision_config）、生成对齐模型（gen_aligner_config）、生成头模型（gen_head_config）及语言模型（language_config）。在初始化时，类会根据传入的关键字参数（kwargs）初始化这些子模块的配置。如果language_config传入的是LlamaConfig实例，则直接使用该实例，否则会创建新的LlamaConfig实例。

```
class MultiModalityConfig(PretrainedConfig):
    model_type = "multi_modality"
    vision_config: VisionConfig
    aligner_config: AlignerConfig

    gen_vision_config: GenVisionConfig
    gen_aligner_config: GenAlignerConfig
    gen_head_config: GenHeadConfig

    language_config: LlamaConfig

    def __init__(self, **kwargs):
        super().__init__(**kwargs)
```

```
        vision_config = kwargs.get("vision_config", {})
        self.vision_config = VisionConfig(**vision_config)

        aligner_config = kwargs.get("aligner_config", {})
        self.aligner_config = AlignerConfig(**aligner_config)

        gen_vision_config = kwargs.get("gen_vision_config", {})
        self.gen_vision_config = GenVisionConfig(**gen_vision_config)

        gen_aligner_config = kwargs.get("gen_aligner_config", {})
        self.gen_aligner_config = GenAlignerConfig(**gen_aligner_config)

        gen_head_config = kwargs.get("gen_head_config", {})
        self.gen_head_config = GenHeadConfig(**gen_head_config)

        language_config = kwargs.get("language_config", {})
        if isinstance(language_config, LlamaConfig):
            self.language_config = language_config
        else:
            self.language_config = LlamaConfig(**language_config)
```

（9）类MultiModalityPreTrainedModel继承自PreTrainedModel，为多模态模型提供了基础配置。它指定了配置类MultiModalityConfig，并为模型设置了前缀multi_modality。此外，类中定义了两个空的列表 _no_split_modules和 _skip_keys_device_placement，后者用于指定某些键（如past_key_values）在设备放置时不被处理。

```
class MultiModalityPreTrainedModel(PreTrainedModel):
    config_class = MultiModalityConfig
    base_model_prefix = "multi_modality"
    _no_split_modules = []
    _skip_keys_device_placement = "past_key_values"
```

（10）类MultiModalityCausalLM继承自MultiModalityPreTrainedModel，实现了一个多模态因果语言模型。类MultiModalityCausalLM通过配置对象初始化多个子模块，包括视觉模型、对齐器、生成视觉模型、生成对齐器、生成头部和语言模型。该模型主要包括两个方法：prepare_inputs_embeds用于将图像和语言输入转换为嵌入表示，并支持图像和语言输入结合；prepare_gen_img_embeds用于处理生成图像的嵌入。该模型能够处理多模态输入，支持图像与文本结合，并能通过多个嵌入层和对齐机制生成输出。

```
class MultiModalityCausalLM(MultiModalityPreTrainedModel):
    def __init__(self, config: MultiModalityConfig):
        super().__init__(config)

        vision_config = config.vision_config
```

```
        vision_cls = model_name_to_cls(vision_config.cls)
        self.vision_model = vision_cls(**vision_config.params)

        aligner_config = config.aligner_config
        aligner_cls = model_name_to_cls(aligner_config.cls)
        self.aligner = aligner_cls(aligner_config.params)

        gen_vision_config = config.gen_vision_config
        gen_vision_cls = model_name_to_cls(gen_vision_config.cls)
        self.gen_vision_model = gen_vision_cls()

        gen_aligner_config = config.gen_aligner_config
        gen_aligner_cls = model_name_to_cls(gen_aligner_config.cls)
        self.gen_aligner = gen_aligner_cls(gen_aligner_config.params)

        gen_head_config = config.gen_head_config
        gen_head_cls = model_name_to_cls(gen_head_config.cls)
        self.gen_head = gen_head_cls(gen_head_config.params)

        self.gen_embed = torch.nn.Embedding(
            gen_vision_config.params.image_token_size,
            gen_vision_config.params.n_embed
        )

        language_config = config.language_config
        self.language_model = LlamaForCausalLM(language_config)

    def prepare_inputs_embeds(
        self,
        input_ids: torch.LongTensor,
        pixel_values: torch.FloatTensor,
        images_seq_mask: torch.LongTensor,
        images_emb_mask: torch.LongTensor,
        **kwargs,
    ):
        """
        准备输入的嵌入

        参数：
            input_ids (torch.LongTensor): [b, T]
            pixel_values (torch.FloatTensor):   [b, n_images, 3, h, w]
            images_seq_mask (torch.BoolTensor): [b, T]
            images_emb_mask (torch.BoolTensor): [b, n_images, n_image_tokens]

            assert torch.sum(images_seq_mask) == torch.sum(images_emb_mask)
```

```
        返回：
            input_embeds (torch.Tensor): [b, T, D]
        """
        bs, n = pixel_values.shape[0:2]
        images = rearrange(pixel_values, "b n c h w -> (b n) c h w")
        # [b x n, T2, D]
        images_embeds = self.aligner(self.vision_model(images))

        # [b x n, T2, D] -> [b, n x T2, D]
        images_embeds = rearrange(images_embeds, "(b n) t d -> b (n t) d",
                                  b=bs, n=n)
        # [b, n, T2] -> [b, n x T2]
        images_emb_mask = rearrange(images_emb_mask, "b n t -> b (n t)")

        # [b, T, D]
        input_ids[input_ids < 0] = 0  # 忽略图像嵌入
        inputs_embeds = self.language_model.get_input_embeddings()(input_ids)

        # 用图像嵌入替换
        inputs_embeds[images_seq_mask] = images_embeds[images_emb_mask]

        return inputs_embeds

    def prepare_gen_img_embeds(self, image_ids: torch.LongTensor):
        return self.gen_aligner(self.gen_embed(image_ids))
```

（11）下面这段代码的功能是使用AutoConfig和AutoModelForCausalLM进行注册，将不同的配置类（如VisionConfig、AlignerConfig等）与对应的模型类型关联起来。通过这些注册操作，系统可以自动加载和实例化相关的配置和模型，使不同模块和模型的配置管理变得更加灵活和可扩展。

```
AutoConfig.register("vision", VisionConfig)
AutoConfig.register("aligner", AlignerConfig)
AutoConfig.register("gen_vision", GenVisionConfig)
AutoConfig.register("gen_aligner", GenAlignerConfig)
AutoConfig.register("gen_head", GenHeadConfig)
AutoConfig.register("multi_modality", MultiModalityConfig)
AutoModelForCausalLM.register(MultiModalityConfig, MultiModalityCausalLM)
```

5.5.7 多模态处理器

在推理和训练过程中，通常需要将用户输入的文本和图像转换为模型可以处理的格式。文件processing_vlm.py定义了一个处理器（VLChatProcessor），用于处理多模态输入（文本和图像），并生成适合模型输入的格式。

文件processing_vlm.py的具体实现流程如下所示。

（1）类DictOutput的作用是提供一个类似于字典的对象接口，使其实例可以像字典一样访问和修改属性。它通过重载keys()、__getitem__() 和 __setitem__() 方法，使对象能够通过obj["key"] 方式读取和赋值，同时保留类的属性访问方式（如obj.key）。这为数据处理和存储提供了更灵活的访问方式。

```
class DictOutput(object):
    def keys(self):
        return self.__dict__.keys()

    def __getitem__(self, item):
        return self.__dict__[item]

    def __setitem__(self, key, value):
        self.__dict__[key] = value
```

（2）类VLChatProcessorOutput继承自DictOutput，并使用 @dataclass装饰器，使其自动生成 __init__ 等方法。类VLChatProcessorOutput用于存储多模态聊天处理后的输出，包括以下几种。

- sft_format：表示处理后的数据格式（如SFT训练格式）。
- input_ids：文本输入的Token ID（torch.Tensor）。
- pixel_values：图像的像素值（torch.Tensor）。
- num_image_tokens：图像对应的Token数量（torch.IntTensor）。
- __len__ 方法返回input_ids的长度，以便快速获取文本序列的长度。

由于继承了DictOutput，类VLChatProcessorOutput的实例可以像字典一样访问和修改属性，兼具面向对象和字典风格的访问方式，方便数据处理。

```
@dataclass
class VLChatProcessorOutput(DictOutput):
    sft_format: str                          # 处理过的格式
    input_ids: torch.Tensor                  # 输入的文本 ID
    pixel_values: torch.Tensor               # 图像的像素值
    num_image_tokens: torch.IntTensor        # 图像标记的数量

    def __len__(self):
        return len(self.input_ids)
```

（3）类BatchedVLChatProcessorOutput继承自DictOutput，使用 @dataclass进行结构化定义，用于存储批量处理后的多模态聊天数据。其主要功能是管理和转换多模态输入，包括文本和图像数据，适用于大模型的批量输入处理。

```
@dataclass
class BatchedVLChatProcessorOutput(DictOutput):
    sft_format: List[str]                    # 处理格式列表
    input_ids: torch.Tensor                  # 输入的文本 ID
```

```
    pixel_values: torch.Tensor             # 图像的像素值
    attention_mask: torch.Tensor           # 注意力掩码
    images_seq_mask: torch.BoolTensor      # 图像序列掩码
    images_emb_mask: torch.BoolTensor      # 图像嵌入掩码

    def to(self, device, dtype=torch.bfloat16):
        self.input_ids = self.input_ids.to(device)
        self.attention_mask = self.attention_mask.to(device)
        self.images_seq_mask = self.images_seq_mask.to(device)
        self.images_emb_mask = self.images_emb_mask.to(device)
        self.pixel_values = self.pixel_values.to(device=device, dtype=dtype)
        return self
```

（4）类VLChatProcessor继承自ProcessorMixin，是一个用于处理多模态对话的处理器，结合文本和图像信息，并支持格式化、多轮对话处理及数据批量化。类VLChatProcessor的主要功能包括：

- 处理输入文本和图像，支持LlamaTokenizerFast分词器和VLMImageProcessor图像处理器。
- 通过特殊标记（如<image_placeholder>）对输入中的图像进行标记和处理。
- 支持多轮对话的SFT（Supervised Fine-Tuning）格式转换。
- 统一管理input_ids（文本）、pixel_values（图像数据）、num_image_tokens（图像标记数）等数据结构。
- 支持将数据批量化，提高训练和推理效率。

类VLChatProcessor包含了如下成员。

- __init__()：初始化VLChatProcessor，设置图像处理器、分词器以及各种特殊标记（如<image_placeholder> <begin_of_image> 等），确保分词器支持图像标记，并调用ProcessorMixin的初始化方法。
- new_chat_template()：创建一个新的对话模板，并设置系统提示词，以标准化对话格式。
- apply_sft_template_for_multi_turn_prompts()：将多轮对话转换为SFT（Supervised Fine-Tuning，监督微调）格式，按照指定模板组织对话，并返回格式化后的文本。
- image_token()：返回图像占位符 <image_placeholder>，用于在文本中表示图像位置。
- image_id()：返回分词器中对应image_tag的token ID，如果不存在，则自动添加该特殊标记。
- image_start_id()：返回图像起始标记 <begin_of_image> 在分词器中的token ID。
- image_end_id()：返回图像结束标记 <end_of_image> 在分词器中的token ID。
- image_start_token()：返回图像起始标记 <begin_of_image> 的文本表示。
- image_end_token()：返回图像结束标记 <end_of_image> 的文本表示。
- pad_id()：返回填充标记 < | __pad__ | > 在分词器中的token ID，用于填充对齐。
- add_image_token()：在输入ID序列中插入图像相关标记，包括 <begin_of_image>、占位符token和 <end_of_image>，并返回更新后的input_ids。
- process_one()：对单个输入（文本或对话 + 图像）进行处理，转换为适合大模型输入的格式，进行tokenization，并添加图像相关标记，最终返回处理后的数据结构。

- __call__()：调用process_one() 处理输入，并根据force_batchify参数决定是否对数据进行批量化处理。
- batchify()：将多个VLChatProcessorOutput结果进行批量处理，统一数据格式，确保batch之间的对齐，适用于批量推理或训练。

```
class VLChatProcessor(ProcessorMixin):
    image_processor_class = "AutoImageProcessor"  # 图像处理类
    tokenizer_class = ("LlamaTokenizer", "LlamaTokenizerFast")  # 分词器类

    attributes = ["image_processor", "tokenizer"] # 需要处理的属性

    system_prompt = (
        "You are a helpful language and vision assistant. "
        "You are able to understand the visual content that the user provides, "
        "and assist the user with a variety of tasks using natural language."
    )

    def __init__(
        self,
        image_processor: VLMImageProcessor,                # 图像处理器
        tokenizer: LlamaTokenizerFast,                     # 分词器
        image_tag: str = "<image_placeholder>",            # 图像标签
        image_start_tag: str = "<begin_of_image>",         # 图像开始标签
        image_end_tag: str = "<end_of_image>",             # 图像结束标签
        pad_tag: str = "<｜▁pad▁｜>",                       # 填充标签
        num_image_tokens: int = 576,                       # 图像的标记数量
        add_special_token: bool = False,                   # 是否添加特殊标记
        sft_format: str = "deepseek",                      # SFT 格式
        mask_prompt: bool = True,                          # 是否遮蔽提示符
        ignore_id: int = -100,                             # 忽略的 ID
        **kwargs,
    ):
        self.image_processor = image_processor
        self.tokenizer = tokenizer

        image_id = self.tokenizer.vocab.get(image_tag)
        if image_id is None:
            special_tokens = [image_tag]
            special_tokens_dict = {"additional_special_tokens": special_tokens}
            self.tokenizer.add_special_tokens(special_tokens_dict)
            print(f"Add image tag = {image_tag} to the tokenizer")

        self.image_tag = image_tag
```

```
        self.image_start_tag = image_start_tag
        self.image_end_tag = image_end_tag
        self.pad_tag = pad_tag

        self.num_image_tokens = num_image_tokens
        self.add_special_token = add_special_token
        self.sft_format = sft_format
        self.mask_prompt = mask_prompt
        self.ignore_id = ignore_id

        super().__init__(
            image_processor,
            tokenizer,
            image_tag,
            num_image_tokens,
            add_special_token,
            sft_format,
            mask_prompt,
            ignore_id,
            **kwargs,
        )

    def new_chat_template(self):
        conv = get_conv_template(self.sft_format)
        conv.set_system_message(self.system_prompt)
        return conv

    def apply_sft_template_for_multi_turn_prompts(
        self,
        conversations: List[Dict[str, str]], # 多轮对话
        sft_format: str = "deepseek",         # SFT 格式
        system_prompt: str = "",              # 系统提示
    ):

        conv = get_conv_template(sft_format)
        conv.set_system_message(system_prompt)
        for message in conversations:
            conv.append_message(message["role"], message["content"].strip())
        sft_prompt = conv.get_prompt().strip()

        return sft_prompt

    @property
    def image_token(self):
        return self.image_tag
```

```
    @property
    def image_id(self):
        image_id = self.tokenizer.vocab.get(self.image_tag)
        return image_id

    @property
    def image_start_id(self):
        image_start_id = self.tokenizer.vocab.get(self.image_start_tag)
        return image_start_id

    @property
    def image_end_id(self):
        image_end_id = self.tokenizer.vocab.get(self.image_end_tag)
        return image_end_id

    @property
    def image_start_token(self):
        return self.image_start_tag

    @property
    def image_end_token(self):
        return self.image_end_tag

    @property
    def pad_id(self):
        pad_id = self.tokenizer.vocab.get(self.pad_tag)
        return pad_id

    def add_image_token(
        self,
        image_indices: List[int],  # 图像索引
        input_ids: torch.LongTensor,  # 输入的 ID
    ):

        input_slices = []

        start = 0
        for index in image_indices:
            if self.add_special_token:
                end = index + 1
            else:
                end = index
            # original text tokens
            input_slices.append(input_ids[start:end])
            # add boi, image tokens, eoi and set the mask as False
            input_slices.append(self.image_start_id * torch.ones((1),
```

```
                                dtype=torch.long))
            input_slices.append(
                self.image_id * torch.ones((self.num_image_tokens,),
                                           dtype=torch.long)
            )
            input_slices.append(self.image_end_id * torch.ones((1),
                                dtype=torch.long))
            start = index + 1

        # the left part
        input_slices.append(input_ids[start:])

        # concat all slices
        input_ids = torch.cat(input_slices, dim=0)
        num_image_tokens = torch.IntTensor([self.num_image_tokens] *
                                           len(image_indices))

        return input_ids, num_image_tokens

    def process_one(
        self,
        prompt: str = None,
        conversations: List[Dict[str, str]] = None,
        images: List[Image] = None,
        **kwargs,
    ):
        assert (
            prompt is None or conversations is None
        ), "prompt and conversations cannot be used at the same time."

        if prompt is None:
            # apply sft format
            sft_format = self.apply_sft_template_for_multi_turn_prompts(
                conversations=conversations,
                sft_format=self.sft_format,
                system_prompt=self.system_prompt,
            )
        else:
            sft_format = prompt

        # tokenize
        input_ids = self.tokenizer.encode(sft_format)
        input_ids = torch.LongTensor(input_ids)

        # add image tokens to the input_ids
        image_token_mask: torch.BoolTensor = input_ids == self.image_id
```

```
        image_indices = image_token_mask.nonzero()
        input_ids, num_image_tokens = self.add_image_token(
            image_indices=image_indices,
            input_ids=input_ids,
        )

        # load images
        images_outputs = self.image_processor(images, return_tensors="pt")

        prepare = VLChatProcessorOutput(
            sft_format=sft_format,
            input_ids=input_ids,
            pixel_values=images_outputs.pixel_values,
            num_image_tokens=num_image_tokens,
        )

        return prepare

    def __call__(
        self,
        *,
        prompt: str = None,
        conversations: List[Dict[str, str]] = None,
        images: List[Image] = None,
        force_batchify: bool = True,
        **kwargs,
    ):

        prepare = self.process_one(
            prompt=prompt, conversations=conversations, images=images
        )

        if force_batchify:
            prepare = self.batchify([prepare])

        return prepare

    def batchify(
        self, prepare_list: List[VLChatProcessorOutput]
    ) -> BatchedVLChatProcessorOutput:
        batch_size = len(prepare_list)
        sft_format = []
        n_images = []
        seq_lens = []
        for prepare in prepare_list:
            n_images.append(len(prepare.num_image_tokens))
```

```
        seq_lens.append(len(prepare))

    input_token_max_len = max(seq_lens)
    max_n_images = max(1, max(n_images))
```

5.6 JanusFlow 模型架构

janus\janusflow\models 目录包含 JanusFlow 模型的核心架构与相关实现。其中，modeling_vlm.py 定义了一个多模态模型，结合视觉理解、视觉生成和语言模型，用于处理图像和文本的联合推理任务；processing_vlm.py 实现了 VLChatProcessor 类，用于预处理图像和文本输入并生成适合模型的输入格式；uvit.py 实现了基于 U-Net Transformer 架构的视觉变换模型，用于图像去噪、超分辨率等任务。janus\janusflow\models 目录中的其他程序文件和前面介绍的“janus/models”目录中的同名文件相同，为节省本书篇幅不再讲解。

5.6.1 多模态模型

文件 modeling_vlm.py 实现了一个多模态模型，结合了视觉理解、视觉生成和语言模型，此文件包含三个主要的配置类：VisionUnderstandEncoderConfig、VisionGenerationEncoderConfig 和 VisionGenerationDecoderConfig，每个类负责不同部分的配置。在类 MultiModalityCausalLM 中，模型集成了视觉理解编码器、视觉生成编码器和解码器以及语言模型。该类通过将视觉输入转化为特征嵌入，并结合语言输入，进行多模态处理。此外，还定义了 prepare_inputs_embeds () 方法，用于处理图像和文本的嵌入，并生成合适的输入格式。

```
def model_name_to_cls(cls_name):
    if "CLIPVisionTower" in cls_name:
        cls = CLIPVisionTower
    elif "ShallowUViTEncoder" in cls_name:
        cls = ShallowUViTEncoder
    elif "ShallowUViTDecoder" in cls_name:
        cls = ShallowUViTDecoder
    else:
        raise ValueError(f"class_name {cls_name} is invalid.")

    return cls


class VisionUnderstandEncoderConfig(PretrainedConfig):
    model_type = "vision_und_enc"
    cls: str = ""
    params: AttrDict = {}
```

```
    def __init__(self, **kwargs):
        super().__init__(**kwargs)

        self.cls = kwargs.get("cls", "")
        if not isinstance(self.cls, str):
            self.cls = self.cls.__name__

        self.params = AttrDict(kwargs.get("params", {}))


class VisionGenerationEncoderConfig(PretrainedConfig):
    model_type = "vision_gen_enc"
    cls: str = ""
    params: AttrDict = {}

    def __init__(self, **kwargs):
        super().__init__(**kwargs)

        self.cls = kwargs.get("cls", "")
        if not isinstance(self.cls, str):
            self.cls = self.cls.__name__

        self.params = AttrDict(kwargs.get("params", {}))


class VisionGenerationDecoderConfig(PretrainedConfig):
    model_type = "vision_gen_dec"
    cls: str = ""
    params: AttrDict = {}

    def __init__(self, **kwargs):
        super().__init__(**kwargs)

        self.cls = kwargs.get("cls", "")
        if not isinstance(self.cls, str):
            self.cls = self.cls.__name__

        self.params = AttrDict(kwargs.get("params", {}))


class MultiModalityConfig(PretrainedConfig):
    model_type = "multi_modality"
    vision_und_enc_config: VisionUnderstandEncoderConfig
    language_config: LlamaConfig
```

```
    def __init__(self, **kwargs):
        super().__init__(**kwargs)
        vision_und_enc_config = kwargs.get("vision_und_enc_config", {})
        self.vision_und_enc_config = VisionUnderstandEncoderConfig(
            **vision_und_enc_config
        )

        vision_gen_enc_config = kwargs.get("vision_gen_enc_config", {})
        self.vision_gen_enc_config = VisionGenerationEncoderConfig(
            **vision_gen_enc_config
        )

        vision_gen_dec_config = kwargs.get("vision_gen_dec_config", {})
        self.vision_gen_dec_config = VisionGenerationDecoderConfig(
            **vision_gen_dec_config
        )

        language_config = kwargs.get("language_config", {})
        if isinstance(language_config, LlamaConfig):
            self.language_config = language_config
        else:
            self.language_config = LlamaConfig(**language_config)


class MultiModalityPreTrainedModel(PreTrainedModel):
    config_class = MultiModalityConfig
    base_model_prefix = "multi_modality"
    _no_split_modules = []
    _skip_keys_device_placement = "past_key_values"


class MultiModalityCausalLM(MultiModalityPreTrainedModel):

    def __init__(self, config: MultiModalityConfig):
        super().__init__(config)

        # 视觉理解编码器
        vision_und_enc_config = config.vision_und_enc_config
        vision_und_enc_cls = model_name_to_cls(vision_und_enc_config.cls)
        self.vision_und_enc_model = vision_und_enc_cls(**vision_und_enc_
config.params)

        # 视觉理解对齐器
        self.vision_und_enc_aligner = nn.Linear(1024, 2048, bias=True)

        # 视觉理解嵌入的开始标志
```

```
        self.beg_of_und_embed = nn.Parameter(torch.zeros(1, 2048))

        # 视觉生成编码器
        vision_gen_enc_config = config.vision_gen_enc_config
        vision_gen_enc_cls = model_name_to_cls(vision_gen_enc_config.cls)
        self.vision_gen_enc_model = vision_gen_enc_cls(**vision_gen_enc_
                                                       config.params)

        # 视觉生成编码器对齐器
        self.vision_gen_enc_aligner = nn.Linear(768, 2048, bias=True)

        # 视觉生成解码器
        vision_gen_dec_config = config.vision_gen_dec_config
        vision_gen_dec_cls = model_name_to_cls(vision_gen_dec_config.cls)
        self.vision_gen_dec_model = vision_gen_dec_cls(**vision_gen_dec_
                                                       config.params)

        # 语言模型
        language_config = config.language_config
        self.language_model = LlamaForCausalLM(language_config)

        # 视觉生成解码器对齐器
        self.vision_gen_dec_aligner_norm = LlamaRMSNorm(
            2048, eps=language_config.rms_norm_eps
        )
        self.vision_gen_dec_aligner = nn.Linear(2048, 768, bias=True)

    def prepare_inputs_embeds(
        self,
        input_ids: torch.LongTensor,
        pixel_values: torch.FloatTensor,
        images_seq_mask: torch.LongTensor,
        images_emb_mask: torch.LongTensor,
        **kwargs,
    ):
        """
        准备输入嵌入
        参数：
            input_ids (torch.LongTensor): [b, T]
            pixel_values (torch.FloatTensor): [b, n_images, 3, h, w]
            images_seq_mask (torch.BoolTensor): [b, T]
            images_emb_mask (torch.BoolTensor): [b, n_images, n_image_tokens]

            assert torch.sum(images_seq_mask) == torch.sum(images_emb_mask)

        返回：
```

```
        input_embeds (torch.Tensor): [b, T, D]
    """

    bs, n = pixel_values.shape[0:2]
    images = rearrange(pixel_values, "b n c h w -> (b n) c h w")
    # [b x n, T2, D]
    images_embeds = self.vision_und_enc_model(images)
    images_embeds = self.vision_und_enc_aligner(images_embeds)
    beg_of_und_embed = self.beg_of_und_embed[0].detach().clone()
    images_embeds = torch.cat(
        [
            beg_of_und_embed.view(1, 1, -1).repeat(images_embeds.
                                                   shape[0], 1, 1),
            images_embeds,
        ],
        dim=1,
    )
    # [b x n, T2, D] -> [b, n x T2, D]
    images_embeds = rearrange(images_embeds, "(b n) t d -> b (n t) d",
                              b=bs, n=n)
    # [b, n, T2] -> [b, n x T2]
    images_emb_mask = rearrange(images_emb_mask, "b n t -> b (n t)")

    # [b, T, D]
    input_ids[input_ids < 0] = 0  # 忽略图像嵌入
    inputs_embeds = self.language_model.get_input_embeddings()(input_ids)

    # 用图像嵌入替换
    inputs_embeds[images_seq_mask] = images_embeds[images_emb_mask]

    return inputs_embeds

AutoConfig.register("vision_und_enc", VisionUnderstandEncoderConfig)
AutoConfig.register("vision_gen_enc", VisionGenerationEncoderConfig)
AutoConfig.register("vision_gen_dec", VisionGenerationDecoderConfig)
AutoConfig.register("multi_modality", MultiModalityConfig)
AutoModelForCausalLM.register(MultiModalityConfig, MultiModalityCausalLM)
```

5.6.2 数据预处理

文件processing_vlm.py实现了一个用于处理视觉和语言多模态输入的VLChatProcessor类。其主要功能是将图像和文本输入进行预处理，为模型生成合适的输入格式。具体来说，文件processing_vlm.py包含以下几个方面的功能。

- 图像和文本的联合处理：通过向文本中添加特定的图像标记（如 <image_placeholder>、

<begin_of_image> 等），结合图像和文本输入，构建包含图像内容的多模态数据。

- 模板管理：提供了多轮对话模板和SFT格式的生成函数，以便于对多轮对话进行处理，适应不同的格式需求。
- 图像和文本的嵌入：通过image_processor处理图像数据，将图像转换为相应的像素值，并将其与文本信息一同编码到输入中。图像在文本中以特殊的标记序列表示。
- 批量处理：支持批量输入的处理，确保不同输入的数据结构和维度一致，方便模型进行并行计算。

总体来说，VLChatProcessor类实现了一个集成图像和文本数据预处理的模块，能够为多模态模型提供经过处理的输入数据，支持图像与文本的联合推理任务。

5.6.3 U-ViT 模型

文件uvit.py是DeepSeek团队基于UViT架构实现的一个视觉变换模型（Vision Transformer），主要用于图像处理任务，比如图像去噪、超分辨率、生成式建模等。代码改编自denoising-diffusion-pytorch，并结合了LLaMA模型的一些规范化方法，如RMSNorm。

文件uvit.py的基本架构如下。

1. UViT 结构

- ShallowUViTEncoder（浅层编码器）：提取图像的特征表示，包含时间步编码（Timesteps和TimestepEmbedding），以及输入卷积层和中间块（UVitBlock）。
- ShallowUViTDecoder（浅层解码器）：根据编码器的输出和时间步信息，进行图像重建。
- UVitBlock（核心U-Net块）：包含ResNet风格的ConvNextBlock，以及可选的Downsample2D（降采样）和Upsample2D（上采样）。

2. Patchify 和 Unpatchify

- Patchify将输入图像转换成嵌入块，用于Transformer计算。
- Unpatchify负责将Transformer处理后的特征图还原回完整的图像。

文件uvit.py的具体实现流程如下所示。

（1）类ImageHead继承自nn.Module，主要功能是对输入的图像特征进行归一化处理，然后通过一个线性层将这些特征映射到指定的输出空间。类ImageHead在多模态模型中用于将视觉特征与语言模型的嵌入空间对齐，以实现图像和文本的联合处理。

```
class ImageHead(nn.Module):
    """
    图像头模块，用于处理图像相关任务。
    """
    def __init__(self, decoder_cfg, gpt_cfg, layer_id=None):
        super().__init__()
        self.layer_id = layer_id
```

```
cfg = (
    AttrDict(
        norm_type="layernorm",
        is_exp_norm=False,
        sequence_parallel=False,
        use_userbuffer=False,
        norm_eps=1e-5,
        norm_bias=True,
        gradient_accumulation_fusion=True,
        use_fp32_head_weight=False,
    )
    + gpt_cfg
)
group = PG.tensor_parallel_group()
assert cfg.norm_type in [
    "layernorm",
    "rmsnorm",
], f"Norm type: {cfg.norm_type} not supported"
if cfg.norm_type == "rmsnorm":
    self.norm = DropoutAddRMSNorm(
        cfg.n_embed,
        prenorm=False,
        eps=cfg.norm_eps,
        is_exp_norm=cfg.is_exp_norm,
        sequence_parallel=cfg.sequence_parallel,
    )
else:
    self.norm = DropoutAddLayerNorm(
        cfg.n_embed,
        prenorm=False,
        eps=cfg.norm_eps,
        is_exp_norm=cfg.is_exp_norm,
        sequence_parallel=cfg.sequence_parallel,
        bias=cfg.norm_bias,
    )

multiple_of = 256
if decoder_cfg.in_channels % multiple_of != 0:
    warnings.warn(
        f"建议把vocab_size设置为{multiple_of}的倍数，否则会影响矩阵乘法
          的性能"
    )

dtype = default_dtype = torch.get_default_dtype()
if cfg.use_fp32_head_weight:
    dtype = torch.float32
```

```
            print(
                "使用 fp32 head weight!!!! 与原来的 bf16 head weight 不兼容\n",
                end="",
                flush=True,
            )
        torch.set_default_dtype(dtype)
        self.head = ColumnParallelLinear(
            cfg.n_embed,
            decoder_cfg.in_channels,
            bias=True,
            group=group,
            sequence_parallel=cfg.sequence_parallel,
            use_userbuffer=cfg.use_userbuffer,
            gradient_accumulation_fusion=cfg.gradient_accumulation_fusion,
            use_fp32_output=False,
        )
        torch.set_default_dtype(default_dtype)

        self.use_fp32_head_weight = cfg.use_fp32_head_weight

    def forward(
        self, input_args,
        images_split_mask: Optional[torch.BoolTensor] = None, **kwargs
    ):
        """
        图像头模块的前向传播。
        """
        residual = None
        if isinstance(input_args, tuple):
            x, residual = input_args
        else:
            x = input_args

        x = self.norm(x, residual)

        if self.use_fp32_head_weight:
            assert (
                self.head.weight.dtype == torch.float32
            ), f"head.weight is {self.head.weight.dtype}"
            x = x.float()

        if images_split_mask is None:
            logits = self.head(x)
        else:
            bs, n_images = images_split_mask.shape[:2]
            n_embed = x.shape[-1]
```

```
        images_embed = torch.masked_select(
            x.unsqueeze(1), images_split_mask.unsqueeze(-1)
        )
        images_embed = images_embed.view((bs * n_images, -1, n_embed))
        logits = self.head(images_embed)

    return logits
```

（2）类GlobalResponseNorm（GRN）是一个全局响应归一化模块，用于标准化输入特征的全局响应。通过计算输入张量在 (H, W) 维度（空间维度）的L2 范数，进一步进行归一化，并应用可学习的weight和bias参数，以增强模型的稳定性和特征表达能力。GRN主要用于计算机视觉任务，特别是在多模态模型和视觉Transformer结构中，用于特征归一化和自适应缩放。

```
class GlobalResponseNorm(nn.Module):
    """
    全局响应归一化（Global Response Normalization, GRN）模块。
    """

    def __init__(self, dim):
        """
        初始化 GRN 模块。

        参数:
        - dim (int): 输入特征的通道数（即特征维度）。
        """
        super().__init__()
        self.weight = nn.Parameter(torch.zeros(1, 1, 1, dim))  # 归一化权重
        self.bias = nn.Parameter(torch.zeros(1, 1, 1, dim))    # 归一化偏置

    def forward(self, x):
        """
        GRN 模块的前向传播。

        参数:
        - x (Tensor): 输入张量，形状为 (B, H, W, C)。

        返回:
        - Tensor: 归一化后的输出张量，形状与输入相同。
        """
        gx = torch.norm(x, p=2, dim=(1, 2), keepdim=True)
                                                # 计算空间维度 (H, W) 的L2 范数
        nx = gx / (gx.mean(dim=-1, keepdim=True) + 1e-6)
                                                      # 进行归一化，避免除零问题
```

```
        return torch.addcmul(self.bias, (self.weight * nx + 1), x, value=1)
                                        # 应用权重和偏置，并返回归一化结果
```

（3）类Downsample2D是一个 二维降采样（下采样）模块，用于对输入的 2D特征图进行尺寸缩减，同时支持可选的卷积（use_conv=True）或均值池化（use_conv=False）。类Downsample2D支持可选的归一化层（LayerNorm或RMSNorm），用于实现标准化输入特征。支持如下两种降采样方式：

- 使用卷积降采样（use_conv=True）：采用 2D卷积层进行降采样，可以学习到更丰富的特征。
- 使用均值池化降采样（use_conv=False）：使用AvgPool2D进行降采样，更适合保留整体信息。

类Downsample2D特别适用于计算机视觉和多模态任务，在CNN和Transformer结构中可用于减少计算量，提高效率。

```
class Downsample2D(nn.Module):
    """
    一个 2D 降采样层，可选使用卷积进行降采样。
    """

    def __init__(
        self,
        channels: int,
        use_conv: bool = False,
        out_channels: Optional[int] = None,
        padding: int = 1,
        name: str = "conv",
        kernel_size=3,
        stride=2,
        norm_type=None,
        eps=None,
        elementwise_affine=None,
        bias=True,
    ):
        """
        初始化 Downsample2D 模块。

        参数:
        - channels (int): 输入通道数。
        - use_conv (bool): 是否使用卷积进行降采样，默认 False（使用均值池化）。
        - out_channels (Optional[int]): 输出通道数，默认为 None（与输入通道数相同）。
        - padding (int): 卷积填充大小，默认为 1。
        - name (str): 卷积层的命名方式，默认 "conv"。
        - kernel_size (int): 卷积核大小，默认 3。
        - stride (int): 步长，默认 2。
        - norm_type (Optional[str]): 归一化类型，可选 `"ln_norm"`（LayerNorm）或
          `"rms_norm"`（RMSNorm）。
```

```
    - eps (Optional[float])：归一化计算中的小常数，避免除零错误。
    - elementwise_affine (Optional[bool])：是否为 LayerNorm 归一化使用可学习参
      数。
    - bias (bool)：是否在卷积层中使用偏置，默认为 True。
    """
    super().__init__()
    self.channels = channels
    self.out_channels = out_channels or channels
    self.use_conv = use_conv
    self.padding = padding
    self.name = name

    # 归一化层（可选）
    if norm_type == "ln_norm":
        self.norm = nn.LayerNorm(channels, eps, elementwise_affine)
    elif norm_type == "rms_norm":
        self.norm = RMSNorm(channels, eps)
    elif norm_type is None:
        self.norm = None
    else:
        raise ValueError(f"未知的归一化类型：{norm_type}")

    # 降采样方式：使用卷积或均值池化
    if use_conv:
        conv = nn.Conv2d(
            self.channels,
            self.out_channels,
            kernel_size=kernel_size,
            stride=stride,
            padding=padding,
            bias=bias,
        )
    else:
        assert self.channels == self.out_channels
        conv = nn.AvgPool2d(kernel_size=stride, stride=stride)

    # 根据 `name` 进行命名
    if name == "conv":
        self.Conv2d_0 = conv
        self.conv = conv
    elif name == "Conv2d_0":
        self.conv = conv
    else:
        self.conv = conv

def forward(self, hidden_states: torch.Tensor, *args,
```

```
        **kwargs) -> torch.Tensor:
"""
Downsample2D 模块的前向传播。

参数:
- hidden_states (Tensor): 输入的特征图，形状为 (B, C, H, W)。

返回:
- Tensor: 降采样后的特征图，形状为 (B, C, H/2, W/2)（假设 stride=2）。
"""
assert hidden_states.shape[1] == self.channels, "输入通道数不匹配! "

# 归一化（如果适用）
if self.norm is not None:
    hidden_states = self.norm(hidden_states.permute(0, 2, 3, 1)).
                              permute(0, 3, 1, 2)

# 当 padding=0 且使用卷积时，手动填充
if self.use_conv and self.padding == 0:
    pad = (0, 1, 0, 1)  # 右侧和底部填充 1
    hidden_states = F.pad(hidden_states, pad, mode="constant", value=0)

assert hidden_states.shape[1] == self.channels,
                                "归一化后通道数发生变化! "

# 进行降采样（卷积或均值池化）
hidden_states = self.conv(hidden_states)

return hidden_states
```

（4）类Upsample2D是一个二维上采样（Upsampling）模块，用于将低分辨率的2D特征图进行上采样，恢复更高的分辨率，同时支持可选的卷积或反卷积（转置卷积）操作。类Upsample2D支持如下三种上采样方式：

- 使用反卷积（use_conv_transpose=True）：通过ConvTranspose2d进行上采样，适用于需要学习放大特征的任务。
- 使用插值 + 卷积（use_conv=True）：先进行最近邻插值（F.interpolate），然后使用Conv2d进一步提取特征。
- 纯插值（use_conv=False）：仅使用最近邻插值进行放大，计算量最小。

另外，类Upsample2D支持归一化操作，通过LayerNorm（ln_norm）或RMSNorm（rms_norm）实现标准化输入特征，提高数值稳定性。

类Upsample2D很适用于计算机视觉和多模态任务，可用于恢复高分辨率特征图，例如Stable Diffusion、DALL·E或JanusFlow中的文本–图像生成任务。

```
class Upsample2D(nn.Module):
    """
    一个 2D 上采样层，可选使用卷积或反卷积进行上采样。
    """

    def __init__(
        self,
        channels: int,
        use_conv: bool = False,
        use_conv_transpose: bool = False,
        out_channels: Optional[int] = None,
        name: str = "conv",
        kernel_size: Optional[int] = None,
        padding=1,
        stride=2,
        norm_type=None,
        eps=None,
        elementwise_affine=None,
        bias=True,
        interpolate=True,
    ):
        """
        初始化 Upsample2D 模块。

        参数:
        - channels (int): 输入通道数。
        - use_conv (bool): 是否在上采样后使用卷积进行特征提取，默认 False。
        - use_conv_transpose (bool): 是否使用转置卷积（反卷积）进行上采样，默认 False。
        - out_channels (Optional[int]): 输出通道数，默认为 None（与输入通道数相同）。
        - name (str): 卷积层的名称，默认为 "conv"。
        - kernel_size (Optional[int]): 卷积核大小，默认 None（如果未指定，会根据
          `use_conv` 或 `use_conv_transpose` 自动选择）。
        - padding (int): 卷积填充大小，默认 1。
        - stride (int): 步长，默认 2。
        - norm_type (Optional[str]): 归一化类型，可选 `"ln_norm"`（LayerNorm）或
          `"rms_norm"`（RMSNorm）。
        - eps (Optional[float]): 归一化计算中的小常数，避免除零错误。
        - elementwise_affine (Optional[bool]): 是否为 LayerNorm 归一化使用可学习参
          数。
        - bias (bool): 是否在卷积层中使用偏置，默认为 True。
        - interpolate (bool): 是否使用最近邻插值进行上采样，默认为 True。
        """
        super().__init__()
        self.channels = channels
        self.out_channels = out_channels or channels
        self.use_conv = use_conv
```

```
        self.use_conv_transpose = use_conv_transpose
        self.name = name
        self.interpolate = interpolate
        self.stride = stride

        # 归一化层（可选）
        if norm_type == "ln_norm":
            self.norm = nn.LayerNorm(channels, eps, elementwise_affine)
        elif norm_type == "rms_norm":
            self.norm = RMSNorm(channels, eps)
        elif norm_type is None:
            self.norm = None
        else:
            raise ValueError(f"未知的归一化类型：{norm_type}")

        conv = None
        # 使用转置卷积（反卷积）进行上采样
        if use_conv_transpose:
            if kernel_size is None:
                kernel_size = 4
            conv = nn.ConvTranspose2d(
                channels,
                self.out_channels,
                kernel_size=kernel_size,
                stride=stride,
                padding=padding,
                bias=bias,
            )
        # 使用普通卷积（先插值后卷积）
        elif use_conv:
            if kernel_size is None:
                kernel_size = 3
            conv = nn.Conv2d(
                self.channels,
                self.out_channels,
                kernel_size=kernel_size,
                padding=padding,
                bias=bias,
            )

        # 赋值给合适的变量
        if name == "conv":
            self.conv = conv
        else:
            self.Conv2d_0 = conv
```

```
def forward(
    self,
    hidden_states: torch.Tensor,
    output_size: Optional[int] = None,
    *args,
    **kwargs,
) -> torch.Tensor:
    """
    Upsample2D 模块的前向传播。

    参数：
    - hidden_states (Tensor): 输入的特征图，形状为 (B, C, H, W)。
    - output_size (Optional[int]): 目标输出大小，如果提供则使用 `F.interpolate`
      进行调整。

    返回：
    - Tensor: 上采样后的特征图，形状为 (B, C, H*2, W*2)（假设 stride=2）。
    """
    assert hidden_states.shape[1] == self.channels, "输入通道数不匹配！"

    # 归一化（如果适用）
    if self.norm is not None:
        hidden_states = self.norm(hidden_states.permute(0, 2, 3, 1)).
                                  permute(0, 3, 1, 2)

    # 直接使用反卷积进行上采样
    if self.use_conv_transpose:
        return self.conv(hidden_states)

    # 由于 `upsample_nearest2d_out_frame` 不支持 bfloat16，因此转换为 float32
    # TODO(Suraj): 一旦 PyTorch 修复该问题，移除此转换
    dtype = hidden_states.dtype
    if dtype == torch.bfloat16:
        hidden_states = hidden_states.to(torch.float32)

    # 大批量数据时，避免 `upsample_nearest_nhwc` 失败
    # 参见：https://github.com/huggingface/diffusers/issues/984
    if hidden_states.shape[0] >= 64:
        hidden_states = hidden_states.contiguous()

    # 如果 `output_size` 被提供，则强制调整到该大小
    # 否则，使用 `scale_factor=2` 进行最近邻插值
    if self.interpolate:
        if output_size is None:
            hidden_states = F.interpolate(
                hidden_states, scale_factor=self.stride, mode="nearest"
```

```
            )
        else:
            hidden_states = F.interpolate(
                hidden_states, size=output_size, mode="nearest"
            )

    # 如果原始数据是 bfloat16，则转换回 bfloat16
    if dtype == torch.bfloat16:
        hidden_states = hidden_states.to(dtype)

    # 进行卷积（如果启用）
    if self.use_conv:
        if self.name == "conv":
            hidden_states = self.conv(hidden_states)
        else:
            hidden_states = self.Conv2d_0(hidden_states)

    return hidden_states
```

（5）类ConvNextBlock是ConvNeXt结构的基础块，用于实现图像特征提取和变换操作，结合深度卷积（Depthwise Convolution）和通道线性变换（Channel-wise Linear Transformation），实现高效的特征处理。该模块包含前馈网络（FFN），通过GELU激活函数和全局响应归一化（GlobalResponseNorm）提高数值稳定性，并使用残差连接以增强梯度传播。此外，该模块支持条件缩放（Scale）+偏移（Shift），使其能够结合外部条件（如文本信息）进行特征调整，在多模态学习和生成任务（如文本到图像生成）中具有重要应用。

```
class ConvNextBlock(nn.Module):
    """
    一个 ConvNeXt 结构的基础块，包含深度卷积和通道线性变换层。
    """

    def __init__(
        self,
        channels,
        norm_eps,
        elementwise_affine,
        use_bias,
        hidden_dropout,
        hidden_size,
        res_ffn_factor: int = 4,
    ):
        """
        初始化 ConvNextBlock 模块。
```

```
        参数：
        - channels (int)：输入通道数。
        - norm_eps (float)：归一化层中的 epsilon 值，避免数值不稳定。
        - elementwise_affine (bool)：是否使用逐元素仿射变换（归一化）。
        - use_bias (bool)：是否在卷积和全连接层中使用偏置。
        - hidden_dropout (float)：隐藏层的 dropout 率。
        - hidden_size (int)：条件嵌入的维度（`cond_embeds` 输入大小）。
        - res_ffn_factor (int)：FFN（前馈网络）扩展倍数，默认为 4。
        """
        super().__init__()

        # 深度卷积层（仅在通道维度内进行卷积操作）
        self.depthwise = nn.Conv2d(
            channels,
            channels,
            kernel_size=7,
            padding=3,
            groups=channels,  # 分组数等于通道数，即 Depthwise Convolution
            bias=use_bias,
        )

        # 归一化层
        self.norm = RMSNorm(channels, norm_eps)

        # 通道变换（前馈网络）
        self.channelwise_linear_1 = nn.Linear(
            channels, int(channels * res_ffn_factor), bias=use_bias
        )  # 通道数放大 res_ffn_factor 倍
        self.channelwise_act = nn.GELU()  # GELU 激活函数
        self.channelwise_norm = GlobalResponseNorm(int(channels *
                                                 res_ffn_factor))
                                                 # 全局归一化
        self.channelwise_linear_2 = nn.Linear(
            int(channels * res_ffn_factor), channels, bias=use_bias
        )  # 还原回原始通道数
        self.channelwise_dropout = nn.Dropout(hidden_dropout)  # Dropout 层

        # 条件嵌入映射（用于可控生成）
       self.cond_embeds_mapper = nn.Linear(hidden_size, channels * 2, use_bias)

    def forward(self, x, cond_embeds):
        """
        ConvNextBlock 模块的前向传播。

        参数：
        - x (Tensor)：输入特征图，形状为 (B, C, H, W)。
```

```
- cond_embeds (Tensor): 条件嵌入向量，形状为 (B, hidden_size)。

返回：
- Tensor: 处理后的特征图，形状与输入相同 (B, C, H, W)。
"""
x_res = x  # 残差连接

# 深度卷积操作
x = self.depthwise(x)

# 变换通道顺序 (B, C, H, W) → (B, H, W, C)
x = x.permute(0, 2, 3, 1)

# 归一化
x = self.norm(x)

# 前馈网络
x = self.channelwise_linear_1(x) # 线性变换（通道数扩展）
x = self.channelwise_act(x)      # GELU 激活
x = self.channelwise_norm(x)     # 全局响应归一化
x = self.channelwise_linear_2(x) # 线性变换（通道数还原）
x = self.channelwise_dropout(x)  # Dropout

# 还原通道顺序 (B, H, W, C) → (B, C, H, W)
x = x.permute(0, 3, 1, 2)

# 残差连接
x = x + x_res

# 计算条件缩放（Scale）和偏移（Shift）
scale, shift = self.cond_embeds_mapper(F.silu(cond_embeds)).chunk(2,
                                                   dim=1)

# 计算最终输出： x = (1 + scale) * x + shift
x = torch.addcmul(
    shift[:, :, None, None], x, (1 + scale)[:, :, None, None], value=1
)

return x
```

（6）下列代码定义了一个名为Patchify的PyTorch模块，其主要功能是将输入的图像数据划分为多个不重叠的patch，并通过卷积操作和归一化处理将这些块转换为嵌入向量（embeddings）。具体来说，Patchify使用一个卷积层（self.patch_conv）来提取图像块，并通过自定义的RMSNorm归一化层（self.norm）对提取的块进行归一化处理。

```
class Patchify(nn.Module):
    def __init__(
        self,
        in_channels,
        block_out_channels,
        patch_size,
        bias,
        elementwise_affine,
        eps,
        kernel_size=None,
    ):
        super().__init__()
        if kernel_size is None:
            kernel_size = patch_size
        self.patch_conv = nn.Conv2d(
            in_channels,
            block_out_channels,
            kernel_size=kernel_size,
            stride=patch_size,
            bias=bias,
        )
        self.norm = RMSNorm(block_out_channels, eps)

    def forward(self, x):
        embeddings = self.patch_conv(x)
        embeddings = embeddings.permute(0, 2, 3, 1)
        embeddings = self.norm(embeddings)
        embeddings = embeddings.permute(0, 3, 1, 2)
        return embeddings
```

（7）类unpatchify是一个图像切片（Patchification）模块，用于将输入图像转换为非重叠的patch，然后进行归一化处理，以适应Transformer或CNN结构。

```
class Unpatchify(nn.Module):
    def __init__(
        self, in_channels, out_channels, patch_size, bias,
        elementwise_affine, eps
    ):
        super().__init__()
        self.norm = RMSNorm(in_channels, eps)
        self.unpatch_conv = nn.Conv2d(
            in_channels,
            out_channels * patch_size * patch_size,
            kernel_size=1,
            bias=bias,
        )
```

```
        self.pixel_shuffle = nn.PixelShuffle(patch_size)
        self.patch_size = patch_size

    def forward(self, x):
        x = x.permute(0, 2, 3, 1)
        x = self.norm(x)
        x = x.permute(0, 3, 1, 2)
        x = self.unpatch_conv(x)
        x = self.pixel_shuffle(x)
        return x
```

（8）类UVitBlock是U-ViT（U-Net结合Vision Transformer）结构中的残差块，用于特征提取、变换和上/下采样。类UVitBlock的主要功能如下：

● 残差学习（Residual Learning）：通过多个ConvNextBlock层构建深度残差网络，提高模型的非线性表达能力。

● 支持可选的上/下采样：通过Downsample2D进行下采样，减少特征尺寸，提高计算效率；通过Upsample2D进行上采样，恢复更高分辨率的特征图。

类UVitBlock适用于多模态学习和生成任务。该结构适用于JanusFlow、Stable Diffusion、DALL·E等模型，在图像理解、生成、编辑任务中发挥关键作用。

```
class UVitBlock(nn.Module):
    """
    一个 U-ViT（U-Net 结合 Vision Transformer）残差块，支持可选的上 / 下采样。
    """

    def __init__(
        self,
        channels,
        out_channels,
        num_res_blocks,
        stride,
        hidden_size,
        hidden_dropout,
        elementwise_affine,
        norm_eps,
        use_bias,
        downsample: bool,
        upsample: bool,
        res_ffn_factor: int = 4,
        seq_len=None,
        concat_input=False,
        original_input_channels=None,
        use_zero=True,
        norm_type="RMS",
```

```
):
    """
    初始化 UVitBlock 模块。

    参数：
    - channels (int)：输入通道数。
    - out_channels (int)：输出通道数。
    - num_res_blocks (int)：该模块包含的残差块数量。
    - stride (int)：上 / 下采样的步长。
    - hidden_size (int)：线性层的隐藏维度。
    - hidden_dropout (float)：dropout 率，防止过拟合。
    - elementwise_affine (bool)：归一化层是否使用逐元素仿射变换。
    - norm_eps (float)：归一化层的小数偏移量，防止除零错误。
    - use_bias (bool)：是否在卷积层和线性层中使用偏置。
    - downsample (bool)：是否启用下采样（降采样）。
    - upsample (bool)：是否启用上采样（升采样）。
    - res_ffn_factor (int)：前馈网络（FFN）通道扩展因子，默认 4 倍。
    - seq_len (Optional[int])：可选的序列长度（用于 Transformer 相关任务）。
    - concat_input (bool)：是否拼接输入（如 U-Net 结构）。
    - original_input_channels (Optional[int])：原始输入的通道数（用于特定应用）。
    - use_zero (bool)：是否初始化为零（影响梯度流动）。
    - norm_type (str)：归一化类型，默认 `"RMS"`（均方根归一化）。
    """
    super().__init__()

    # 初始化多个残差块（ConvNext 结构）
    self.res_blocks = nn.ModuleList()
    for i in range(num_res_blocks):
        conv_block = ConvNextBlock(
            channels,
            norm_eps,
            elementwise_affine,
            use_bias,
            hidden_dropout,
            hidden_size,
            res_ffn_factor=res_ffn_factor,
        )
        self.res_blocks.append(conv_block)

    # 下采样（可选）
    if downsample:
        self.downsample = Downsample2D(
            channels=channels,
            out_channels=out_channels,
            use_conv=True,
            name="Conv2d_0",
```

```
                kernel_size=3,
                padding=1,
                stride=stride,
                norm_type="rms_norm",
                eps=norm_eps,
                elementwise_affine=elementwise_affine,
                bias=use_bias,
            )
        else:
            self.downsample = None

        # 上采样（可选）
        if upsample:
            self.upsample = Upsample2D(
                channels=channels,
                out_channels=out_channels,
                use_conv_transpose=False,
                use_conv=True,
                kernel_size=3,
                padding=1,
                stride=stride,
                name="conv",
                norm_type="rms_norm",
                eps=norm_eps,
                elementwise_affine=elementwise_affine,
                bias=use_bias,
                interpolate=True,
            )
        else:
            self.upsample = None

    def forward(self, x, emb, recompute=False):
        """
        UVitBlock 模块的前向传播。

        参数：
        - x (Tensor)：输入特征图，形状为 (B, C, H, W)。
        - emb (Tensor)：额外的条件嵌入向量（如文本、时间步信息等）。
        - recompute (bool)：是否在训练时使用重计算策略（降低显存占用）。

        返回：
        - Tensor：经过残差块和上/下采样处理后的特征图，形状可能变化。
        """
        # 通过多个残差块处理特征
        for res_block in self.res_blocks:
            x = res_block(x, emb)
```

```
        # 下采样（如果启用）
        if self.downsample is not None:
            x = self.downsample(x)

        # 上采样（如果启用）
        if self.upsample is not None:
            x = self.upsample(x)

        return x
```

（9）类ShallowUViTEncoder是U-ViT（U-Net + Vision Transformer）结构中的浅层（Shallow）编码器，用于提取图像特征并结合时间步嵌入。类ShallowUViTEncoder的主要功能如下。

- 时间步嵌入（Timestep Embedding）：通过Timesteps和TimestepEmbedding对时间信息进行编码，使得模型能够理解动态变化（如扩散模型中的时间步）。
- 卷积特征提取：in_conv采用大核卷积（7×7）+步长4进行下采样，捕获全局信息并减少计算量。
- 中间处理块（可选）：通过UVitBlock进一步处理特征，增强视觉表征能力。

类ShallowUViTEncoder适用于多模态任务，可以用于图像理解、生成任务，如Stable Diffusion、DALL · E 、JanusFlow，能够融合时间和视觉特征。

```
class ShallowUViTEncoder(nn.Module):
    """
    一个浅层 U-ViT（U-Net + Vision Transformer）编码器。
    """

    def __init__(
        self,
        input_channels=3,
        stride=4,
        kernel_size=7,
        padding=None,
        block_out_channels=(768,),
        layers_in_middle=2,
        hidden_size=2048,
        elementwise_affine=True,
        use_bias=True,
        norm_eps=1e-6,
        dropout=0.0,
        use_mid_block=True,
        **kwargs,
    ):
        """
        初始化 ShallowUViTEncoder 模块。
```

```
参数：
- input_channels (int): 输入图像的通道数（如RGB图像为 3）。
- stride (int): 下采样步长，默认为 4。
- kernel_size (int): 卷积核大小，默认为 7。
- padding (Optional[int]): 填充大小，若未指定，则自动计算。
- block_out_channels (Tuple[int]): 每个层的输出通道数，默认 768。
- layers_in_middle (int): `UVitBlock` 处理的层数，默认 2 层。
- hidden_size (int): 线性层的隐藏维度。
- elementwise_affine (bool): 归一化层是否使用逐元素仿射变换。
- use_bias (bool): 是否在卷积层中使用偏置。
- norm_eps (float): 归一化层中的epsilon，避免除零错误。
- dropout (float): Dropout概率，默认 0。
- use_mid_block (bool): 是否使用 `UVitBlock` 进行中间处理。
"""
super().__init__()

# 时间步投影（用于扩散模型）
self.time_proj = Timesteps(
    block_out_channels[0], flip_sin_to_cos=True,
    downscale_freq_shift=0
)
self.time_embed = TimestepEmbedding(
    block_out_channels[0], hidden_size, sample_proj_bias=use_bias
)

# 自动计算填充大小
if padding is None:
    padding = math.ceil(kernel_size - stride)

# 输入卷积层（用于特征提取 + 下采样）
self.in_conv = nn.Conv2d(
    in_channels=input_channels,
    out_channels=block_out_channels[0],
    kernel_size=kernel_size,
    stride=stride,
    padding=padding,
)

# 可选的中间处理块
if use_mid_block:
    self.mid_block = UVitBlock(
        block_out_channels[-1],              # 输入通道数
        block_out_channels[-1],              # 输出通道数
        num_res_blocks=layers_in_middle,     # 残差块数量
        hidden_size=hidden_size,             # 线性层隐藏大小
```

```
                hidden_dropout=dropout,              # Dropout 率
                elementwise_affine=elementwise_affine,
                norm_eps=norm_eps,
                use_bias=use_bias,
                downsample=False,
                upsample=False,
                stride=1,
                res_ffn_factor=4,
            )
        else:
            self.mid_block = None

    def get_num_extra_tensors(self):
        """
        获取额外张量的数量（用于计算或调试）。
        """
        return 2

    def forward(self, x, timesteps):
        """
        ShallowUViTEncoder 的前向传播。

        参数：
        - x (Tensor)：输入图像，形状为 (B, C, H, W)。
        - timesteps (Tensor)：时间步张量，表示当前图像的时间信息。

        返回：
        - x_emb (Tensor)：经过编码器处理后的特征图。
        - t_emb (Tensor)：经过时间步嵌入后的特征。
        - hs (List[Tensor])：经过处理的所有特征图。
        """
        bs = x.shape[0]   # 获取批次大小
        dtype = x.dtype   # 记录数据类型

        # 计算时间步嵌入
        t_emb = self.time_proj(timesteps.flatten()).view(bs, -1).to(dtype)
        t_emb = self.time_embed(t_emb)

        # 计算图像特征
        x_emb = self.in_conv(x)

        # 通过 UVitBlock 进行额外处理（如果启用）
        if self.mid_block is not None:
            x_emb = self.mid_block(x_emb, t_emb)

        # 存储处理后的特征图
```

```
        hs = [x_emb]

        return x_emb, t_emb, hs
```

（10）类ShallowUViTDecoder是一个浅层的U-ViT（U型视觉变换器）解码器模块，主要用于处理输入的特征并通过多个层级进行处理，最终生成输出。该解码器支持可配置的中间块（mid_block）和上采样操作（out_convs）。在实现中，mid_block是一个可选的块，可以在解码过程的中间阶段进行额外处理，处理后的特征经过上采样层逐步生成输出图像。

类ShallowUViTDecoder的具体功能如下。

- 输入归一化：对输入特征进行标准化处理，以提高模型的训练稳定性。
- 中间块（mid_block）：可选的中间处理块，使用U-ViT块对特征进行进一步的处理。该模块支持配置残差块、隐藏层大小、dropout、是否使用偏置等参数。
- 上采样卷积层（out_convs）：根据设置的上采样次数，使用卷积层逐步恢复输入特征的空间分辨率。最终生成输出图像的通道数由out_channels指定。

ShallowUViTDecoder解码器可以应用于视觉变换器（Vision Transformer）架构，特别是在图像生成、分割等任务中，作为解码模块与编码器部分进行结合。

```
class ShallowUViTDecoder(nn.Module):
    """
    一个浅层的 U-ViT 解码器。
    """
    def __init__(
        self,
        in_channels=768,
        out_channels=3,
        block_out_channels: Tuple[int] = (768,),
        upsamples=2,
        layers_in_middle=2,
        hidden_size=2048,
        elementwise_affine=True,
        norm_eps=1e-6,
        use_bias=True,
        dropout=0.0,
        use_mid_block=True,
        **kwargs,
    ):
        super().__init__()
        if use_mid_block:
            self.mid_block = UVitBlock(
                in_channels + block_out_channels[-1],
                block_out_channels[
                    -1
                ], # 实际上，这个参数没有使用，因为在禁用下采样和上采样时它没有任何效果。
```

```
                num_res_blocks=layers_in_middle,
                hidden_size=hidden_size,
                hidden_dropout=dropout,
                elementwise_affine=elementwise_affine,
                norm_eps=norm_eps,
                use_bias=use_bias,
                downsample=False,
                upsample=False,
                stride=1,
                res_ffn_factor=4,
            )
        else:
            self.mid_block = None
        self.out_convs = nn.ModuleList()
        for rank in range(upsamples):
            if rank == upsamples - 1:
                curr_out_channels = out_channels
            else:
                curr_out_channels = block_out_channels[-1]
            if rank == 0:
                curr_in_channels = block_out_channels[-1] + in_channels
            else:
                curr_in_channels = block_out_channels[-1]
            self.out_convs.append(
                Unpatchify(
                    curr_in_channels,
                    curr_out_channels,
                    patch_size=2,
                    bias=use_bias,
                    elementwise_affine=elementwise_affine,
                    eps=norm_eps,
                )
            )
        self.input_norm = RMSNorm(in_channels, norm_eps)

    def forward(self, x, hs, t_emb):
        """
        ShallowUViTDecoder 模块的前向传递。
        """
        x = x.permute(0, 2, 3, 1)
        x = self.input_norm(x)
        x = x.permute(0, 3, 1, 2)

        x = torch.cat([x, hs.pop()], dim=1)
        if self.mid_block is not None:
            x = self.mid_block(x, t_emb)
```

```
        for out_conv in self.out_convs:
            x = out_conv(x)
        assert len(hs) == 0
        return x
```

5.7 模型推理

在DeepSeek的Janus项目中提供了一个统一的多模态框架，支持多模态理解和文本到图像生成。模型推理过程包括加载预训练的Janus或JanusFlow模型，通过VLChatProcessor处理输入的文本和图像数据，生成相应的文本回答或图像输出。推理脚本分为多模态理解（inference.py）、文本到图像生成（generation_inference.py）和交互式图像生成（interactivechat.py），分别用于不同的应用场景。

5.7.1 多模态推理测试

文件inference.py的功能是通过使用Janus-1.3B模型进行推理，处理一个多模态的对话任务。在这个例子中，用户提供了一张包含数学公式的图像，并要求助手将该公式转换为LaTeX代码。

```
# 指定模型路径
model_path = "deepseek-ai/Janus-1.3B"
vl_chat_processor: VLChatProcessor = VLChatProcessor.from_pretrained(model_
path)
tokenizer = vl_chat_processor.tokenizer

vl_gpt: MultiModalityCausalLM = AutoModelForCausalLM.from_pretrained(
    model_path, trust_remote_code=True
)
vl_gpt = vl_gpt.to(torch.bfloat16).cuda().eval()

conversation = [
    {
        "role": "User",
        "content": "<image_placeholder>\nConvert the formula into latex 
code.",
        "images": ["images/equation.png"],
    },
    {"role": "Assistant", "content": ""},
]

# 加载图像并为输入做好准备
pil_images = load_pil_images(conversation)
prepare_inputs = vl_chat_processor(
    conversations=conversation, images=pil_images, force_batchify=True
```

```
).to(vl_gpt.device)

# 运行图像编码器以获取图像嵌入
inputs_embeds = vl_gpt.prepare_inputs_embeds(**prepare_inputs)

# 运行模型以生成响应
outputs = vl_gpt.language_model.generate(
    inputs_embeds=inputs_embeds,
    attention_mask=prepare_inputs.attention_mask,
    pad_token_id=tokenizer.eos_token_id,
    bos_token_id=tokenizer.bos_token_id,
    eos_token_id=tokenizer.eos_token_id,
    max_new_tokens=512,
    do_sample=False,
    use_cache=True,
)

answer = tokenizer.decode(outputs[0].cpu().tolist(), skip_special_
tokens=True)
print(f"{prepare_inputs['sft_format'][0]}", answer)
```

执行后会输出输出一个LaTeX代码字符串，表示用户提供的数学公式，例如：

```
<image_placeholder>
Convert the formula into latex code.
\frac{a}{b} = c
```

5.7.2 文生图推理

文件generation_inference.py的功能是使用Janus-1.3B多模态语言模型根据文本提示生成图像，输入文本描述场景或对象后，模型会基于这些描述生成相应的图像。生成过程包括编码输入提示、生成表示图像的token、将这些token解码成图像，并将生成的图像保存为JPEG文件。文件generation_inference.py使用了MultiModalityCausalLM模型和VLChatProcessor来处理文本和图像的多模态输入。

```
model_path = "deepseek-ai/Janus-1.3B"
vl_chat_processor: VLChatProcessor = VLChatProcessor.from_pretrained(model_
path)
tokenizer = vl_chat_processor.tokenizer

vl_gpt: MultiModalityCausalLM = AutoModelForCausalLM.from_pretrained(
    model_path, trust_remote_code=True
)
vl_gpt = vl_gpt.to(torch.bfloat16).cuda().eval()
```

```
conversation = [
    {
        "role": "User",
        "content": "A close-up high-contrast photo of Sydney Opera House
                    sitting next to Eiffel tower, under a blue night sky of
                    roiling energy, exploding yellow stars, and radiating
                    swirls of blue.",
    },
    {"role": "Assistant", "content": ""},
]

sft_format = vl_chat_processor.apply_sft_template_for_multi_turn_prompts(
    conversations=conversation,
    sft_format=vl_chat_processor.sft_format,
    system_prompt="",
)
prompt = sft_format + vl_chat_processor.image_start_tag


@torch.inference_mode()
def generate(
    mmgpt: MultiModalityCausalLM,
    vl_chat_processor: VLChatProcessor,
    prompt: str,
    temperature: float = 1,
    parallel_size: int = 16,
    cfg_weight: float = 5,
    image_token_num_per_image: int = 576,
    img_size: int = 384,
    patch_size: int = 16,
):
    input_ids = vl_chat_processor.tokenizer.encode(prompt)
    input_ids = torch.LongTensor(input_ids)

    tokens = torch.zeros((parallel_size*2, len(input_ids)),
                        dtype=torch.int).cuda()
    for i in range(parallel_size*2):
        tokens[i, :] = input_ids
        if i % 2 != 0:
            tokens[i, 1:-1] = vl_chat_processor.pad_id

    inputs_embeds = mmgpt.language_model.get_input_embeddings()(tokens)

    generated_tokens = torch.zeros((parallel_size, image_token_num_per_
                        image), dtype=torch.int).cuda()
```

```
    for i in range(image_token_num_per_image):
        outputs = mmgpt.language_model.model(inputs_embeds=inputs_embeds,
                   use_cache=True, past_key_values=outputs.past_key_values if
                   i != 0 else None)
        hidden_states = outputs.last_hidden_state

        logits = mmgpt.gen_head(hidden_states[:, -1, :])
        logit_cond = logits[0::2, :]
        logit_uncond = logits[1::2, :]

        logits = logit_uncond + cfg_weight * (logit_cond-logit_uncond)
        probs = torch.softmax(logits / temperature, dim=-1)

        next_token = torch.multinomial(probs, num_samples=1)
        generated_tokens[:, i] = next_token.squeeze(dim=-1)

        next_token = torch.cat([next_token.unsqueeze(dim=1), next_token.
                                unsqueeze(dim=1)], dim=1).view(-1)
        img_embeds = mmgpt.prepare_gen_img_embeds(next_token)
        inputs_embeds = img_embeds.unsqueeze(dim=1)


    dec = mmgpt.gen_vision_model.decode_code(generated_tokens.to(dtype=torch.
        int), shape=[parallel_size, 8, img_size//patch_size, img_size//
        patch_size])
    dec = dec.to(torch.float32).cpu().numpy().transpose(0, 2, 3, 1)

    dec = np.clip((dec + 1) / 2 * 255, 0, 255)

    visual_img = np.zeros((parallel_size, img_size, img_size, 3),
                          dtype=np.uint8)
    visual_img[:, :, :] = dec

    os.makedirs('generated_samples', exist_ok=True)
    for i in range(parallel_size):
        save_path = os.path.join('generated_samples',
                                 "img_{}.jpg".format(i))
        PIL.Image.fromarray(visual_img[i]).save(save_path)


generate(
    vl_gpt,
    vl_chat_processor,
    prompt,
)
```

执行后将会生成与用户提供的文本描述相对应的图像，并将其保存在generated_samples文件夹下。每个生成的图像将保存为img_x.jpg格式，其中x是图像的索引。

5.7.3 交互式文生图推理

文件interactivechat.py实现了一个交互式图像生成器，允许用户通过输入描述来生成图像。用户可以指定每次生成多少张图像，程序根据用户的描述生成相应的图像，并将这些图像保存在本地的generated_samples文件夹中。图像生成的过程依赖预训练的Janus-1.3B模型，并结合了多模态的生成方法，其中输入不仅包含文本描述，还可以包括图像信息。

```
# 指定模型的路径
model_path = "deepseek-ai/Janus-1.3B"
vl_chat_processor: VLChatProcessor = VLChatProcessor.from_pretrained(model_path)
tokenizer = vl_chat_processor.tokenizer

vl_gpt: MultiModalityCausalLM = AutoModelForCausalLM.from_pretrained(
    model_path, trust_remote_code=True
)
vl_gpt = vl_gpt.to(torch.bfloat16).cuda().eval()


# 创建提示文本
def create_prompt(user_input: str) -> str:
    conversation = [
        {
            "role": "User",
            "content": user_input,
        },
        {"role": "Assistant", "content": ""},  # 助手的回答为空
    ]

    # 应用 SFT 模板进行多轮对话提示格式化
    sft_format = vl_chat_processor.apply_sft_template_for_multi_turn_prompts(
        conversations=conversation,
        sft_format=vl_chat_processor.sft_format,
        system_prompt="",
    )
    prompt = sft_format + vl_chat_processor.image_start_tag  # 拼接开始标记
    return prompt


# 生成图像
@torch.inference_mode()
def generate(
```

```
    mmgpt: MultiModalityCausalLM,
    vl_chat_processor: VLChatProcessor,
    prompt: str,
    short_prompt: str,
    parallel_size: int = 16,
    temperature: float = 1,
    cfg_weight: float = 5,
    image_token_num_per_image: int = 576,
    img_size: int = 384,
    patch_size: int = 16,
):
    input_ids = vl_chat_processor.tokenizer.encode(prompt)
    input_ids = torch.LongTensor(input_ids)

    # 初始化并设置张量
    tokens = torch.zeros((parallel_size * 2, len(input_ids)),
                         dtype=torch.int).cuda()
    for i in range(parallel_size * 2):
        tokens[i, :] = input_ids
        if i % 2 != 0:
            tokens[i, 1:-1] = vl_chat_processor.pad_id  # 填充标记

    inputs_embeds = mmgpt.language_model.get_input_embeddings()(tokens)

    generated_tokens = torch.zeros((parallel_size,
                                    image_token_num_per_image),
                                    dtype=torch.int).cuda()
    outputs = None  # 初始化输出以便在循环中使用

    # 循环生成图像的 tokens
    for i in range(image_token_num_per_image):
        outputs = mmgpt.language_model.model(
            inputs_embeds=inputs_embeds,
            use_cache=True,
            past_key_values=outputs.past_key_values if i != 0 else None
                                                                  # 使用缓存
        )
        hidden_states = outputs.last_hidden_state

        logits = mmgpt.gen_head(hidden_states[:, -1, :])
        logit_cond = logits[0::2, :]
        logit_uncond = logits[1::2, :]

        logits = logit_uncond + cfg_weight * (logit_cond - logit_uncond)
                                                        # 加权条件 logits
        probs = torch.softmax(logits / temperature, dim=-1)
```

```
            next_token = torch.multinomial(probs, num_samples=1)
            generated_tokens[:, i] = next_token.squeeze(dim=-1)

            # 准备图像嵌入
            next_token = torch.cat([next_token.unsqueeze(dim=1),
                                    next_token.unsqueeze(dim=1)], dim=1).view(-1)
            img_embeds = mmgpt.prepare_gen_img_embeds(next_token)
            inputs_embeds = img_embeds.unsqueeze(dim=1)

        # 解码生成的图像
        dec = mmgpt.gen_vision_model.decode_code(
            generated_tokens.to(dtype=torch.int),
            shape=[parallel_size, 8, img_size // patch_size, img_size // patch_size]
        )
        dec = dec.to(torch.float32).cpu().numpy().transpose(0, 2, 3, 1)

        # 归一化并调整图像数据
        dec = np.clip((dec + 1) / 2 * 255, 0, 255)

        visual_img = np.zeros((parallel_size, img_size, img_size, 3), dtype=np.uint8)
        visual_img[:, :, :] = dec

        # 创建保存图像的文件夹
        os.makedirs('generated_samples', exist_ok=True)

        # 创建时间戳
        timestamp = time.strftime("%Y%m%d-%H%M%S")

        # 清理短提示文本以确保文件名安全
        short_prompt = re.sub(r'\W+', '_', short_prompt)[:50]

        # 使用时间戳和部分用户提示生成图像文件名
        for i in range(parallel_size):
            save_path = os.path.join('generated_samples',
                                     f"img_{timestamp}_{short_prompt}_{i}.jpg")
            PIL.Image.fromarray(visual_img[i]).save(save_path)

# 交互式图像生成器
def interactive_image_generator():
    print("欢迎使用交互式图像生成器！")

    # 询问生成的图像数量
    while True:
        num_images_input = input("请输入每个提示生成多少张图像？（请输入正整数）：")
        if num_images_input.isdigit() and int(num_images_input) > 0:
```

```
            parallel_size = int(num_images_input)
            break
        else:
            print(" 无效输入，请输入正整数。")

    while True:
        user_input = input(" 请输入您希望生成的图像描述（或输入 'exit' 退出）: ")

        if user_input.lower() == 'exit':
            print(" 退出图像生成器，再见！ ")
            break

        prompt = create_prompt(user_input)

        # 创建用户输入的简化版本作为文件名
        short_prompt = re.sub(r'\W+', '_', user_input)[:50]

        print(f" 为以下内容生成 {parallel_size} 张图像: '{user_input}'")
        generate(
            mmgpt=vl_gpt,
            vl_chat_processor=vl_chat_processor,
            prompt=prompt,
            short_prompt=short_prompt,
            parallel_size=parallel_size  # 传递用户指定的图像数量
        )

        print(" 图像生成完成！请检查 'generated_samples' 文件夹中的输出。\n")

if __name__ == "__main__":
    interactive_image_generator()
```

5.8 Web 交互测试

本项目的“demo”目录实现了Web交互功能，通过Gradio和FastAPI提供了用户友好的界面和API接口，方便用户与多模态模型进行交互。

5.8.1 FastAPI 测试

（1）文件fastapi_app.py提供了一种自动化的方法来测试和展示DeepSeek Janus模型的多模态理解和图像生成能力，用于调用本项目中的两个主要API。

- 图像理解（Multimodal Understanding）：通过understand_image_and_question()函数向http://localhost:8000/understand_image_and_question/ 端点发送图像和问题，并接收模型的回答，从而实

现对图像内容的理解。

● 文本生成图像（Text-to-Image Generation）：通过generate_images()函数向http://localhost:8000/generate_images/ 端点发送文本提示（prompt），并接收模型生成的图像，最终将其保存为本地文件。

```
# 图像理解 API 端点
understand_image_url = "http://localhost:8000/understand_image_and_question/"

# 文本生成图像 API 端点
generate_images_url = "http://localhost:8000/generate_images/"

# 你的图片文件路径
image_path = "images/equation.png"

# 调用图像理解 API 的函数
def understand_image_and_question(image_path, question, seed=42, top_p=0.95,
                                  temperature=0.1):
    """
    发送图像和问题到图像理解端点，并打印模型的回答。

    参数:
    - image_path: str，图像文件的路径
    - question: str，用户提出的问题
    - seed: int，随机种子，确保生成的答案一致性
    - top_p: float，核采样参数，控制生成的多样性
    - temperature: float，温度参数，控制输出的随机性
    """
    files = {'file': open(image_path, 'rb')}
    data = {
        'question': question,
        'seed': seed,
        'top_p': top_p,
        'temperature': temperature
    }
    response = requests.post(understand_image_url, files=files, data=data)
    response_data = response.json()
    print(" 图像理解模型回答 :", response_data['response'])

# 调用文本生成图像 API 的函数
def generate_images(prompt, seed=None, guidance=5.0):
    """
    发送文本提示到图像生成端点，并保存返回的图片。

    参数:
    - prompt: str，文本提示，用于描述要生成的图像内容
```

```
    - seed: int，可选，随机种子，以确保生成的图像一致性
    - guidance: float，控制提示词的引导强度，数值越大，模型对提示词的遵循程度越高
    """
    data = {
        'prompt': prompt,
        'seed': seed,
        'guidance': guidance
    }
    response = requests.post(generate_images_url, data=data, stream=True)

    if response.ok:
        img_idx = 1
        # 创建字节流存储多个图像数据
        buffers = {}

        try:
            for chunk in response.iter_content(chunk_size=1024):
                if chunk:
                    # 使用边界检测来确定新图像的起始位置
                    if img_idx not in buffers:
                        buffers[img_idx] = io.BytesIO()
                    buffers[img_idx].write(chunk)
                    # 尝试打开图像
                    try:
                        buffer = buffers[img_idx]
                        buffer.seek(0)
                        image = Image.open(buffer)
                        img_path = f"generated_image_{img_idx}.png"
                        image.save(img_path)
                        print(f"已保存：{img_path}")

                        # 关闭当前缓冲区，准备下一个图像
                        buffer.close()
                        img_idx += 1

                    except Exception as e:
                        # 继续加载当前缓冲区中的数据
                        continue
        except Exception as e:
            print("处理图像时发生错误：", e)
    else:
        print("生成图像失败。")

# 示例用法
if __name__ == "__main__":
    # 调用图像理解 API
```

```
understand_image_and_question(image_path, "这张图片的内容是什么？")

# 调用文本生成图像 API
generate_images("一幅美丽的日落山脉数字艺术画。")
```

（2）文件fastapi_client.py作为fastapi_app.py的客户端，用于测试API服务器的功能。执行后调用项目的两个API端点，以测试和展示模型的多模态能力

5.8.2 Gradio 交互

文件app.py使用Gradio框架实现了一个交互式界面，基于DeepSeek的普通版多模态模型（Janus版）实现了一个图片理解和文本生成图像（Text-to-Image）系统，允许用户实现如下功能。

（1）多模态理解（Multimodal Understanding）。

- 用户上传图片并输入问题，模型会基于图像和文本进行推理并返回答案。
- 通过multimodal_understanding函数，处理图像和文本输入，并利用DeepSeek的VLChatProcessor和AutoModelForCausalLM进行推理。

（2）文本生成图像（Text-to-Image Generation）。

- 用户输入文本描述（prompt），模型生成相应的图像。
- generate_image函数使用DeepSeek的VLChatProcessor处理输入文本，并通过generate()函数生成图像的token，再由unpack进行解码，最终输出一组高质量图像。

```
# 加载模型和处理器
model_path = "deepseek-ai/Janus-1.3B"
config = AutoConfig.from_pretrained(model_path)
language_config = config.language_config
language_config._attn_implementation = 'eager'  # 设置注意力机制实现方式
vl_gpt = AutoModelForCausalLM.from_pretrained(model_path,
                                              language_config=language_config,
                                              trust_remote_code=True)
vl_gpt = vl_gpt.to(torch.bfloat16).cuda()          # 将模型转换为半精度并移动到 GPU
vl_chat_processor = VLChatProcessor.from_pretrained(model_path)
tokenizer = vl_chat_processor.tokenizer
cuda_device = 'cuda' if torch.cuda.is_available() else 'cpu'

# --------------- 多模态理解(Multimodal Understanding)---------------
@torch.inference_mode()  # 关闭梯度计算，优化推理性能
def multimodal_understanding(image, question, seed, top_p, temperature):
    """
    处理多模态输入（图片 + 文字），输出回答。
    :param image: 输入图片（NumPy 数组格式）
    :param question: 用户输入的问题
    :param seed: 随机种子，保证生成一致性
    :param top_p: 采样参数，控制生成多样性
```

```
    :param temperature: 生成温度，影响随机性
    :return: 生成的文本答案
    """
    torch.cuda.empty_cache()  # 清理 CUDA 缓存
    torch.manual_seed(seed)   # 设置随机种子
    np.random.seed(seed)
    torch.cuda.manual_seed(seed)

    conversation = [
        {"role": "User", "content": f"<image_placeholder>\n{question}",
         "images": [image]},
        {"role": "Assistant", "content": ""},
    ]
    pil_images = [Image.fromarray(image)]  # 将 NumPy 数组转换为 PIL 图像格式

    # 预处理输入
    prepare_inputs = vl_chat_processor(
        conversations=conversation, images=pil_images, force_batchify=True
    ).to(cuda_device, dtype=torch.bfloat16 if torch.cuda.is_available() else
         torch.float16)

    # 生成嵌入表示
    inputs_embeds = vl_gpt.prepare_inputs_embeds(**prepare_inputs)
    outputs = vl_gpt.language_model.generate(
        inputs_embeds=inputs_embeds,
        attention_mask=prepare_inputs.attention_mask,
        pad_token_id=tokenizer.eos_token_id,
        bos_token_id=tokenizer.bos_token_id,
        eos_token_id=tokenizer.eos_token_id,
        max_new_tokens=512,
        do_sample=False if temperature == 0 else True,
        use_cache=True,
        temperature=temperature,
        top_p=top_p,
    )
    answer = tokenizer.decode(outputs[0].cpu().tolist(), skip_special_tokens=True)
    return answer

# --------------- 生成图像（Text-to-Image Generation）---------------
@torch.inference_mode()
def generate_image(prompt, seed=None, guidance=5):
    """
    生成基于文本的图像。
    :param prompt: 文本输入（描述希望生成的图像）
    :param seed: 随机种子
    :param guidance: CFG（Classifier-Free Guidance）引导系数
```

```
    :return: 生成的图像列表
    """
    torch.cuda.empty_cache()  # 清理 CUDA 缓存
    if seed is not None:
        torch.manual_seed(seed)
        torch.cuda.manual_seed(seed)
        np.random.seed(seed)

    width = 384
    height = 384
    parallel_size = 5      # 生成多个样本以提高多样性

    with torch.no_grad():
        messages = [{'role': 'User', 'content': prompt},
                    {'role': 'Assistant', 'content': ''}]
        text = vl_chat_processor.apply_sft_template_for_multi_turn_prompts(
            conversations=messages,
            sft_format=vl_chat_processor.sft_format,
            system_prompt=''
        )
        text = text + vl_chat_processor.image_start_tag
        input_ids = torch.LongTensor(tokenizer.encode(text))

        # 调用图像生成函数
        output, patches = generate(
            input_ids,
            width // 16 * 16,
            height // 16 * 16,
            cfg_weight=guidance,
            parallel_size=parallel_size
        )

        images = unpack(patches, width // 16 * 16, height // 16 * 16)
        return [Image.fromarray(images[i]).resize((1024, 1024),
                Image.LANCZOS) for i in range(parallel_size)]

# --------------- Gradio 交互界面 ---------------
with gr.Blocks() as demo:
    gr.Markdown(value="# 多模态理解与文本生成图像系统 ")

    # 多模态理解界面
    gr.Markdown(value="## 多模态理解 ")
    with gr.Row():
        image_input = gr.Image()
        with gr.Column():
            question_input = gr.Textbox(label=" 请输入问题 ")
```

```
            und_seed_input = gr.Number(label="随机种子", precision=0,
                                       value=42)
            top_p = gr.Slider(minimum=0, maximum=1, value=0.95, step=0.05,
                              label="Top_p")
            temperature = gr.Slider(minimum=0, maximum=1, value=0.1,
                                    step=0.05, label="Temperature")

    understanding_button = gr.Button("解析")
    understanding_output = gr.Textbox(label="回答")

    # 绑定多模态理解功能
    understanding_button.click(
        multimodal_understanding,
        inputs=[image_input, question_input, und_seed_input, top_p,
                temperature],
        outputs=understanding_output
    )

    # 文本生成图像界面
    gr.Markdown(value="## 文本生成图像")
    prompt_input = gr.Textbox(label="请输入文本描述")
    seed_input = gr.Number(label="随机种子（可选）", precision=0, value=12345)
    cfg_weight_input = gr.Slider(minimum=1, maximum=10, value=5, step=0.5,
                                 label="CFG 权重")

    generation_button = gr.Button("生成图像")
    image_output = gr.Gallery(label="生成的图像", columns=2, rows=2,
                              height=300)

    # 绑定文本生成图像功能
    generation_button.click(
        fn=generate_image,
        inputs=[prompt_input, seed_input, cfg_weight_input],
        outputs=image_output
    )

# 启动 Gradio 服务器
demo.launch(share=True)
```

除了上述代码，DeepSeek还提供了如下两个Web测试文件。

- app_januspro.py：使用Gradio框架实现了一个交互式界面，基于DeepSeek的Janus-Pro版多模态模型实现。
- app_janusflow.py：使用Gradio框架实现了一个交互式界面，基于DeepSeek的JanusFlow版多模态模型实现。

执行后可以通过Web页面调用多模态模型，使用Janus-Pro版多模态模型实现文生图测试的效果如图5-2所示。

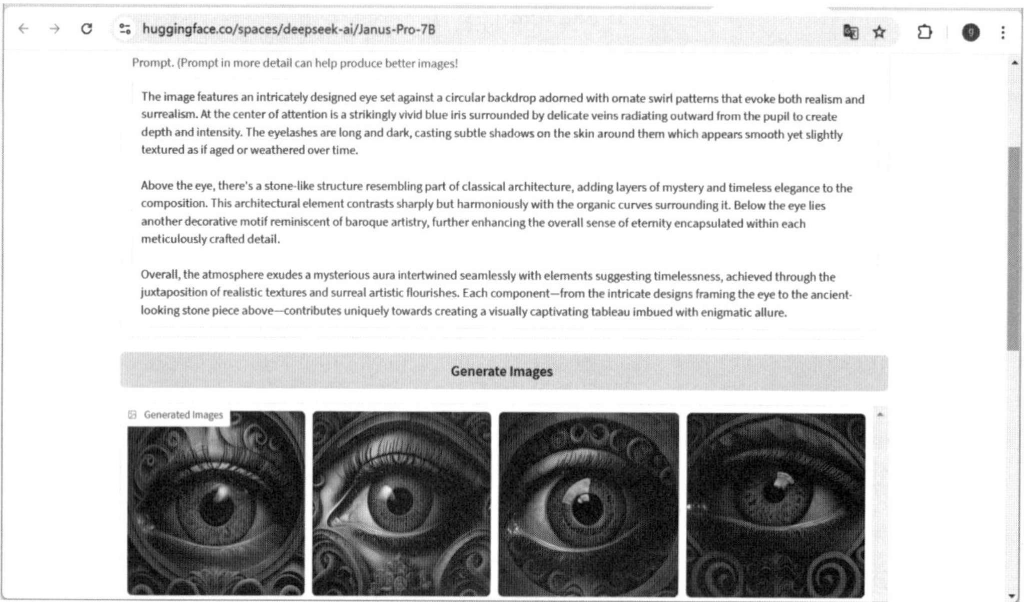

图5-2　文生图测试

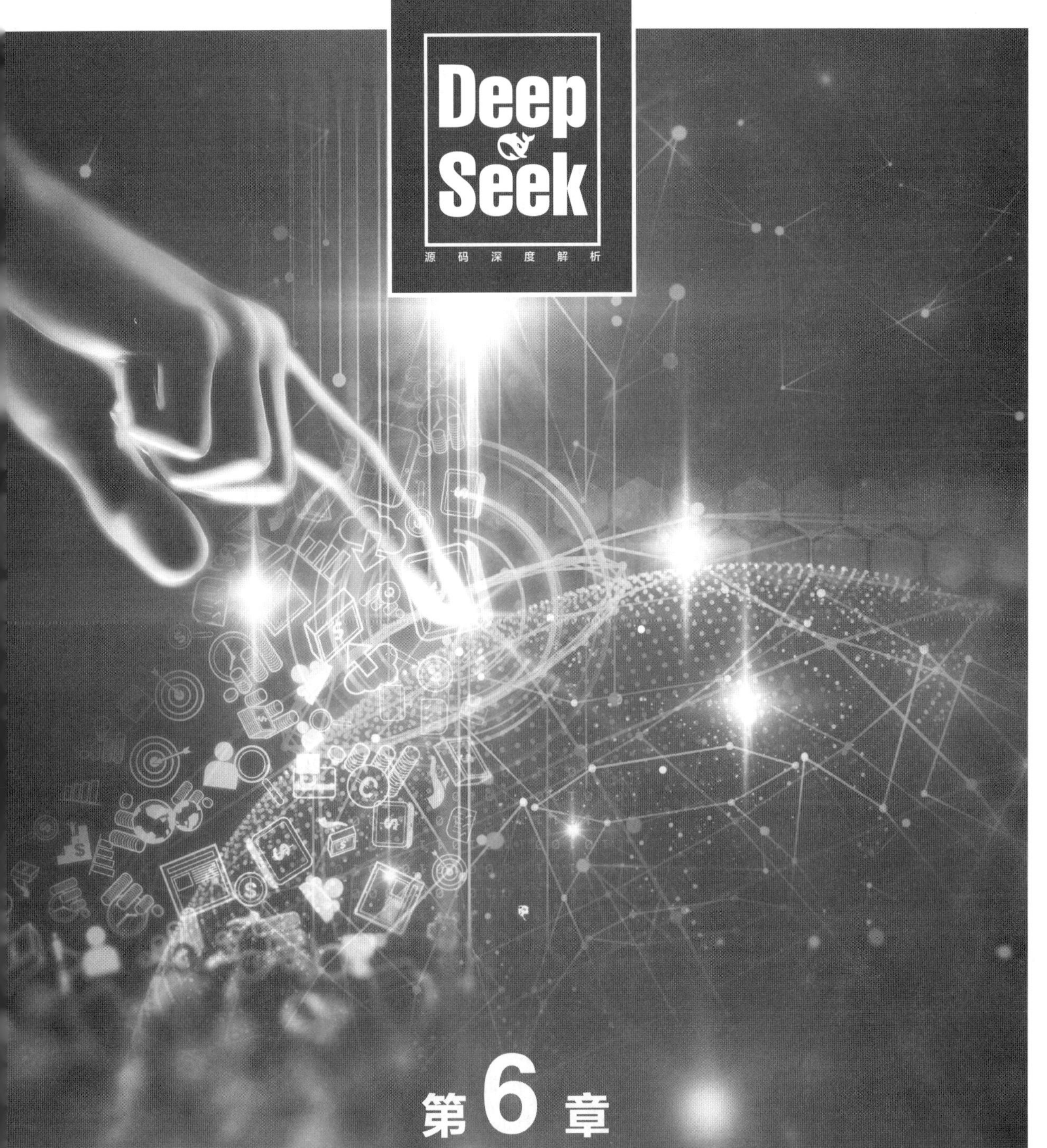

第 6 章 适用于高分辨率图像的多模态模型

DeepSeek-VL2 是一个多模态视觉语言模型，结合了高度优化的视觉编码器（SigLIP ViT）和强大的语言生成能力，支持文本与图像的联合推理和生成。通过模块化设计，项目包括视觉编码、视觉—语言特征对齐、语言生成及稀疏激活的MoE（混合专家）机制，提升了计算效率。DeepSeek-VL2 提供了完整的推理和部署代码，并支持命令行与Gradio Web界面的实时交互部署，是一个高效且灵活的多模态推理解决方案。

6.1 项目介绍

DeepSeek-VL2 是由DeepSeek AI团队开源的大型Vision-Language模型系列，通过MoE架构显著增强了多模态理解能力。相较于前代DeepSeek-VL，该模型在视觉和语言两个核心组件上进行了重大升级。

6.1.1 模型架构

DeepSeek-VL2 的架构包括6个核心模块，具体说明如下所示。

1. 视觉编码器

DeepSeek-VL2 引入了动态平铺（dynamic tiling）视觉编码策略，能够处理不同分辨率和宽高比的高分辨率图像。它将高分辨率图像分割成局部平铺，使用共享的SigLIP-SO400M-384 视觉变换器处理每个平铺，从而提取丰富的视觉特征。

具体来说，DeepSeek-VL2 的视觉编码器采用了一种动态平铺的策略，用于处理不同分辨率和宽高比的高分辨率图像。其主要步骤如下：

（1）图像分割与平铺：对于输入图像，模型首先根据长边和短边计算合适的候选分辨率，并选择一个最小化填充区域的分辨率。其次将图像调整到该分辨率后，分割成多个 384×384 的局部平铺，同时添加一个全局缩略图平铺。

（2）共享视觉编码器：所有（$1 + m_i \times n_i$）个平铺均由共享的SigLIP-SO400M-384 视觉变换器处理。每个平铺经过编码后，输出固定数量的视觉嵌入，即每个平铺产生 27×27 = 729个嵌入，每个嵌入的维度为 1152。

这种设计既保留了图像的局部细节，也能够高效地生成固定长度的视觉token序列，便于后续与语言模型进行融合。

2. 视觉—语言适配器

为了将视觉特征与语言模型的嵌入空间对齐，DeepSeek-VL2 引入了一个视觉—语言适配器。在处理每个平铺的视觉特征后，模型通过像素洗牌操作将其压缩到 14×14=196 个视觉标记。对于全局缩略图平铺，添加 14 个 <tile_newline> 标记，形成 14×15=210 个视觉标记。局部平铺被排列成一个二维网格，形状为（i×14，i×14），并在最终列末尾添加 i×14 个 <tile_newline> 标记。此外，在全局缩略图平铺和局部平铺之间插入一个 <view_separator> 标记。处理后的完整视觉序列包含 210 +1 + i×14×（i×14 +1）个视觉标记，这些标记随后被投影到语言模型的嵌入空间，使用两层MLP。

这种方法使模型能够有效地处理高分辨率图像，并将视觉信息与语言模型紧密结合，增强了多模态任务的性能。

3. MoE 语言模型

● DeepSeekMoE语言模型：模型在语言部分采用了改进的Transformer架构，基于DeepSeek-V2语言模型实现，并集成了MoE架构。这种设计允许模型在保持较高容量的同时，通过稀疏激活只使用部分专家，从而提高推理效率。

● 多头潜在注意力：该机制通过将Key-Value缓存压缩到潜在向量中，实现了高效的注意力计算和更快的推理速度。

● 专家选择与辅助损失：MoE模块中的门控机制（MoEGate）会根据输入特征计算门控得分，并采用贪婪或分组等策略选择Top-K专家。辅助损失机制用于优化专家选择过程，使得专家的路由更加合理。

DeepSeek-VL2 提供了三个不同大小的模型变体：DeepSeek-VL2-Tiny（1.0B激活参数）、DeepSeek-VL2-Small（2.8B激活参数）和DeepSeek-VL2（4.5B激活参数）。这些模型通过MoE架构实现了高效的稀疏计算，可以根据输入选择不同的专家网络来处理任务。

4. 数据预处理与对话管理

● 多模态数据预处理：通过专门的处理模块（如processing_deepseek_vl_v2.py和utils.py中的函数），模型能够将文本与图像数据转换为统一格式的输入，包括对话模板格式化、特殊token的处理和图像转换（如PIL到张量或Base64 编码）。

● 对话与上下文管理：在对话生成过程中，系统将用户输入、图像和历史对话结合在一起，构造适合模型推理的输入序列，实现多轮对话交互。

5. 训练与微调策略

● 多阶段训练流程：DeepSeek-VL2 的训练采用了三阶段策略实现，分别是视觉语言对齐、大规模预训练、监督微调和指令调优。

● 数据多样性：训练数据涵盖了VQA、OCR、文档/表格理解、视觉推理等多种任务，确保模型具备广泛的应用能力。

6. 部署与推理

● 推理接口与Gradio部署：项目不仅开源了模型的核心架构和预训练模型，还提供了完整的推理和部署代码（如inference.py和web_demo.py），支持通过命令行和Gradio Web界面进行实时推理和多模态交互。

● 流式生成与增量预填充：在推理时，模型支持流式文本输出和增量预填充技术，这些技术提高了响应速度和系统交互的流畅性。

总之，DeepSeek-VL2 项目采用了模块化设计，将视觉编码、视觉语言融合、语言生成和专家选择机制有机结合。

6.1.2 技术创新与亮点

DeepSeek-VL2 通过混合专家架构、动态分辨率支持和高效训练策略，实现了多模态任务的高性能与高效率。其技术原理和架构设计为AI视觉领域的创新与应用提供了强大支持，其技术创新与亮点如下。

1. 动态切片视觉编码策略（Dynamic Tiling Strategy）

（1）背景问题：传统视觉编码器通常处理固定分辨率图像，对于高分辨率且长宽比例变化较大的图像（如文档、图表、复杂场景）会面临计算资源浪费和细节捕捉不足的问题。

（2）创新方案：通过将高分辨率图像拆分为多个局部切片，每个切片经过共享的视觉Transformer处理，从而在不增加计算复杂度的前提下实现对图像细节的高效解析。适用于文档、表格、图表分析及视觉定位任务。

（3）动态切片策略的设计原理。

- 候选分辨率选择：定义一组候选分辨率 $CR = \{(m\times384, n\times384) \mid m, n \in N, 1 \leqslant m, n, mn \leqslant 9\}$（在某些任务中可扩展到 $mn \leqslant 18$），对于输入图像 (H, W) 计算每个候选分辨率下需要的填充面积，然后选取填充面积最小的候选分辨率。
- 图像切片与处理：将图像调整至选定分辨率后，划分为 $m\_i \times n\_i$ 个局部切片（每个切片尺寸为 384×384），同时额外生成一个全局缩略图切片。所有切片均由同一预训练SigLIP模型进行处理，提取出的视觉嵌入用于后续任务。

这种方法既能捕捉图像的局部细节，又能避免因分辨率增加带来的计算复杂度呈二次增长的问题，从而兼顾了效率与精度。

2. 多头潜在注意力（MLA）

（1）背景问题：传统注意力机制的计算复杂度较高，特别是在处理大型视觉—语言模型时。

（3）优化方法：引入MLA机制，将Key-Value缓存压缩为潜在向量，显著提高推理效率和吞吐量。通过DeepSeek MoE框架实现稀疏计算，有效提升计算效率。

3. 视觉语言适配器（Vision-Language Adaptor）

视觉语言适配器承担了将视觉信息和语言模型有效对接的任务，其主要流程包括：

（1）特征压缩与重排：利用 2×2 像素混洗操作将每个切片输出的视觉token从原始的 27×27 压缩至 14×14，这样既减少了token数量，又保留了局部空间结构信息。

（2）添加特殊标记：在全局缩略图切片后，为每一行添加 <tile_newline> 标记，在局部切片的2D网格结构中添加行结束标记，并在全局与局部切片之间插入 <view_separator> 标记。这些特殊标记帮助模型理解视觉token序列的层次结构和分隔信息。

（3）映射到语言空间：最终使用一个两层的多层感知器 (MLP) 将整合后的视觉token序列映射

到与语言模型嵌入空间一致的维度，保证后续语言模型处理时能够融合视觉与文本信息。

总之，视觉语言适配器通过像素混洗和特定标记（如 <tile_newline>、<view_separator> 等）将视觉特征序列映射到语言模型的嵌入空间，实现视觉与文本信息的无缝对接。

4. 优化算法和超参数

DeepSeek-VL2 使用AdamW优化器进行训练，采用cosine和step调度策略调整学习率。模型的权重衰减因子为 0.1，梯度裁剪的值为 1.0。在训练过程中，模型的超参数根据不同的任务和数据集进行调整，以获得最佳的性能。

总之，这一系列的创新设计和实现使得DeepSeek-VL2 成为一个强大且灵活的多模态大模型，为视觉与语言的联合理解与生成提供了全新的解决方案。

6.1.3 模型训练

1. 训练数据准备

DeepSeek-VL2 的训练数据包括视觉—语言对齐数据、视觉—语言预训练数据和监督微调数据。视觉—语言对齐数据用于初始阶段的MLP连接训练，视觉—语言预训练数据用于提升模型的多模态理解能力，监督微调数据则用于进一步优化模型的性能。

这些数据来自多个公开数据集和内部收集的数据，涵盖多种任务，如视觉问答、光学字符识别、文档理解、视觉定位等。数据的多样性和丰富性确保了模型的泛化能力和适应性。

2. 模型训练流程

DeepSeek-VL2 的训练流程分为三个主要阶段：模型初始化、视觉—语言预训练和监督微调（SFT），具体说明如下所示。

（1）在模型初始化阶段，视觉编码器和语言模型被分别预训练，以获取基本的视觉和语言特征表示。

（2）在视觉—语言预训练阶段，使用大规模的多模态数据来训练模型的视觉—语言对齐能力和多模态融合能力。在这个阶段，模型通过多任务学习和对比学习等技术，不断提高其对不同任务的理解和处理能力。

（3）在监督微调阶段，使用标注的多模态数据进一步优化模型性能。在这个阶段，模型通过优化任务特定的损失函数，学习如何更好地完成特定任务，如视觉问答、文档理解等。总之，通过监督微调和指令调优提升了模型的对话和多模态推理能力。

3. 并行训练与资源优化

● 并行策略：在模型训练中采用了模型并行、数据并行和专家并行等多种策略，充分利用GPU集群（如NVIDIA A100）进行大规模分布式训练。

● 负载均衡与切片调度：针对动态切片策略下的视觉编码器，由于不同图像切片数量存在差异，所以采用细粒度层分割和数据并行策略，可以实现在前向与反向过程中各GPU负载的均衡调度。

6.1.4 对比 Janus 项目

DeepSeek-VL2 和Janus都是DeepSeek开源的多模态项目，但它们在设计侧重点和实现细节上有所不同。DeepSeek-VL2 主要聚焦于通过动态平铺策略和SigLIP视觉编码器实现高分辨率图像的细粒度特征提取，并利用视觉—语言适配器和MoE语言模型（结合Multi-head Latent Attention）来实现视觉问答、OCR、文档解析和视觉定位等任务；Janus则致力于构建一个统一的多模态理解与生成框架，其核心思想在于将视觉编码解耦成独立通路，以便在同一模型中同时优化理解与生成任务，并进一步通过JanusFlow融合自回归与rectified flow技术，实现了高效地从文本到图像的生成。

1. 设计侧重点

DeepSeek-VL2：侧重于高分辨率图像的精细特征提取和视觉—语言信息的深度融合，适用于视觉问答、OCR、文档理解和视觉定位等任务。

Janus：侧重于构建统一的多模态平台，解耦视觉编码以同时支持理解和生成任务，并通过JanusFlow进一步提升从文本到图像生成的质量。

2. 视觉编码方法

- DeepSeek-VL2：采用动态平铺策略结合SigLIP视觉编码器，对图像进行分块处理，然后通过视觉—语言适配器将图像特征转换为语言模型输入。
- Janus：通过解耦视觉编码，将视觉特征分别提取后统一输入Transformer架构中，从而降低视觉编码与生成任务之间的冲突。

3. 语言模型架构

- DeepSeek-VL2：基于改进的DeepSeek-V2 Transformer，部分变体引入MoE机制和MLA机制，以提高参数利用率和推理效率。
- Janus：采用单一统一的Transformer架构处理多模态输入，重点在于通过解耦设计实现更好的任务适应性和交互性，同时在JanusFlow中融合自回归与rectified flow技术实现图像生成。

4. 训练与推理优化

- DeepSeek-VL2：采用多阶段训练策略（视觉—语言对齐、预训练、监督微调）和增量预填充技术，确保在高分辨率、多模态场景下能够高效推理。
- Janus：强调数据和模型规模的扩展，通过优化训练策略提升多模态理解和生成能力，确保在统一框架下兼顾多种任务性能。

总之，虽然这两个项目都是多模态模型，但DeepSeek-VL2 更侧重于细粒度视觉信息与语言模型的深度融合，而Janus通过解耦视觉编码和集成生成技术，追求更广泛的多模态统一性和灵活性。

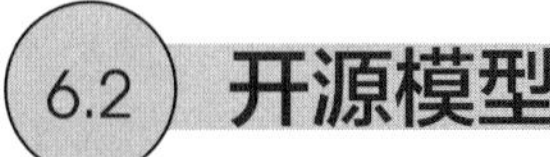

6.2 开源模型

目前DeepSeek-VL2 提供了三个不同规模的模型变体，在Hugging Face平台公开发布。这些模型通过MoE架构实现了高效的稀疏计算，能够根据输入选择不同的专家网络来处理任务。

（1）DeepSeek-VL2-Tiny：拥有约 10 亿个激活参数，适合资源受限场景。如图 6-1 所示，大家可以从Hugging Face获取并部署。

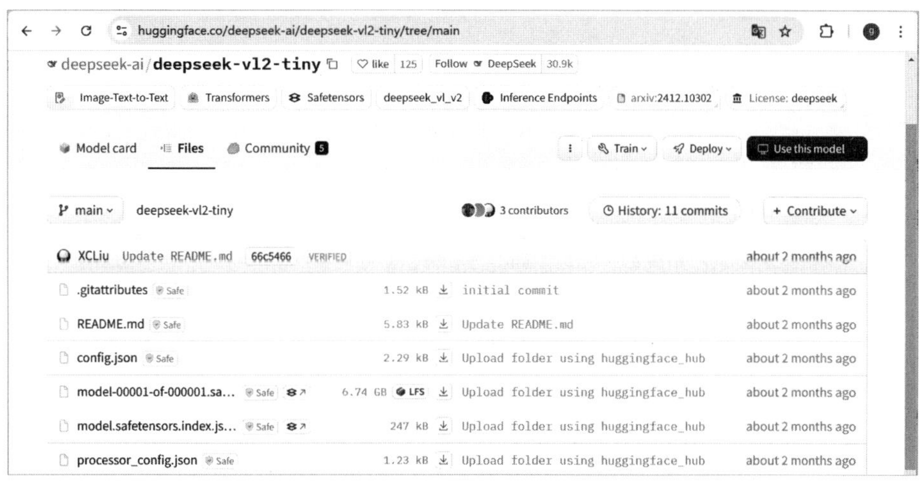

图 6-1　DeepSeek-VL2-Tiny模型

（2）DeepSeek-VL2-Small：拥有约 28 亿个激活参数，实现性能与资源消耗之间的较好平衡。如图 6-2 所示，大家可以从Hugging Face获取并部署。

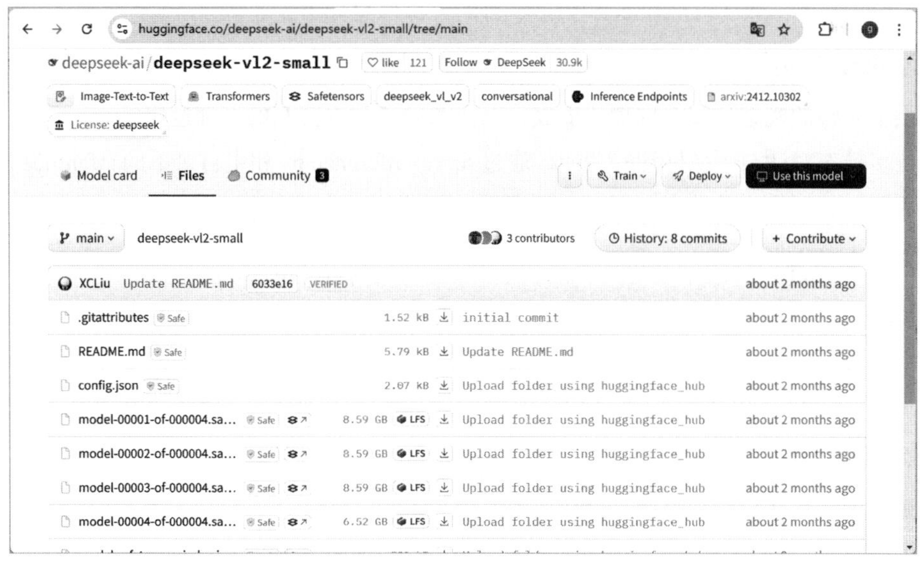

图 6-2　DeepSeek-VL2-Small模型

（3）DeepSeek-VL2：拥有约 45 亿个激活参数，适用于追求更高性能和复杂任务的场景。如图6-3所示，大家可以从Hugging Face获取并部署。

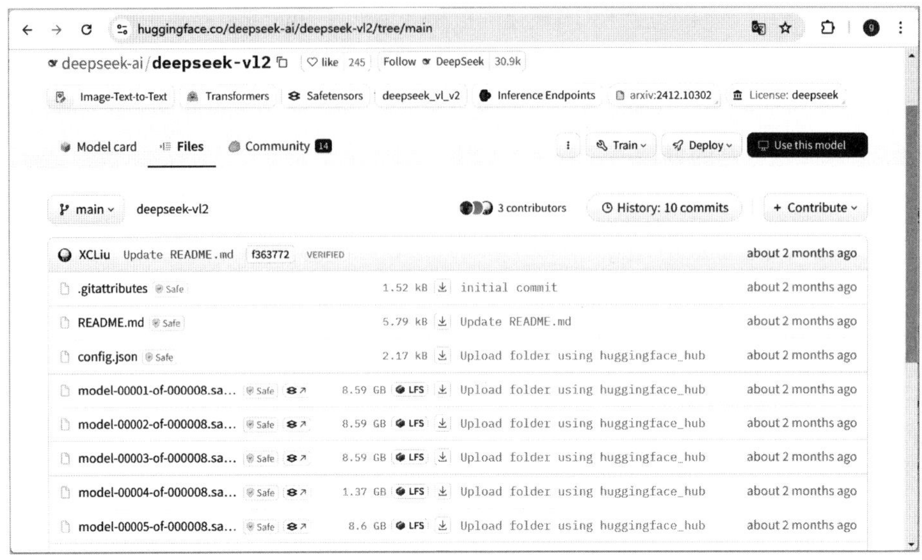

图6-3　DeepSeek-VL2模型

我们可以直接从Hugging Face上获取这些预训练好的模型权重，而无须重新训练模型。

6.3 开源信息介绍

目前（作者写本书时），DeepSeek-VL2项目的开源信息如下。

1. 模型推理与部署

- 提供了完整的模型加载、推理、生成，以及图像和文本交互的推理逻辑。
- 支持多模态对话与图像标注等功能，通过serve/inference.py和主目录下的inference.py实现。
- 集成了Gradio作为前端交互界面，允许用户通过Web界面直接调用模型功能。

2. 数据预处理与格式化

- processing_deepseek_vl_v2.py文件中包含对多模态输入的处理与格式化逻辑，支持图像与文本的联合输入。
- 自定义Tokenizer和Processor，用于将原始文本与图像数据转换为模型可处理的输入格式。

3. 自定义模型组件

- 开源了模型的部分核心组件，包括多头注意力机制、MLP层、MoE门控层等模块。
- 提供自定义的DeepseekVLV2ForCausalLM模型类，支持多模态输入的推理。

4. Gradio 前端 Web 调用

● 提供了完整的Gradio Web Demo实现，通过该脚本可以直接在本地或服务器上运行模型推理服务，并通过Web界面与模型交互。

● 开源了Gradio组件封装与事件处理逻辑，构建了用户输入输出交互界面，实现了文本和图像的联合输入与输出功能。

总之，DeepSeek-VL2 项目在Gradio Web部署方面提供了较为完整的开源实现，包含从后端模型推理到前端界面交互的全流程。用户可以借助该项目实现高效的多模态推理服务部署与Web交互调用。

6.4 配置文件

在DeepSeek-VL2项目的"utils"目录中提供一些通用工具和辅助函数，文件io.py提供了加载预训练模型、加载图像及读取JSON文件的功能。具体来说，文件io.py包含了从指定路径加载预训练的DeepSeek-VL2 模型、从对话数据加载图像，以及从指定文件路径加载JSON数据的函数。

```
import json
from typing import Dict, List
import PIL.Image
import torch
from transformers import AutoModelForCausalLM

def load_pretrained_model(model_path: str):
    """
    加载预训练的 DeepSeek-VL2 模型。

    参数：
        model_path (str): 模型的路径。

    返回：
        tokenizer: 模型的分词器。
        vl_chat_processor: 视觉语言处理器。
        vl_gpt: 预训练的视觉语言模型。
    """
    from deepseek_vl2.models.processing_deepseek_vl_v2 import
        DeepseekVLV2Processor
    from deepseek_vl2.models.modeling_deepseek_vl_v2 import
        DeepseekVLV2ForCausalLM

    vl_chat_processor = DeepseekVLV2Processor.from_pretrained(model_path)
    tokenizer = vl_chat_processor.tokenizer
```

```
    vl_gpt: DeepseekVLV2ForCausalLM = AutoModelForCausalLM.from_pretrained(
        model_path, trust_remote_code=True
    )
    vl_gpt = vl_gpt.to(torch.bfloat16).cuda().eval()

    return tokenizer, vl_chat_processor, vl_gpt

def load_pil_images(conversations: List[Dict[str, str]]) -> List[PIL.Image.Image]:
    """
    从对话列表中加载图像。

    参数：
        conversations (List[Dict[str, str]]): 包含消息的对话列表。示例:
            [
                {
                    "role": "User",
                    "content": "<image>\nExtract all information from this
                            image and convert them into markdown format.",
                    "images": ["./examples/table_datasets.png"]
                },
                {"role": "Assistant", "content": ""},
            ]

    返回：
        pil_images (List[PIL.Image.Image]): PIL 图像的列表。
    """
    pil_images = []

    for message in conversations:
        if "images" not in message:
            continue

        for image_path in message["images"]:
            pil_img = PIL.Image.open(image_path)
            pil_img = pil_img.convert("RGB")
            pil_images.append(pil_img)

    return pil_images

def load_json(filepath):
    """
    从指定文件路径加载 JSON 数据。

    参数：
        filepath (str): JSON 文件的路径。
```

```
返回：
    data: 加载的 JSON 数据。
"""
with open(filepath, "r") as f:
    data = json.load(f)
    return data
```

在上述代码中包含以下函数。

- load_pretrained_model()：加载预训练的DeepSeek-VL2 模型。该函数从指定路径加载模型的处理器和语言模型，并将模型设置为评估模式。
- load_pil_images()：从对话列表中加载图像。该函数遍历对话中的消息，检查是否包含图像路径，并将这些图像加载为PIL图像对象。
- load_json()：从指定文件路径加载JSON数据。

6.5 模型架构

在DeepSeek-VL2 项目的 "models" 目录中，汇聚了DeepSeek-VL2 模型的核心架构和实现代码，包括模型配置、对话逻辑、视觉编码器、语言模型（如MoE模块及其辅助组件）、数据预处理和权重初始化等多个部分。这些文件共同构成了多模态视觉语言模型的整体实现框架，支持模型的初始化、训练和推理，并为模型的各项高级功能（如动态平铺、Flash Attention、多头注意力及辅助损失计算）提供了完整的技术支撑。

6.5.1 模型配置

文件configuration_deepseek.py定义了DeepSeek-V2 模型的配置类DeepseekV2Config，该类继承自PretrainedConfig，用于存储和管理模型的配置信息。它包含了模型的各种超参数和结构设置，如词汇表大小、隐藏层维度、注意力头数量等，并提供了默认值。通过这个配置类，可以方便地实例化DeepSeek-V2 模型，并根据需要调整模型的参数。

```
from transformers.configuration_utils import PretrainedConfig
from transformers.utils import logging

logger = logging.get_logger(__name__)

DEEPSEEK_PRETRAINED_CONFIG_ARCHIVE_MAP = {}

class DeepseekV2Config(PretrainedConfig):
    """
    这是存储 [`DeepseekV2Model`] 配置的配置类。它用于根据指定的参数实例化DeepSeek模型，
定义模型架构。
```

```
    使用默认值实例化配置将生成一个类似于具有多潜在注意力的 DeepSeek-V2 的配置。

    配置对象继承自 [`PretrainedConfig`]，可以用于控制模型输出。有关更多信息，请参阅
[`PretrainedConfig`] 的文档。

    参数：
        vocab_size (`int`, 可选, 默认值为 102400):
            Deep 模型的词汇表大小。定义了在调用 [`DeepseekV2Model`] 时传递的
            `inputs_ids` 可以表示的不同标记的数量。
        hidden_size (`int`, 可选, 默认值为 4096):
            隐藏表示的维度。
        intermediate_size (`int`, 可选, 默认值为 11008):
            MLP 表示的维度。
        moe_intermediate_size (`int`, 可选, 默认值为 1407):
            MoE 表示的维度。
        num_hidden_layers (`int`, 可选, 默认值为 32):
            Transformer 解码器中的隐藏层数量。
        num_attention_heads (`int`, 可选, 默认值为 32):
            Transformer 解码器中每个注意力层的注意力头数量。
        n_shared_experts (`int`, 可选, 默认值为 None):
            共享专家的数量，None 表示密集模型。
        n_routed_experts (`int`, 可选, 默认值为 None):
            路由专家的数量，None 表示密集模型。
        routed_scaling_factor (`float`, 可选, 默认值为 1.0):
            路由专家的缩放因子。
        topk_method (`str`, 可选, 默认值为 `gready`):
            路由门中使用的 Topk 方法。
        n_group (`int`, 可选, 默认值为 None):
            路由专家的组数。
        topk_group (`int`, 可选, 默认值为 None):
            每个 token 选择的组数（确保每个 token 选择的专家仅在 `topk_group` 组内）。
        num_experts_per_tok (`int`, 可选, 默认值为 None):
            每个 token 选择的专家数量，None 表示密集模型。
        moe_layer_freq (`int`, 可选, 默认值为 1):
            MoE 层的频率：每 `moe_layer_freq - 1` 个密集层有一个专家层。
        first_k_dense_replace (`int`, 可选, 默认值为 0):
            浅层中密集层的数量（嵌入 -> 密集 -> 密集 -> ... -> 密集 -> moe ->
                            moe... -> lm_head）。
        norm_topk_prob (`bool`, 可选, 默认值为 False):
            是否归一化路由专家的权重。
        scoring_func (`str`, 可选, 默认值为 'softmax'):
            计算专家权重的方法。
        aux_loss_alpha (`float`, 可选, 默认值为 0.001):
            辅助损失权重系数。
        seq_aux (`bool`, 可选, 默认值为 True):
            是否为每个单独的样本计算辅助损失。
```

```
num_key_value_heads (`int`, 可选):
    用于实现分组查询注意力的 key_value 头的数量。如果 `num_key_value_
    heads=num_attention_heads`，模型将使用多头注意力（MHA）；如果 `num_
    key_value_heads=1`，模型将使用多查询注意力（MQA），否则使用 GQA。当将多
    头检查点转换为 GQA 检查点时，每个组的 key 和 value 头应通过平均池化该组内的
    所有原始头来构建。有关详细信息，请参阅 [这篇论文](https://arxiv.org/
    pdf/2305.13245.pdf)。如果未指定，将默认为 `num_attention_heads`。
hidden_act (`str` 或 `function`, 可选, 默认值为 `"silu"`):
    解码器中的非线性激活函数（函数或字符串）。
max_position_embeddings (`int`, 可选, 默认值为 2048):
    该模型可能使用的最大序列长度。
initializer_range (`float`, 可选, 默认值为 0.02):
    用于初始化所有权重矩阵的截断正态分布的标准差。
rms_norm_eps (`float`, 可选, 默认值为 1e-06):
    rms 归一化层中使用的 epsilon。
use_cache (`bool`, 可选, 默认值为 `True`):
    模型是否应返回最后的 key/values 注意力（并非所有模型都使用）。
    仅在 `config.is_decoder=True` 时相关。
pad_token_id (`int`, 可选):
    填充标记 id。
bos_token_id (`int`, 可选, 默认值为 1):
    流开始标记 id。
eos_token_id (`int`, 可选, 默认值为 2):
    流结束标记 id。
pretraining_tp (`int`, 可选, 默认值为 1):
    实验功能。预训练期间使用的张量并行秩。有关更多信息，请参阅 [这份文档]
    (https://huggingface.co/docs/transformers/parallelism)。此值对于确
    保预训练结果的精确可重复性是必要的。有关详细信息，请参阅 [这个问题](https://
    github.com/pytorch/pytorch/issues/76232)。
tie_word_embeddings (`bool`, 可选, 默认值为 `False`):
    是否绑定权重嵌入。
rope_theta (`float`, 可选, 默认值为 10000.0):
    RoPE 嵌入的基本周期。
rope_scaling (`Dict`, 可选):
    包含 RoPE 嵌入缩放配置的字典。目前支持两种缩放策略：线性和动态。它们的缩放因
    子必须是一个大于 1 的浮点数。预期格式为 `{"type": 策略名称, "factor": 缩
    放因子}`。使用此标志时，不要更新 `max_position_embeddings` 为预期的新最
    大值。
attention_bias (`bool`, 默认为 `False`, 可选, 默认值为 `False`):
    在自注意力期间的查询、键、值和输出投影层中是否使用偏置。
attention_dropout (`float`, 可选, 默认值为 0.0):
    注意力概率的 dropout 比率。
use_mla (`bool`, 可选, 默认值为 `True`): 使用多潜在注意力或多头注意力。如果
                                         为 True，模型将使用多潜在注意力，否
                                         则将使用多头注意力。
```

```
示例:
```python
>>> from transformers import DeepseekV2Model, DeepseekV2Config

>>> # 初始化一个 Deepseek-V2 风格的配置
>>> configuration = DeepseekV2Config()

>>> # 访问模型配置
>>> configuration = model.config
"""
model_type = "deepseek_v2"
keys_to_ignore_at_inference = ["past_key_values"]

def __init__(
    self,
    vocab_size=102400,
    hidden_size=4096,
    intermediate_size=11008,
    moe_intermediate_size = 1407,
    num_hidden_layers=30,
    num_attention_heads=32,
    num_key_value_heads=32,
    n_shared_experts = None,
    n_routed_experts = None,
    ep_size = 1,
    routed_scaling_factor = 1.0,
    kv_lora_rank = 512,
    q_lora_rank = 1536,
    qk_rope_head_dim = 64,
    v_head_dim = 128,
    qk_nope_head_dim = 128,
    topk_method = 'gready',
    n_group = None,
    topk_group = None,
    num_experts_per_tok = None,
    moe_layer_freq = 1,
    first_k_dense_replace = 0,
    norm_topk_prob = False,
    scoring_func = 'softmax',
    aux_loss_alpha = 0.001,
    seq_aux = True,
    hidden_act="silu",
    max_position_embeddings=2048,
    initializer_range=0.02,
    rms_norm_eps=1e-6,
    use_cache=True,
```
```

```
    pad_token_id=None,
    bos_token_id=100000,
    eos_token_id=100001,
    pretraining_tp=1,
    tie_word_embeddings=False,
    rope_theta=10000.0,
    rope_scaling=None,
    attention_bias=False,
    attention_dropout=0.0,
    use_mla=True,
    **kwargs,
):
    self.vocab_size = vocab_size
    self.max_position_embeddings = max_position_embeddings
    self.hidden_size = hidden_size
    self.intermediate_size = intermediate_size
    self.moe_intermediate_size = moe_intermediate_size
    self.num_hidden_layers = num_hidden_layers
    self.num_attention_heads = num_attention_heads
    self.n_shared_experts = n_shared_experts
    self.n_routed_experts = n_routed_experts
    self.ep_size = ep_size
    self.routed_scaling_factor = routed_scaling_factor
    self.kv_lora_rank = kv_lora_rank
    self.q_lora_rank = q_lora_rank
    self.qk_rope_head_dim = qk_rope_head_dim
    self.v_head_dim = v_head_dim
    self.qk_nope_head_dim = qk_nope_head_dim
    self.topk_method = topk_method
    self.n_group = n_group
    self.topk_group = topk_group
    self.num_experts_per_tok = num_experts_per_tok
    self.moe_layer_freq = moe_layer_freq
    self.first_k_dense_replace = first_k_dense_replace
    self.norm_topk_prob = norm_topk_prob
    self.scoring_func = scoring_func
    self.aux_loss_alpha = aux_loss_alpha
    self.seq_aux = seq_aux
    # 为了向后兼容
    if num_key_value_heads is None:
        num_key_value_heads = num_attention_heads

    self.num_key_value_heads = num_key_value_heads
    self.hidden_act = hidden_act
    self.initializer_range = initializer_range
    self.rms_norm_eps = float(rms_norm_eps)
```

```
        self.pretraining_tp = pretraining_tp
        self.use_cache = use_cache
        self.rope_theta = rope_theta
        self.rope_scaling = rope_scaling
        self.attention_bias = attention_bias
        self.attention_dropout = attention_dropout
        self.use_mla = use_mla

        super().__init__(
            pad_token_id=pad_token_id,
            bos_token_id=bos_token_id,
            eos_token_id=eos_token_id,
            tie_word_embeddings=tie_word_embeddings,
            **kwargs,
        )
```

6.5.2 多模态模型架构

文件modeling_deepseek_vl_v2.py定义了DeepSeek-VL2模型的核心组件，包括视觉编码器、多层感知机投影器和语言模型。此文件实现了多模态融合，能够处理视觉和文本输入，并生成相应的输出。具体来说，文件modeling_deepseek_vl_v2.py实现如下核心功能。

- 视觉编码器（VisionTransformer）：用于提取输入图像的视觉特征。
- 多层感知机投影器（MlpProjector）：将视觉特征投影到语言模型的嵌入空间，实现视觉和文本特征的对齐。
- 语言模型（DeepseekV2ForCausalLM）：基于Transformer的语言模型，能够生成文本输出。
- 多模态融合：将视觉特征和文本特征融合，实现对复杂多模态任务的理解和处理。

在接下来的内容中，将详细讲解文件modeling_deepseek_vl_v2.py的实现流程。

（1）类类是DeepSeek-VL2用于将视觉特征投影到语言模型的嵌入空间，此类支持多种类型的投影器，包括恒等投影、线性投影、带有GELU激活函数的多层感知机投影，以及下采样多层感知机投影等。

```
class MlpProjector(nn.Module):

    def __init__(self, cfg):

        super().__init__()

        self.cfg = cfg

        if cfg.projector_type == "identity":
            modules = nn.Identity()
```

```
        elif cfg.projector_type == "linear":
            modules = nn.Linear(cfg.input_dim, cfg.n_embed)

        elif cfg.projector_type == "mlp_gelu":
            mlp_depth = cfg.depth
            modules = [nn.Linear(cfg.input_dim, cfg.n_embed)]
            for _ in range(1, mlp_depth):
                modules.append(nn.GELU())
                modules.append(nn.Linear(cfg.n_embed, cfg.n_embed))
            modules = nn.Sequential(*modules)

        elif cfg.projector_type == "downsample_mlp_gelu":
            mlp_depth = cfg.depth
            mlp_ratio = cfg.mlp_ratio
            modules = [nn.Linear(cfg.input_dim * cfg.downsample_ratio *
                        cfg.downsample_ratio, cfg.n_embed * mlp_ratio)]
            for _ in range(1, mlp_depth - 1):
                modules.append(nn.GELU())
                modules.append(nn.Linear(cfg.n_embed * mlp_ratio,
                                cfg.n_embed * mlp_ratio))
            modules.append(nn.GELU())
            modules.append(nn.Linear(cfg.n_embed * mlp_ratio, cfg.n_embed))
            modules = nn.Sequential(*modules)

        else:
            raise ValueError(f"Unknown projector type: {cfg.projector_type}")

        if cfg.token_pooling:
            self.token_pooling_layer = nn.Linear(cfg.input_dim * 4,
                                                 cfg.input_dim)

        self.layers = modules

    def forward(self, x):
        if self.cfg.token_pooling:
            batch_size, wxh, channels = x.shape
            w = h = int(wxh ** 0.5)
            x = x.view(batch_size, w, h, channels)
            x = x.permute(0, 3, 1, 2)
            # import ipdb; ipdb.set_trace()
            patches = x.unfold(2, 2, 2).unfold(3, 2, 2)
            batch_size, channels, h_patches, w_patches, _, _ = patches.size()
            # 在通道维度上拼接
            patches = patches.contiguous().view(batch_size, channels,
                                            h_patches * w_patches, -1)
```

```
        # 通过线性层
        patches = patches.permute(0, 2, 1, 3).contiguous()
        patches = patches.view(batch_size, h_patches * w_patches, channels * 4)

        x = self.token_pooling_layer(patches)

    elif self.cfg.projector_type == 'downsample_mlp_gelu':
        bs, hw, input_dim = x.shape
        h = w = int((hw) ** 0.5)

        """compute padding"""
        if h % self.cfg.downsample_ratio:
            pad = self.cfg.downsample_ratio - h % self.cfg.downsample_ratio
        else:
            pad = 0
        x = x.reshape(bs, h, w, input_dim)
        if pad > 0:
            x = F.pad(x, (0, 0, 0, pad, 0, pad), "constant", 0)

        """4 to 1 concat"""
        x = x.permute(0, 3, 1, 2)  # B, C, H, W
        x = F.unfold(x, kernel_size=self.cfg.downsample_ratio,
                     stride=self.cfg.downsample_ratio,
                     padding=0)  # B, C*4, HW // 4
        x = x.permute(0, 2, 1)

    return self.layers(x)
```

（2）下列代码定义了视觉编码器的配置类VisionEncoderConfig，该类继承自PretrainedConfig。它用于存储模型的超参数和结构设置，如模型名称、图像尺寸、注意力头数量等，以便根据这些参数实例化模型并进行训练。

```
class VisionEncoderConfig(PretrainedConfig):
    model_type: str = "vision"

    model_name: str = "siglip_large_patch16_384"
    image_size: int = 384
    patch_size: int = 16
    width: int = 1024
    layers: int = 24
    heads: int = 16
    mlp_ratio: int = 4
    global_pool: str = "map"
    ignore_head: bool = True
    class_token: bool = False
```

```
    num_classes: int = 0
    use_checkpoint: bool = False
    weight_init: str = "skip"
    deterministic: bool = False
    num_recomputing_layers: int = 0

    def __init__(
            self,
            model_name: str = "siglip_large_patch16_384",
            image_size: int = 384,
            patch_size: int = 16,
            width: int = 1024,
            layers: int = 24,
            heads: int = 16,
            mlp_ratio: int = 4,
            global_pool: str = "map",
            ignore_head: bool = True,
            class_token: bool = False,
            num_classes: int = 0,
            use_checkpoint: bool = False,
            **kwargs
    ):
        self.model_name = model_name
        self.image_size = image_size
        self.patch_size = patch_size
        self.width = width
        self.layers = layers
        self.heads = heads
        self.mlp_ratio = mlp_ratio
        self.global_pool = global_pool
        self.ignore_head = ignore_head
        self.class_token = class_token
        self.num_classes = num_classes
        self.use_checkpoint = use_checkpoint

        super().__init__(**kwargs)
```

类DeepseekVLV2ForCausalLM是一个多模态模型，旨在处理视觉和语言输入。DeepseekVLV2ForCausalLM结合了视觉编码器、视觉—语言投影器和语言模型，以实现图像和文本的联合处理。其中，视觉编码器使用VisionTransformer对图像进行特征提取，投影器将视觉特征映射到与语言模型兼容的嵌入空间，语言模型则基于处理后的视觉和文本输入生成输出。该类还包含处理图像和文本输入的函数，如prepare_inputs_embeds和incremental_prefilling，以确保模型能够有效地处理多模态数据。

（3）MlpProjectorConfig是一个继承自PretrainedConfig的配置类，用于定义多层感知器（MLP）

投影器的参数。类MlpProjectorConfig包含了模型类型、投影器类型、输入维度、嵌入维度、深度、MLP比例、下采样比例等参数。通过初始化方法，用户可以自定义这些参数，以配置特定的MLP投影器结构。这使得模型的配置更加灵活和可定制。

```
class MlpProjectorConfig(PretrainedConfig):
    model_type = "mlp_projector"
    projector_type: str = "downsample_mlp_gelu"
    input_dim: int = 1152
    n_embed: int = 2048
    depth: int = 2
    mlp_ratio: int = 1
    downsample_ratio: int = 2
    token_pooling: bool = False

    def __init__(
            self,
            projector_type: str = "downsample_mlp_gelu",
            input_dim: int = 1152,
            n_embed: int = 2048,
            depth: int = 2,
            mlp_ratio: int = 1,
            downsample_ratio: int = 2,
            **kwargs
    ):
        self.projector_type = projector_type
        self.input_dim = input_dim
        self.n_embed = n_embed
        self.depth = depth
        self.mlp_ratio = mlp_ratio
        self.downsample_ratio = downsample_ratio

        super().__init__(**kwargs)
```

（4）下列代码定义了一个名为DeepSeekVLV2CausalLMOutputWithPast的数据类，该类继承自ModelOutput，用于表示DeepSeek-VL2 自回归语言模型的输出结果。该数据类包含多个可选属性，如loss、logits、past_key_values、hidden_states、attentions和rope_deltas，分别用于存储语言模型的损失值、预测得分、过去的键值对、隐藏状态、注意力权重及序列长度与多模态旋转位置编码（rope）之间的索引差异。这些属性有助于在模型训练和推理过程中获取模型的输出信息和内部状态。

```
@dataclass
class DeepSeekVLV2CausalLMOutputWithPast(ModelOutput):
    loss: Optional[torch.FloatTensor] = None
    logits: torch.FloatTensor = None
    past_key_values: Optional[List[torch.FloatTensor]] = None
```

```
    hidden_states: Optional[Tuple[torch.FloatTensor]] = None
    attentions: Optional[Tuple[torch.FloatTensor]] = None
    rope_deltas: Optional[torch.LongTensor] = None
```

（5）类DeepseekVLV2Config是一个配置类，继承自PretrainedConfig，用于存储和管理DeepseekVLV2 模型的配置参数。DeepseekVLV2Config包含视觉编码器 (vision_config)、多层感知器投影器 (projector_config) 和语言模型 (language_config) 的配置。此外，该类还定义了图像标记方式 (tile_tag)、全局视图位置 (global_view_pos) 和候选分辨率 (candidate_resolutions) 等参数。在初始化时，DeepseekVLV2Config会从传入的关键字参数中提取各个子配置的参数，并实例化相应的配置对象。

```
class DeepseekVLV2Config(PretrainedConfig):
    model_type = "deepseek_vl_v2"
    vision_config: VisionEncoderConfig
    projector_config: MlpProjectorConfig
    language_config: DeepseekV2Config

    tile_tag: str = "2D"
    global_view_pos: str = "head"
    candidate_resolutions: Tuple[Tuple[int, int]] = ((384, 384),)

    def __init__(
            self,
            tile_tag: str = "tile_tag",
            global_view_pos: str = "head",
            candidate_resolutions: Tuple[Tuple[int, int]] = ((384, 384),),
            **kwargs
    ):
        super().__init__(**kwargs)

        vision_config = kwargs.get("vision_config", {})
        self.vision_config = VisionEncoderConfig(**vision_config)

        projector_config = kwargs.get("projector_config", {})
        self.projector_config = MlpProjectorConfig(**projector_config)

        language_config = kwargs.get("language_config", {})
        if isinstance(language_config, DeepseekV2Config):
            self.language_config = language_config
        else:
            self.language_config = DeepseekV2Config(**language_config)

        self.tile_tag = tile_tag
        self.global_view_pos = global_view_pos
```

```
        self.candidate_resolutions = candidate_resolutions
```

（6）下列代码定义的DeepseekVLV2PreTrainedModel是一个预训练模型的基类，继承自PreTrainedModel。它指定了模型的配置类为DeepseekVLV2Config，并将模型的前缀设置为"deepseek_vl_v2"。此外，它定义了 _no_split_modules和 _skip_keys_device_placement等属性，以控制模型在保存、加载和分布式训练时的行为。该类为DeepSeek-VL2 模型的具体实现提供了基础设施。

```
class DeepseekVLV2PreTrainedModel(PreTrainedModel):
    config_class = DeepseekVLV2Config
    base_model_prefix = "deepseek_vl_v2"
    _no_split_modules = []
    _skip_keys_device_placement = "past_key_values"
```

（7）类DeepseekVLV2ForCausalLM实现了一个多模态自回归语言模型，集成了视觉编码器、视觉投影器和语言模型，从而能够将图像信息和文本输入联合处理。具体来说，类DeepseekVLV2ForCausalLM首先利用VisionTransformer对输入图像进行特征提取，再通过MLP投影器将视觉特征映射到与文本嵌入空间一致的表示，之后将这些视觉嵌入与文本标记结合起来，传递给语言模型进行生成。此外，该类还支持增量预填充机制，通过分块前向计算和缓存past key values来加速自回归生成过程，同时提供辅助函数以管理CUDA内存和调整缓存状态，实现高效的多模态推理和生成任务。

- 方法__init__()：用于初始化DeepseekVLV2ForCausalLM类的实例。在初始化过程中，它根据传入的配置构建各个子模块，包括使用VisionTransformer构建视觉编码器、使用MlpProjector构建视觉投影器，并根据配置确定图像标记的格式（如 2D或 1D）。同时，它还实例化语言模型部分（DeepseekV2ForCausalLM），为多模态输入构建整体模型结构，并设置一些用于图像token格式化的特殊参数（例如 <|view_separator|> 和 <tile_newline>）。

```
class DeepseekVLV2ForCausalLM(DeepseekVLV2PreTrainedModel):

    def __init__(self, config: DeepseekVLV2Config):
        super().__init__(config)

        self._use_flash_attention_2 = config._attn_implementation == "flash_
            attention_2"

        # ----------- vision encoder ------------
        vision_config = config.vision_config
        self.vision = VisionTransformer(
            img_size=vision_config.image_size,
            patch_size=vision_config.patch_size,
            embed_dim=vision_config.width,
            depth=vision_config.layers,
```

```
        num_heads=vision_config.heads,
        mlp_ratio=vision_config.mlp_ratio,
        class_token=vision_config.class_token,
        global_pool=vision_config.global_pool,
        ignore_head=vision_config.ignore_head,
        weight_init=vision_config.weight_init,
        num_classes=0,
        deterministic=vision_config.deterministic,
        num_recomputing_layers=vision_config.num_recomputing_layers
    )

    # ----------- vl projector ------------
    projector_config = config.projector_config
    self.projector = MlpProjector(projector_config)

    # image token format 形式
    # FIXME 目前 tile tag & global_view_pos 的默认取值都是之前的实验策略;
    # 后续应当去掉默认取值，改为没有取值就 raise error
    self.tile_tag = config.tile_tag
    self.global_view_pos = config.global_view_pos

    # 用于 format image token sequence 的特殊 token
    embed_std = 1 / torch.sqrt(torch.tensor(projector_config.n_embed,
                            dtype=torch.float32))
    if self.tile_tag == "2D":
        # <|view_separator|>, <|\n|>
        self.image_newline = nn.Parameter(torch.randn(projector_config.
                                         n_embed) * embed_std)
        # fix the typo: view_seperater
        self.view_seperator = nn.Parameter(torch.randn(projector_config.
                                         n_embed) * embed_std)
    elif self.tile_tag == "1D":
        # <|tile_x|>, <|tile_global|>
        candidate_resolutions = config.candidate_resolutions
        if len(candidate_resolutions) == 0:
            raise ValueError(
                f"len(candidate_resolutions) should be larger than 0,
                      but got {len(candidate_resolutions)}")
        tile_variants_num = len(candidate_resolutions)
        self.tile_indicators = nn.Parameter(
            torch.randn(size=(tile_variants_num + 1, config.aligner.
                        params.n_embed)) * embed_std
        )
    else:
        raise ValueError(f"tile tag should be either 1D or 2D, but got
                      {self.tile_tag}")
```

```
    # ----------- language model ------------
    language_config = config.language_config
    self.language = DeepseekV2ForCausalLM(language_config)
```

● 方法prepare_inputs_embeds()：负责生成输入嵌入，将文本token嵌入与经过视觉编码器和投影器处理后的图像token嵌入结合起来。首先，检查是否提供图像输入；若未提供，则直接返回文本嵌入。否则，根据图像空间切分信息计算每个batch中的平铺数量，从图像中提取视觉特征，并通过投影器获得图像嵌入。其次，根据预设的tile格式（2D或1D）对全局和局部视觉特征进行格式化，并利用masked_scatter_ 方法将图像token嵌入填充到对应的文本嵌入序列中，最终返回融合了视觉信息的输入嵌入。

```
def prepare_inputs_embeds(
        self,
        input_ids: torch.LongTensor,
        images: Optional[torch.FloatTensor] = None,
        images_seq_mask: Optional[torch.LongTensor] = None,
        images_spatial_crop: Optional[torch.LongTensor] = None,
        **ignore_kwargs
):

    if images is None or images_spatial_crop.sum() == 0:
        return self.language.get_input_embeddings()(input_ids)

    bs, max_n_images, _ = images_spatial_crop.shape
    batch_num_tiles = [0 for _ in range(bs)]
    total_tiles = []
    for idx in range(bs):
        for jdx in range(max_n_images):
            num_width_tiles, num_height_tiles = images_spatial_crop[idx, jdx]
            if num_width_tiles == 0 or num_height_tiles == 0:
                break
            batch_num_tiles[idx] += (1 + num_width_tiles * num_height_tiles)

        total_tiles.append(images[idx, :batch_num_tiles[idx]])

    # [batch_all_tiles, 3, height, width]
    total_tiles = torch.cat(total_tiles, dim=0)
    assert total_tiles.shape[0] == sum(batch_num_tiles)
    if total_tiles.shape[0] == 0:
        return self.language.get_input_embeddings()(input_ids)

    # [batch_all_tiles, vit_seq_len, c]
    images_feature = self.vision(total_tiles)
```

```
# [batch_all_tiles, hw, D]
images_embeds = self.projector(images_feature)
_, hw, n_dim = images_embeds.shape
h = w = int(hw ** 0.5)

# put image tokens into the input_embeds, [b, T, D]
input_embeds = self.language.get_input_embeddings()(input_ids)

# 根据 self.tile_tag & self.global_view_pos 填充 image token sequence
tile_index = 0
for idx in range(images_spatial_crop.shape[0]):
    images_in_this_batch = []
    for jdx in range(images_spatial_crop.shape[1]):

        # extra global & local features
        num_width_tiles, num_height_tiles = images_spatial_crop[idx, jdx]
        if num_width_tiles == 0 or num_height_tiles == 0:
            break

        num_tiles_in_image = num_width_tiles * num_height_tiles

        # [hw, D]
        global_features = images_embeds[tile_index]

        # [num_height_tiles * num_width_tiles, hw, D]
        local_features = images_embeds[tile_index + 1: tile_index +
                                       1 + num_tiles_in_image]

        tile_index += num_tiles_in_image + 1

        # format global and local features
        if self.tile_tag == "2D":

            # ------------- global view add newline ---------------
            # [hw, D] -> [h, w, D]
            global_features = global_features.view(h, w, n_dim)
            # [D]     -> [h, 1, D]
            new_lines_in_global = repeat(self.image_newline,
                                         "d -> h 1 d", h=h)
            # cat([h, w, D], [h, 1, D], dim=1) -> [h, w + 1, D]
            global_features = torch.cat([global_features,
                                         new_lines_in_global], dim=1)
            # [h, w + 1, D] -> [h * (w + 1), D]
            global_features = global_features.view(-1, n_dim)
```

```
            # ----------------- local view add newline -------------
----
            # [num_height_tiles * num_width_tiles, h * w, D] -> [num_
              height_tiles * h, num_width_tiles * w, D]
            local_features = rearrange(
                local_features,
                "(th tw) (h w) d -> (th h) (tw w) d",
                th=num_height_tiles,
                tw=num_width_tiles,
                h=h,
                w=w
            )

            # [D] -> [num_height_tiles * h, 1, D]
            new_lines_in_local = repeat(
                self.image_newline,
                "d -> (th h) 1 d",
                th=num_height_tiles,
                h=h
            )

            # [num_height_tiles * h, num_width_tiles * w + 1, D]
            local_features = torch.cat([local_features,
                                        new_lines_in_local], dim=1)

            # [num_height_tiles * h, num_width_tiles * w + 1, D]
            #   --> [(num_height_tiles * h) *
                         (num_width_tiles * w + 1), D]
            local_features = local_features.view(-1, n_dim)

            # ----------- merge global and local tiles -----------
            if self.global_view_pos == "head":
                global_local_features = torch.cat(
                    [global_features, self.view_seperator[None, :],
                    local_features], dim=0)
            else:
                global_local_features = torch.cat(
                    [local_features, self.view_seperator[None, :],
                    global_features], dim=0)

        else:
            # abandoned，实际上不会走这个逻辑
            global_features = torch.cat(
                [self.tile_indicators[0:1], global_features], dim=0
            )
            local_features = torch.cat(
```

```
                    [self.tile_indicators[1:num_tiles_in_image +
                     1].unsqueeze(1), local_features], dim=1
                )
                local_features = rearrange(local_features, 'crop_num hw
                                           d -> (crop_num hw) d')

                if self.global_view_pos == "head":
                    global_local_features = torch.cat([global_features,
                                            local_features], dim=0)
                else:
                    global_local_features = torch.cat([local_features,
                                            global_features], dim=0)

            images_in_this_batch.append(global_local_features)

        if len(images_in_this_batch) > 0:
            images_in_this_batch = torch.cat(images_in_this_batch, dim=0)
            input_embeds[idx].masked_scatter_(
                images_seq_mask[idx].unsqueeze(-1),
                images_in_this_batch)

    return input_embeds
```

● 方法incremental_prefilling()：用于实现增量预填充，以减少显存占用并提高推理效率。在该方法中，如果未提供inputs_embeds，则调用prepare_inputs_embeds生成融合视觉信息的嵌入；然后，将输入嵌入按指定的块大小分块处理，逐块进行前向计算，并逐步更新和缓存过去键值对（past key values）。该方法最终返回处理后的输入嵌入以及对应的预填充的过去键值对，供后续自回归生成使用。

```
@torch.no_grad()
def incremental_prefilling(
        self,
        input_ids: Optional[torch.LongTensor] = None,
        attention_mask: Optional[torch.Tensor] = None,
        inputs_embeds: Optional[torch.FloatTensor] = None,

        images: Optional[torch.FloatTensor] = None,
        images_seq_mask: Optional[torch.LongTensor] = None,
        images_spatial_crop: Optional[torch.LongTensor] = None,
        chunk_size: int = 1024
):
    if inputs_embeds is None:
        inputs_embeds = self.prepare_inputs_embeds(
            input_ids=input_ids,
            images=images,
```

```
            images_seq_mask=images_seq_mask,
            images_spatial_crop=images_spatial_crop,
        )

        del images
        del images_seq_mask
        del images_spatial_crop

        if attention_mask is not None:
            attention_mask = attention_mask.to(inputs_embeds.device)

        self._clear_cuda_cache()

    bzs, seq_len, _ = inputs_embeds.shape
    past_key_values = None

    # remain the last token for the next forward
    prefilling_len = seq_len - 1
    for i in range(0, prefilling_len, chunk_size):
        chunk_start = i
        chunk_end = min(i + chunk_size, prefilling_len)
        chunk_inputs_embeds = inputs_embeds[:, chunk_start: chunk_end]
        chunk_attention_mask = attention_mask[:, 0: chunk_end]
        # print(f"start = {chunk_start}, end = {chunk_end},
                prefilling_len = {prefilling_len}, seq_len = {seq_len}")

        # compute position_ids
        if past_key_values is not None:
            position_ids = torch.arange(
                chunk_start,
                chunk_end,
                dtype=torch.long,
                device=inputs_embeds.device
            ).unsqueeze(0)
            past_key_values = self._move_past_key_values_to_gpu(past_
                                  key_values, inputs_embeds.device)
        else:
            position_ids = None

        # chunk-forward
        with torch.no_grad():
            outputs = self.forward(
                inputs_embeds=chunk_inputs_embeds,
                attention_mask=chunk_attention_mask,
                past_key_values=past_key_values,
                position_ids=position_ids,
```

```
                use_cache=True,
            )
            # update past_key_values
            past_key_values = outputs.past_key_values
            past_key_values = self._move_past_key_values_to_cpu(past_
                                key_values)

            del outputs, position_ids
            self._clear_cuda_cache()

    prefilling_key_values = []
    for layer_past in past_key_values:
        prefilling_key_values.append(
            (
                layer_past[0][:, :, 0: prefilling_len, ...].to(inputs_
                    embeds.device),
                layer_past[1][:, :, 0: prefilling_len, ...].to(inputs_
                    embeds.device),
            )
        )

    return inputs_embeds, prefilling_key_values
```

● 方法forward()：模型的标准前向传播方法，用于处理输入数据并生成模型输出。该方法首先检查是否提供了融合视觉信息的嵌入，如未提供则调用prepare_inputs_embeds生成；接着，将输入（包括attention_mask、position_ids、past_key_values、labels等）传递给语言模型的forward方法进行处理，最终返回包含logits、隐藏状态、注意力权重等信息的输出结果，支持自回归生成和条件生成任务。

```
def forward(
        self,
        input_ids: Optional[torch.LongTensor] = None,

        attention_mask: Optional[torch.Tensor] = None,
        position_ids: Optional[torch.LongTensor] = None,
        past_key_values: Optional[List[torch.FloatTensor]] = None,
        inputs_embeds: Optional[torch.FloatTensor] = None,

        images: Optional[torch.FloatTensor] = None,
        images_seq_mask: Optional[torch.LongTensor] = None,
        images_spatial_crop: Optional[torch.LongTensor] = None,

        labels: Optional[torch.LongTensor] = None,
        use_cache: Optional[bool] = None,
        output_attentions: Optional[bool] = None,
```

```
        output_hidden_states: Optional[bool] = None,
        return_dict: Optional[bool] = None,
        cache_position: Optional[torch.LongTensor] = None,
):

    output_attentions = (
        output_attentions
        if output_attentions is not None
        else self.config.output_attentions
    )
    output_hidden_states = (
        output_hidden_states
        if output_hidden_states is not None
        else self.config.output_hidden_states
    )
    use_cache = use_cache if use_cache is not None else self.config.use_cache

    return_dict = (
        return_dict if return_dict is not None else self.config.use_
            return_dict
    )
    if inputs_embeds is None:
        inputs_embeds = self.prepare_inputs_embeds(
            input_ids=input_ids,
            images=images,
            images_seq_mask=images_seq_mask,
            images_spatial_crop=images_spatial_crop,
        )

        if attention_mask is not None:
            attention_mask = attention_mask.to(inputs_embeds.device)

    # print(inputs_embeds.shape)
    outputs = self.language.forward(
        input_ids=None,
        attention_mask=attention_mask,
        position_ids=position_ids,
        past_key_values=past_key_values,
        inputs_embeds=inputs_embeds,
        labels=labels,
        use_cache=use_cache,
        output_attentions=output_attentions,
        output_hidden_states=output_hidden_states,
        return_dict=return_dict,
        cache_position=cache_position
    )
```

```
        return outputs
```

● 方法_clear_cuda_cache()：用于清理CUDA内存缓存，以确保显存得到及时释放。它通过调用Python垃圾回收机制［gc.collect()］、torch.cuda.empty_cache()和torch.cuda.synchronize()清空和同步GPU缓存，从而降低显存占用，保证长时间运行时的稳定性。

```
    def _clear_cuda_cache(self):
        gc.collect()
        if torch.cuda.is_available():
            torch.cuda.empty_cache()
            torch.cuda.synchronize()
```

● 方法_move_past_key_values_to_cpu()：将模型中缓存的past key values从GPU移动到CPU，以释放GPU内存。它接收一个包含past key values的数据结构，对每一层中的每个tensor调用 .cpu()方法，将其转移到CPU上，然后返回转移后的缓存数据，常用于增量预填充或分布式训练过程中内存管理。

```
    def _move_past_key_values_to_cpu(self, past_key_values):
        # print(f"past_key_values -> cpu")
        if past_key_values is None:
            return None
        return tuple(tuple(t.cpu() for t in layer) for layer in past_key_values)
```

● 方法_move_past_key_values_to_gpu()：与上面的方法_move_past_key_values_to_cpu()相反，该方法将缓存的past key values从CPU移动到GPU，以便在模型前向传播时使用。它对传入的past_key_values中的每个tensor调用 .to(device) 方法，将数据转移到指定的GPU上，确保后续计算在GPU上高效进行。

```
    def _move_past_key_values_to_gpu(self, past_key_values, device="cuda:0"):
        # print(f"past_key_values -> gpu")
        if past_key_values is None:
            return None
        return tuple(tuple(t.to(device) for t in layer) for layer in past_
            key_values)
```

● 方法prepare_inputs_for_generation()：用于为生成任务准备输入数据。它首先调用语言模型对应的prepare_inputs_for_generation方法生成基础输入，其次根据当前的缓存位置（cache_position）判断是否需要传递图像相关输入（如images、images_seq_mask、images_spatial_crop）。在缓存解码阶段，如果当前缓存位置为0，则将图像信息添加到模型输入中，最终返回包含所有必要输入数据的字典，确保生成过程能够融合视觉信息。

```
    def prepare_inputs_for_generation(
```

```
        self,
        input_ids,
        past_key_values=None,
        inputs_embeds=None,

        images: Optional[torch.FloatTensor] = None,
        images_seq_mask: Optional[torch.LongTensor] = None,
        images_spatial_crop: Optional[torch.LongTensor] = None,

        attention_mask=None,
        cache_position=None,

        pixel_values=None,
        image_sizes=None,
        num_logits_to_keep=None,
        **kwargs,
):
    model_inputs = self.language.prepare_inputs_for_generation(
        input_ids,
        past_key_values=past_key_values,
        inputs_embeds=inputs_embeds,
        attention_mask=attention_mask,
        cache_position=cache_position,
        num_logits_to_keep=num_logits_to_keep,
        **kwargs,
    )

    cache_position = model_inputs["cache_position"]
    if cache_position[0] == 0:
        model_inputs["images"] = images
        model_inputs["images_seq_mask"] = images_seq_mask
        model_inputs["images_spatial_crop"] = images_spatial_crop

    return model_inputs
```

- 静态方法_reorder_cache()：用于在Beam Search解码过程中对cached past key values进行重新排序。它根据提供的beam索引（beam_idx）对每一层的缓存进行index_select操作，重新排列缓存中的tensor，使其顺序与当前beam的排列一致，从而保证解码过程中缓存数据的正确性和一致性。

```
@staticmethod
def _reorder_cache(past_key_values, beam_idx):
    reordered_past = ()
    for layer_past in past_key_values:
        reordered_past += (
            tuple(
```

```
                past_state.index_select(0, beam_idx.to(past_state.device))
                for past_state in layer_past
            ),
        )
    return reordered_past
```

（8）下面这段代码利用Transformers库提供的注册机制，将自定义的配置类和模型类注册到AutoConfig和AutoModelForCausalLM中，从而使得用户可以通过Transformers的自动化API来加载和使用DeepSeek-VL2 模型。

```
AutoConfig.register("vision", VisionEncoderConfig)
AutoConfig.register("mlp_projector", MlpProjectorConfig)
AutoConfig.register("deepseek_vl_v2", DeepseekVLV2Config)
AutoModelForCausalLM.register(DeepseekVLV2Config, DeepseekVLV2ForCausalLM)
```

6.5.3 数据处理

文件processing_deepseek_vl_v2.py是实现DeepSeek-VL2 模型的数据处理模块，负责将图像与文本进行多模态融合处理，为模型提供输入格式化、批处理、分辨率选择等功能。此外，还包含了自定义的图像处理转换、对话格式化及特殊token管理等功能模块，以确保数据在训练与推理中能够高效、规范地进入模型。

文件processing_deepseek_vl_v2.py的具体实现流程如下所示。

（1）函数select_best_resolution()的功能是，在给定的候选分辨率中选择一个最佳的分辨率，以在尽量减少分辨率浪费的前提下，最大限度地保留原始图像的有效分辨率。具体来说，函数会计算每个候选分辨率的缩放比例、有效分辨率和浪费的分辨率，然后选择最佳分辨率。

```
def select_best_resolution(image_size, candidate_resolutions):
    # 用于裁剪
    original_width, original_height = image_size
    best_fit = None
    max_effective_resolution = 0
    min_wasted_resolution = float("inf")

    for width, height in candidate_resolutions:
        scale = min(width / original_width, height / original_height)
        downscaled_width, downscaled_height = int(original_width * scale),
                                              int(original_height *
                                              scale)
        effective_resolution = min(downscaled_width * downscaled_height,
                                   original_width * original_height)
        wasted_resolution = (width * height) - effective_resolution

        if effective_resolution > max_effective_resolution or (effective_
```

```
resolution == max_effective_resolution and wasted_resolution < min_wasted_resolution):
                max_effective_resolution = effective_resolution
                min_wasted_resolution = wasted_resolution
                best_fit = (width, height)

    return best_fit
```

（2）DictOutput是一个简单的封装类，用于将对象的属性以字典的形式进行访问和操作。此类通过重写keys、__getitem__和__setitem__方法，使得对象的属性可以通过字典的接口进行访问和修改。这使得对象在某些场景下可以像字典一样使用，提供了更灵活的属性访问方式。

```
class DictOutput(object):
    def keys(self):
        return self.__dict__.keys()

    def __getitem__(self, item):
        return self.__dict__[item]

    def __setitem__(self, key, value):
        self.__dict__[key] = value
```

（3）VLChatProcessorOutput是一个使用dataclass装饰器定义的数据类，用于存储与视觉语言处理相关的输出数据。VLChatProcessorOutput继承自DictOutput类，从而可以像字典一样访问其属性。

```
@dataclass
class VLChatProcessorOutput(DictOutput):
    sft_format: str
    input_ids: torch.LongTensor
    target_ids: torch.LongTensor
    images: torch.Tensor
    images_seq_mask: torch.BoolTensor
    images_spatial_crop: torch.LongTensor
    num_image_tokens: List[int]

    def __len__(self):
        return len(self.input_ids)
```

（4）BatchCollateOutput是一个使用dataclass装饰器定义的数据类，用于封装批量处理后的多模态数据。BatchCollateOutput继承自DictOutput类，从而可以像字典一样访问其属性。该类的主要功能是存储和管理批量数据，包括文本和图像的处理结果，以及相关的掩码和元数据。此外，它还提供了一个to方法，用于将所有张量移动到指定的设备（如GPU）上，并可以指定数据类型（如torch.bfloat16），以便进行高效的计算。

```
@dataclass
class BatchCollateOutput(DictOutput):
```

```
    sft_format: List[str]
    input_ids: torch.LongTensor
    labels: torch.LongTensor
    images: torch.Tensor
    attention_mask: torch.Tensor
    images_seq_mask: torch.BoolTensor
    images_spatial_crop: torch.LongTensor
    seq_lens: List[int]

    def to(self, device, dtype=torch.bfloat16):
        self.input_ids = self.input_ids.to(device)
        self.labels = self.labels.to(device)
        self.attention_mask = self.attention_mask.to(device)
        self.images_seq_mask = self.images_seq_mask.to(device)
        self.images_spatial_crop = self.images_spatial_crop.to(device)
        self.images = self.images.to(device=device, dtype=dtype)
        return self
```

（5）类ImageTransform是一个图像处理工具类，用于对输入的PIL图像执行一系列预定义的转换操作。ImageTransform支持将图像转换为张量，并可选地对图像进行归一化处理。归一化操作会使用指定的均值和标准差对图像的每一个通道进行标准化。这个类的主要功能是将图像数据转换为适合深度学习模型输入的格式。

```
class ImageTransform(object):
    def __init__(
            self,
            mean: Optional[Tuple[float, float, float]] = (0.5, 0.5, 0.5),
            std: Optional[Tuple[float, float, float]] = (0.5, 0.5, 0.5),
            normalize: bool = True
    ):
        self.mean = mean
        self.std = std
        self.normalize = normalize

        transform_pipelines = [
            T.ToTensor()
        ]

        if normalize:
            transform_pipelines.append(T.Normalize(mean, std))

        self.transform = T.Compose(transform_pipelines)

    def __call__(self, pil_img: Image.Image):
        x = self.transform(pil_img)
```

```
        return x
```

（6）DeepseekVLV2Processor是一个用于处理视觉语言模型输入的处理器类，它集成了文本和图像数据的预处理功能。该类的主要功能包括：

- 初始化：设置图像处理的参数（如候选分辨率、块大小、归一化参数等），并初始化图像变换工具和文本分词器。
- 文本和图像的预处理：将文本和图像数据转换为模型所需的格式，包括文本的标记化、图像的裁剪和归一化。
- 格式化会话数据：将会话数据（如用户和助手的对话）格式化为模型能够理解的结构化格式。
- 批量处理：将多个样本的数据进行批量处理，以便模型能够一次性处理多个输入。
- 设备和数据类型转换：提供方法将数据移动到指定的设备（如GPU）并转换为指定的数据类型（如torch.bfloat16）。

在DeepseekVLV2Processor类中，通过巧妙地结合视觉和文本数据，实现了模型输入的准备。类DeepseekVLV2Processor能够灵活地处理多种多模态任务，如视觉问答、文档理解等，并且支持多种格式的输入数据。DeepseekVLV2Processor类的实现流程如下：

- 方法__init__()：功能是初始化DeepseekVLV2Processor类实例，配置候选图像分辨率、补丁大小、图像均值和标准差，以及是否归一化等参数。同时，它创建图像变换对象ImageTransform和初始化LlamaTokenizerFast分词器，并添加pad_token作为特殊符号来支持填充。

```
class DeepseekVLV2Processor(ProcessorMixin):
    tokenizer_class = ("LlamaTokenizer", "LlamaTokenizerFast")
    attributes = ["tokenizer"]

    def __init__(
            self,
            tokenizer: LlamaTokenizerFast,
            candidate_resolutions: Tuple[Tuple[int, int]],
            patch_size: int,
            downsample_ratio: int,
            image_mean: Tuple[float, float, float] = (0.5, 0.5, 0.5),
            image_std: Tuple[float, float, float] = (0.5, 0.5, 0.5),
            normalize: bool = True,
            image_token: str = "<image>",
            pad_token: str = "<｜▁pad▁｜>",
            add_special_token: bool = False,
            sft_format: str = "deepseek",
            mask_prompt: bool = True,
            ignore_id: int = -100,
            **kwargs,
    ):
```

```
self.candidate_resolutions = candidate_resolutions
self.image_size = candidate_resolutions[0][0]
self.patch_size = patch_size
self.image_mean = image_mean
self.image_std = image_std
self.normalize = normalize
self.downsample_ratio = downsample_ratio

self.image_transform = ImageTransform(mean=image_mean, std=image_
                                       std, normalize=normalize)
self.tokenizer = tokenizer
self.tokenizer.padding_side = 'left'
                                        # 必须设置这个，填充侧在批量推理中会产生影响

# 添加 pad_token 作为特殊符号以使用 'tokenizer.pad_token' 和 'tokenizer.
  pad_token_id'
if tokenizer.pad_token is None:
    self.tokenizer.add_special_tokens({'pad_token': pad_token})
print(f"将 pad token = ['{pad_token}'] 添加到 tokenizer\n"
      f"{pad_token}:{tokenizer.encode(pad_token, add_special_
                     tokens=False)[0]}")

# 添加图像 token
image_token_id = self.tokenizer.vocab.get(image_token)
if image_token_id is None:
    special_tokens = [image_token]
    special_tokens_dict = {"additional_special_tokens": special_tokens}
    self.tokenizer.add_special_tokens(special_tokens_dict)
self.image_token_id = self.tokenizer.vocab.get(image_token)
print(f"将 image token = ['{image_token}'] 添加到 tokenizer\n"
      f"{image_token}:{tokenizer.encode(image_token,
                                        add_special_tokens=False)[0]}")

# 为与 grounding 相关的任务添加 5 个特殊符号
# <|ref|>, <|/ref|>, <|det|>, <|/det|>, <|grounding|>
special_tokens = ['<|ref|>', '<|/ref|>', '<|det|>', '<|/det|>',
                  '<|grounding|>']
special_tokens_dict = {"additional_special_tokens": special_tokens}
self.tokenizer.add_special_tokens(special_tokens_dict)
print(f"将 grounding-related tokens = {special_tokens} 添加到 tokenizer
        的 input_ids 中\n"
      f"<|ref|>:{tokenizer.encode('<|ref|>', add_special_
        tokens=False)[0]}\n"
      f"<|/ref|>:{tokenizer.encode('<|/ref|>', add_special_
        tokens=False)[0]}\n"
      f"<|det|>:{tokenizer.encode('<|det|>', add_special_
```

```
              tokens=False)[0]}\n"
            f"<|/det|>:{tokenizer.encode('<|/det|>', add_special_
              tokens=False)[0]}\n"
            f"<|grounding|>:{tokenizer.encode('<|grounding|>', add_special_
              tokens=False)[0]}")

        # 为 SFT 数据添加特殊符号
        special_tokens = ["<|User|>", "<|Assistant|>"]
        special_tokens_dict = {"additional_special_tokens": special_tokens}
        self.tokenizer.add_special_tokens(special_tokens_dict)
        print(f"将 chat tokens = {special_tokens} 添加到 tokenizer 的 input_ids
              中 \n"
            f"<|User|>:{tokenizer.encode('<|User|>', add_special_
              tokens=False)[0]}\n"
            f"<|Assistant|>:{tokenizer.encode('<|Assistant|>', add_special_
              tokens=False)[0]}\n")

        self.image_token = image_token
        self.pad_token = pad_token
        self.add_special_token = add_special_token
        self.sft_format = sft_format
        self.mask_prompt = mask_prompt
        self.ignore_id = ignore_id

        super().__init__(
            tokenizer,
            **kwargs,
        )
```

- 方法 new_chat_template()：通过调用 get_conv_template 函数返回一个新的对话模板实例，供后续格式化对话时使用。

```
    def new_chat_template(self):
        conv = get_conv_template(self.sft_format)
        return conv
```

- 方法 format_messages()：将提供的多轮对话数据按照 SFT（Supervised Fine-Tuning，监督微调）模板进行格式化。它通过设置系统提示信息并逐条添加用户与助手消息，生成一个格式化后的对话字符串，用于模型输入。

```
    def format_messages(
            self,
            conversations: List[Dict[str, str]],
            sft_format: str = "deepseek",
            system_prompt: str = "",
```

```
):
    """
    将 SFT 模板应用于会话。

    Args:
        conversations (List[Dict]): 信息列表。
        sft_format (str, optional): 使用的 SFT 模板的格式。默认为 "deepseek"。
        system_prompt (str, optional): 在 SFT 模板中使用的系统提示。默认为 ""。

    Returns:
        sft_prompt (str): 格式化的文本。
    """

    conv = get_conv_template(sft_format)
    conv.set_system_message(system_prompt)
    for message in conversations:
        conv.append_message(message["role"], message["content"].strip())
    sft_prompt = conv.get_prompt().strip()

    return sft_prompt
```

- 方法 format_messages_v2()：对多模态会话进行格式化操作，同时处理文本和图像信息。该方法将图像与对话内容关联起来，生成包含文本、图像标记和图像裁剪信息的嵌入序列，确保模型能够正确解析多模态输入。

```
def format_messages_v2(self, messages, pil_images, systems=None):
    """扮演 format_messages_v2 和 get_images_info 的角色"""
    tokenized_data = []
    masked_tokenized_data = []  # labels
    images_list = []
    images_seq_mask = []
    images_spatial_crop = []
    num_image_tokens = []

    image_index = 0

    conv = get_conv_template(self.sft_format)
    conv_system_message = conv.system_message

    for idx, message in enumerate(messages):
        if idx == 0:
            tokenized_data += [self.bos_id]
            masked_tokenized_data += [self.bos_id]
            images_seq_mask += [False]
            conv.system_message = conv_system_message
```

```
        else:
            conv.system_message = ''

        if message['role'] == conv.roles[0] or message['role'] ==
                                "user":
            conv.reset_message()
            conv.append_message(conv.roles[0],
                                str(message['content']).strip())
            conv.append_message(conv.roles[1], '')
            formatted_question = conv.get_prompt()
            tokenized_str, images, seq_mask, spatial_crop,
            n_image_tokens = self.tokenize_with_images(
                formatted_question,
                pil_images[image_index: image_index + formatted_question.
                           count(self.image_token)],
                bos=False,
                eos=False,
                cropping=len(pil_images) <= 2
            )
            image_index += formatted_question.count(self.image_token)

            tokenized_data += tokenized_str
            if self.mask_prompt:
                masked_tokenized_data += [self.ignore_id] *
                                         len(tokenized_str)
            else:
                masked_tokenized_data += tokenized_str
            images_list += images
            images_seq_mask += seq_mask
            images_spatial_crop += spatial_crop
            num_image_tokens += n_image_tokens

        elif message['role'] == conv.roles[1] or message['role'] ==
                                "assistant":
            formatted_answer = message['content'].strip()
            assert formatted_answer.count(
                self.image_token) == 0, f"在助理的回复中不应该有{self.
                                    image_token}, 但得到 {messages}"
            tokenized_str, images, seq_mask, spatial_crop,
            n_image_tokens = self.tokenize_with_images(
                formatted_answer,
                [],
                bos=False,
                eos=True,
                cropping=len(pil_images) <= 2)
```

```
            tokenized_data += tokenized_str
            masked_tokenized_data += tokenized_str
            images_seq_mask += seq_mask

        elif message['role'] == 'system' or message['role'] ==
                                'deepseekapi-sys':
            # 如果 message 中有 system，则它只能出现在会话的第一句，
              同时 conv 原来的 system 就会失效
            assert idx == 0, 'system information should only exist in
                         the begining of the conversation'
            formatted_system = message['content'].strip()
            tokenized_str = self.encode(formatted_system, bos=False, eos=False)
            tokenized_data += tokenized_str
            if self.mask_prompt:
                masked_tokenized_data += [self.ignore_id] * len(tokenized_str)
            else:
                masked_tokenized_data += tokenized_str
            seq_mask = [False] * len(tokenized_str)
            images_seq_mask += seq_mask

        else:
            assert False, f"Unknown role: {message['role']}"

    assert len(tokenized_data) == len(
        images_seq_mask), f"tokenized_str 的长度 {len(tokenized_str)} 不等
            于 imags_seq_mask 的长度 {len(images_seq_mask)}"
    assert len(images_spatial_crop) == len(num_image_tokens), f"image
        number should be compatible"

    return tokenized_data, masked_tokenized_data, images_list, images_
        seq_mask, images_spatial_crop, num_image_tokens
```

● 方法 format_prompts()：将提供的提示字符串格式化为 SFT 模板格式，添加用户和助手的角色标签，方便模型直接使用该格式进行推理。

```
def format_prompts(
        self,
        prompts: str,
        sft_format: str = "deepseek",
        system_prompt: str = "",
):
    """
    将 SFT 模板应用于提示。

    Args:
        prompts (str): 未格式化的提示；
```

```
        sft_format (str, optional): 使用的 SFT 模板的格式。默认为 "deepseek"。
        system_prompt (str, optional): 在 SFT 模板中使用的系统提示。默认为 ""。

    Returns:
        sft_prompt (str): 格式化的文本。
    """

    conv = get_conv_template(sft_format)
    conv.set_system_message(system_prompt)
    conv.append_message(conv.roles[0], prompts.strip())
    conv.append_message(conv.roles[1], "")

    sft_prompt = conv.get_prompt().strip()

    return sft_prompt
```

● 方法encode()：将输入字符串编码为token序列，可选地在序列开头添加bos标记和结尾添加eos标记，便于模型理解序列边界。方法decode()用于将token序列解码为字符串，支持参数化选项以处理特殊标记或裁剪输出。

```
def encode(self, text: str, bos: bool = True, eos: bool = False):
    t = self.tokenizer.encode(text, add_special_tokens=False)

    if bos:
        t = [self.bos_id] + t
    if eos:
        t = t + [self.eos_id]

    return t

def decode(self, t: List[int], **kwargs) -> str:
    return self.tokenizer.decode(t, **kwargs)
```

● 方法process_one()：用于处理单次对话输入，将文本和图像数据转换为模型需要的格式，包括生成token序列、图像嵌入、掩码等信息，并将其封装为VLChatProcessorOutput结构返回。

```
def process_one(
        self,
        prompt: str = None,
        conversations: List[Dict[str, str]] = None,
        images: List[Image.Image] = None,
        apply_sft_format: bool = False,
        inference_mode: bool = True,
        system_prompt: str = "",
        **kwargs,
```

```
):
    """

    Args:
        prompt (str): 格式化的提示;
        conversations (List[Dict]): 包含消息列表的会话;
        images (List[ImageType]): 图像列表;
        apply_sft_format (bool): 如果 prompt 不为 None，则将 SFT 格式应用于提示;
            如果 conversations 不为 None，则始终将 SFT 格式应用于会话;
        inference_mode (bool): 如果为 True，则移除最后一个 eos token;
        system_prompt (str): 系统提示;
        **kwargs:

    Returns:
        outputs (BaseProcessorOutput): 处理器的输出,
            - input_ids (torch.LongTensor): [N + image tokens]
            - target_ids (torch.LongTensor): [N + image tokens]
            - images (torch.FloatTensor): [n_images, 3, H, W]
            - image_id (int): image token 的 id
            - num_image_tokens (List[int]): image tokens 的数量
    """

    assert (
            prompt is None or conversations is None
    ), "prompt 和 conversations 不能同时使用。"

    if prompt is None:
        # 使用 sft 格式
        sft_format = self.format_messages(
            conversations=conversations,
            sft_format=self.sft_format,
            system_prompt=system_prompt,
        )
        tokenized_str, masked_tokenized_str, images_list,
            images_seq_mask, images_spatial_crop,
            num_image_tokens = self.format_messages_v2(
            conversations, images)
    else:
        if apply_sft_format:
            sft_format = self.format_prompts(
                prompts=prompt,
                sft_format=self.sft_format,
                system_prompt=system_prompt
            )
        else:
            sft_format = prompt
```

```
        tokenized_str, images_list, images_seq_mask, images_spatial_
            crop, num_image_tokens = self.tokenize_with_images(
            sft_format, images, bos=True, eos=True,
            cropping=len(images) <= 2)
        masked_tokenized_str = []
        for token_index in tokenized_str:
            if token_index != self.image_token_id:
                masked_tokenized_str.append(token_index)
            else:
                masked_tokenized_str.append(self.ignore_id)

    assert len(tokenized_str) == len(images_seq_mask) == len(masked_
            tokenized_str), \
        (f"tokenized_str 的长度 {len(tokenized_str)}, input_ids 的长度
                        {len(masked_tokenized_str)}, "
         f"imags_seq_mask 的长度 {len(images_seq_mask)}，不相等 ")

    input_ids = torch.LongTensor(tokenized_str)
    target_ids = torch.LongTensor(masked_tokenized_str)
    images_seq_mask = torch.tensor(images_seq_mask, dtype=torch.bool)

    # 将 input_ids < 0 | input_ids == self.image_token_id 设置为 ignore_id
    target_ids[(input_ids < 0) | (input_ids == self.image_token_id)] =
                        self.ignore_id
    input_ids[input_ids < 0] = self.pad_id

    if inference_mode:
        # 移除结尾的 eos token
        assert input_ids[-1] == self.eos_id
        input_ids = input_ids[:-1]
        target_ids = target_ids[:-1]
        images_seq_mask = images_seq_mask[:-1]

    if len(images_list) == 0:
        images = torch.zeros((1, 3, self.image_size, self.image_size))
        images_spatial_crop = torch.zeros((1, 2), dtype=torch.long)
    else:
        images = torch.stack(images_list, dim=0)
        images_spatial_crop = torch.tensor(images_spatial_crop,
                                    dtype=torch.long)

    prepare = VLChatProcessorOutput(
        sft_format=sft_format,
        input_ids=input_ids,
        target_ids=target_ids,
        images=images,
```

```
            images_seq_mask=images_seq_mask,
            images_spatial_crop=images_spatial_crop,
            num_image_tokens=num_image_tokens
        )

        return prepare
```

- 方法__call__(): 类的主入口方法，支持直接调用对象来处理多模态输入。它调用process_one进行单次处理，并根据force_batchify参数决定是否将输出转换为批次输入。

```
    def __call__(
            self,
            *,
            prompt: str = None,
            conversations: List[Dict[str, str]] = None,
            images: List[Image.Image] = None,
            apply_sft_format: bool = False,
            force_batchify: bool = True,
            inference_mode: bool = True,
            system_prompt: str = "",
            **kwargs,
    ):
        """

        Args:
            prompt (str): 格式化的提示;
            conversations (List[Dict]): 包含消息列表的会话;
            images (List[ImageType]): 图像列表;
            apply_sft_format (bool): 如果 prompt 不为 None, 则将 SFT 格式应用于提示;
                如果 conversations 不为 None, 则始终将 SFT 格式应用于会话;
            force_batchify (bool): 强制 batchify 输入;
            inference_mode (bool): 如果为 True, 则移除最后一个 eos token;
            system_prompt (str): 系统提示;
            **kwargs:

        Returns:
            outputs (BaseProcessorOutput): 处理器的输出,
                - input_ids (torch.LongTensor): [N + image tokens]
                - images (torch.FloatTensor): [n_images, 3, H, W]
                - image_id (int): image token 的 id
                - num_image_tokens (List[int]): image tokens 的数量
        """

        prepare = self.process_one(
            prompt=prompt,
            conversations=conversations,
```

```
            images=images,
            apply_sft_format=apply_sft_format,
            inference_mode=inference_mode,
            system_prompt=system_prompt
        )

        if force_batchify:
            prepare = self.batchify([prepare])

        return prepare
```

- 方法tokenize_with_images()：用于处理包含图像标记的对话，将图像和文本联合编码为token序列，同时生成图像裁剪信息和掩码。该方法确保模型能够理解图像位置及其在序列中的顺序。

```
    def tokenize_with_images(
            self,
            conversation: str,
            images: List[Image.Image],
            bos: bool = True,
            eos: bool = True,
            cropping: bool = True,
    ):
        """ 为带有 <image> 标签的文本进行标记化。"""
        assert conversation.count(self.image_token) == len(images)
        text_splits = conversation.split(self.image_token)
        images_list, images_seq_mask, images_spatial_crop = [], [], []
        num_image_tokens = []
        tokenized_str = []
        for text_sep, image in zip(text_splits, images):
            """ 编码 text_sep"""
            tokenized_sep = self.encode(text_sep, bos=False, eos=False)
            tokenized_str += tokenized_sep
            images_seq_mask += [False] * len(tokenized_sep)

            """ 为 anyres 选择最佳分辨率 """
            if cropping:
                best_width, best_height = select_best_resolution(image.size,
                                                  self.candidate_resolutions)
            else:
                best_width, best_height = self.image_size, self.image_size
            # print(image.size, (best_width, best_height))
                                                  # 检查 select_best_resolutions 函数

            """ 处理全局视图 """
            global_view = ImageOps.pad(image, (self.image_size, self.image_size),
                                   color=tuple(int(x * 255) for x in self.
```

```
                        image_transform.mean))
    images_list.append(self.image_transform(global_view))

    """ 处理局部视图 """
    local_view = ImageOps.pad(image, (best_width, best_height),
                             color=tuple(int(x * 255) for x in
                             self.image_transform.mean))
    for i in range(0, best_height, self.image_size):
        for j in range(0, best_width, self.image_size):
            images_list.append(
                self.image_transform(local_view.crop((j, i, j +
                                    self.image_size,
                                    i + self.image_size))))

    """ 记录宽度 / 高度裁剪数量 """
    num_width_tiles, num_height_tiles = best_width // self.image_size,
                                        best_height // self.image_size
    images_spatial_crop.append([num_width_tiles, num_height_tiles])

    """ 添加 image tokens"""
    h = w = math.ceil((self.image_size // self.patch_size) / self.
                    downsample_ratio)
    # 全局视图 tokens 的数量为 h*(w+1), 1 是用于行分隔符
    tokenized_image = [self.image_token_id] * h * (w + 1)
    # 添加一个分隔符，用于将全局视图和局部视图分开
    tokenized_image += [self.image_token_id]
    # 局部视图 tokens 的数量为 (num_height_tiles * h)*(num_width_tiles * w + 1)
    tokenized_image += [self.image_token_id] * (num_height_tiles * h)
                    * (num_width_tiles * w + 1)

    tokenized_str += tokenized_image
    images_seq_mask += [True] * len(tokenized_image)
    num_image_tokens.append(len(tokenized_image))
    # print(width_crop_num, height_crop_num, len(tokenized_image))
    # 测试 image tokens 数量的正确性

""" 处理最后一个 text_splits"""
tokenized_sep = self.encode(text_splits[-1], bos=False, eos=False)
tokenized_str += tokenized_sep
images_seq_mask += [False] * len(tokenized_sep)

""" 添加 bos 和 eos tokens"""
if bos:
    tokenized_str = [self.bos_id] + tokenized_str
    images_seq_mask = [False] + images_seq_mask
if eos:
```

```
            tokenized_str = tokenized_str + [self.eos_id]
            images_seq_mask = images_seq_mask + [False]

        assert len(tokenized_str) == len(
            images_seq_mask), f"tokenize_with_images 函数中 tokenized_str 的长度
                                {len(tokenized_str)} 不等于 imags_seq_mask 的长
                                度 {len(images_seq_mask)}"

        return tokenized_str, images_list, images_seq_mask, images_spatial_
            crop, num_image_tokens
```

● 方法batchify()：功能是将多个VLChatProcessorOutput样本转换为批次数据，为模型的多模态推理准备输入。该方法对文本序列、图像、裁剪信息进行填充与对齐，确保不同长度的样本能够在同一批次中处理。

```
    def batchify(
            self,
            sample_list: List[VLChatProcessorOutput],
            padding: Literal["left", "right"] = "left"
    ) -> BatchCollateOutput:
        """
        为多模态推理准备输入。

        Args:
            sample_list (List[VLChatProcessorOutput]): VLChatProcessorOutput
                                                        的列表。
            padding (str): 补充方法。默认为 "left"。

        Returns:
            BatchCollateOutput: 用于多模态推理的输入。
        """

        batched_sft_format = [sample.sft_format for sample in sample_list]
        batched_input_ids = [sample.input_ids for sample in sample_list]
        batched_labels = [sample.target_ids for sample in sample_list]
        batched_images_seq_mask = [sample["images_seq_mask"] for sample in
                                    sample_list]
        seq_lens = [len(sample) for sample in sample_list]

        """ 填充 input_ids 和 images_seq_mask"""
        if padding == "left":
            # 默认 tokenizer 填充到左边
            # 如果使用 LlamaTokenizerFast,
            # 则强烈建议使用 `__call__` 方法而不是手动编码和填充。
            padded_input_ids = self.tokenizer.pad({"input_ids": batched_
```

```
                                        input_ids})
        batched_input_ids, batched_attention_mask = padded_input_
            ids["input_ids"], padded_input_ids[
            "attention_mask"].bool()
        batched_labels = self.tokenizer.pad({"input_ids": batched_
            labels})["input_ids"]
        batched_labels[batched_labels == self.pad_id] = self.ignore_id
            # labels 正常不会出现 pad_id，无需额外保护
        batched_images_seq_mask = self.tokenizer.pad({"input_ids":
            batched_images_seq_mask})["input_ids"]
        batched_images_seq_mask[batched_images_seq_mask == self.pad_id]
            = False
    else:
        batched_input_ids = pad_sequence(batched_input_ids, batch_
            first=True, padding_value=self.pad_id)
        batched_labels = pad_sequence(batched_labels, batch_first=True,
            padding_value=self.ignore_id)
        batched_images_seq_mask = pad_sequence(batched_images_seq_mask,
            batch_first=True, padding_value=0)
        batched_attention_mask = batched_input_ids != self.pad_id

    """ 填充图片到 max_patch_num"""
    max_n_patches = max(sample["images"].shape[0] for sample in sample_
        list)
    batched_images = []
    for sample in sample_list:
        images = sample["images"]
        n_pads = max_n_patches - images.shape[0]
        if n_pads > 0:
            pad_images = torch.zeros((n_pads, *images.shape[1:]),
                dtype=images.dtype)
            images = torch.cat([images, pad_images], dim=0)
        batched_images.append(images)
    batched_images = torch.stack(batched_images, dim=0)

    """ 填充 images_spatial_crop 到 max_n_images"""
    max_n_images = max(sample["images_spatial_crop"].shape[0] for sample
        in sample_list)
    batched_images_spatial_crop = []
    for sample in sample_list:
        images_spatial_crop = sample["images_spatial_crop"]
        n_pads = max_n_images - sample["images_spatial_crop"].shape[0]
        if n_pads > 0:
            pad_images_spatial_crop = torch.full((n_pads, 2), 0,
                dtype=images_spatial_crop.dtype)
            images_spatial_crop = torch.cat([images_spatial_crop, pad_
```

```
                images_spatial_crop], dim=0)
            batched_images_spatial_crop.append(images_spatial_crop)
        batched_images_spatial_crop = torch.stack(batched_images_spatial_
            crop, dim=0)

        batched_samples = BatchCollateOutput(
            input_ids=batched_input_ids,
            attention_mask=batched_attention_mask,
            labels=batched_labels,
            images=batched_images,
            images_seq_mask=batched_images_seq_mask,
            images_spatial_crop=batched_images_spatial_crop,
            sft_format=batched_sft_format,
            seq_lens=seq_lens
        )

        return batched_samples
```

上述方法共同确保DeepSeek-VL2 模型能够高效、准确地处理多模态输入数据，为训练与推理提供一致的数据接口。

6.5.4 DeepSeek 模型架构

在本项目中，文件modeling_deepseek.py定义了DeepSeekV2 模型及其组件的实现，主要包括模型配置、解码层、注意力机制、MoE模块等。该文件的目标是提供与DeepSeekV2 和DeepSeekV3 兼容的PyTorch模型，并实现了多种优化策略，如Rotary Position Embedding、Flash Attention和混合专家模块。

文件modeling_deepseek.py的具体实现流程如下所示。

（1）函数 _get_unpad_data()的主要功能是处理输入的attention_mask，用于提取与非填充部分相关的索引、累积序列长度及最大序列长度。

```
def _get_unpad_data(attention_mask):
    seqlens_in_batch = attention_mask.sum(dim=-1, dtype=torch.int32)
    indices = torch.nonzero(attention_mask.flatten(), as_tuple=False).flatten()
    max_seqlen_in_batch = seqlens_in_batch.max().item()
    cu_seqlens = F.pad(
        torch.cumsum(seqlens_in_batch, dim=0, dtype=torch.torch.int32), (1, 0)
    )
    return (
        indices,
        cu_seqlens,
        max_seqlen_in_batch,
    )
```

（2）DeepseekV2RMSNorm是一个基于PyTorch的归一化层，主要功能是对输入的隐藏状态（hidden_states）进行归一化处理。

```
class DeepseekV2RMSNorm(nn.Module):
    def __init__(self, hidden_size, eps=1e-6):
        """
        DeepseekV2RMSNorm is equivalent to T5LayerNorm[^1^]
        """
        super().__init__()
        self.weight = nn.Parameter(torch.ones(hidden_size)) # 初始化权重参数为全 1
        self.variance_epsilon = eps  # 方差的平滑项，防止除以零

    def forward(self, hidden_states):
        # 保存输入的数据类型
        input_dtype = hidden_states.dtype
        # 将输入转换为 float32，以确保计算精度
        hidden_states = hidden_states.to(torch.float32)
        # 计算方差：对输入的平方取均值
        variance = hidden_states.pow(2).mean(-1, keepdim=True)
        # 使用方差的倒数平方根进行归一化
        hidden_states = hidden_states * torch.rsqrt(variance +
                                             self.variance_epsilon)
        # 将归一化后的结果乘以权重参数，并恢复到原始数据类型
        return self.weight * hidden_states.to(input_dtype)
```

（3）DeepseekV2RotaryEmbedding是一个实现旋转位置嵌入（Rotary Position Embedding, RoPE）的PyTorch模块。旋转位置嵌入是一种用于处理序列数据（如自然语言处理中的文本）的位置编码方法，特别适用于Transformer架构。DeepseekV2RotaryEmbedding的主要功能包括：

- 初始化：计算频率（inv_freq），并预先缓存最大序列长度的余弦和正弦值。
- 动态缓存更新：如果输入序列长度超过预先缓存的最大长度，则动态更新缓存。
- 前向传播：根据输入序列长度，返回对应的余弦和正弦值，用于后续的旋转位置嵌入计算。

```
ALL_LAYERNORM_LAYERS.append(DeepseekV2RMSNorm)
                                    # 将 DeepseekV2RMSNorm 添加到归一化层列表中


class DeepseekV2RotaryEmbedding(nn.Module):
    def __init__(self, dim, max_position_embeddings=2048, base=10000,
device=None):
        """
        初始化旋转位置嵌入模块。
        :param dim: 每个头的维度。
        :param max_position_embeddings: 最大位置嵌入长度。
        :param base: 用于计算频率的基础值。
        :param device: 设备（可选）。
```

```
        """
        super().__init__()

        self.dim = dim  # 每个头的维度
        self.max_position_embeddings = max_position_embeddings  # 最大位置嵌入长度
        self.base = base  # 基础值
        # 计算频率
        inv_freq = 1.0 / (
            self.base ** (torch.arange(0, self.dim, 2).float().to(device) /
                          self.dim)
        )
        self.register_buffer("inv_freq", inv_freq, persistent=False)
                                                            # 注册频率为缓冲区

        # 为了使 `torch.jit.trace` 正常工作，这里预先构建缓存
        self._set_cos_sin_cache(
            seq_len=max_position_embeddings,
            device=self.inv_freq.device,
            dtype=torch.get_default_dtype(),
        )
        self.max_seq_len_cached = None  # 缓存的最大序列长度

    def _set_cos_sin_cache(self, seq_len, device, dtype):
        """
        动态设置余弦和正弦缓存。
        :param seq_len: 序列长度。
        :param device: 设备。
        :param dtype: 数据类型。
        """
        self.max_seq_len_cached = seq_len  # 更新缓存的最大序列长度
        t = torch.arange(
            self.max_seq_len_cached, device=device, dtype=self.inv_freq.dtype
        )  # 创建时间序列

        freqs = torch.outer(t, self.inv_freq.to(t.device))  # 计算频率矩阵
        # 将频率矩阵扩展为两倍维度（用于余弦和正弦）
        emb = torch.cat((freqs, freqs), dim=-1)
        self.register_buffer("cos_cached", emb.cos().to(dtype),
                             persistent=False)  # 缓存余弦值
        self.register_buffer("sin_cached", emb.sin().to(dtype),
                             persistent=False)  # 缓存正弦值

    def forward(self, x, seq_len=None):
        """
        前向传播。
        :param x: 输入张量，形状为 [bs, num_attention_heads, seq_len, head_size]。
```

```
        :param seq_len: 输入序列长度（可选）。
        :return: 返回对应的余弦和正弦值。
        """
        # 如果缓存的最大序列长度为空，或者输入序列长度超过缓存的最大长度，则更新缓存
        if self.max_seq_len_cached is None or seq_len > self.max_seq_len_cached:
            self._set_cos_sin_cache(seq_len=seq_len, device=x.device,
                                    dtype=x.dtype)

        # 返回对应的余弦和正弦值
        return (
            self.cos_cached[:seq_len].to(dtype=x.dtype),
            self.sin_cached[:seq_len].to(dtype=x.dtype),
        )
```

（4）DeepseekV2LinearScalingRotaryEmbedding是继承自DeepseekV2RotaryEmbedding的线性缩放位置嵌入类，借鉴了transformers库中LlamaLinearScalingRotaryEmbedding的实现。它引入了一个线性缩放因子scaling_factor，以控制位置嵌入的频率缩放。主要功能包括生成和缓存旋转嵌入的cos值和sin值，用于优化Transformer模型的自注意力计算。

```
# Copied from transformers.models.llama.modeling_llama.LlamaLinearScalingRot-
  aryEmbedding with Llama->DeepseekV2
class DeepseekV2LinearScalingRotaryEmbedding(DeepseekV2RotaryEmbedding):
    """DeepseekV2RotaryEmbedding 扩展了线性缩放功能。致谢 Reddit 用户 /u/kaiokendev"""

    def __init__(
        self,
        dim,
        max_position_embeddings=2048,
        base=10000,
        device=None,
        scaling_factor=1.0,
    ):
        # 线性缩放因子
        self.scaling_factor = scaling_factor
        # 调用父类初始化
        super().__init__(dim, max_position_embeddings, base, device)

    def _set_cos_sin_cache(self, seq_len, device, dtype):
        """
        设置旋转嵌入的 cos 和 sin 缓存。

        Args:
            seq_len (int): 序列长度。
            device (torch.device): 缓存的设备。
            dtype (torch.dtype): 数据类型。
```

```
        生成频率嵌入矩阵并缓存 cos 和 sin 值，以加速模型的嵌入计算。
        """
        # 缓存的最大序列长度
        self.max_seq_len_cached = seq_len

        # 计算时间步（位置索引）并应用线性缩放因子
        t = torch.arange(
            self.max_seq_len_cached, device=device, dtype=self.inv_freq.dtype
        )
        t = t / self.scaling_factor

        # 计算频率嵌入
        freqs = torch.outer(t, self.inv_freq)

        # 与论文中不同的方式，通过不同排列得到相同计算结果
        emb = torch.cat((freqs, freqs), dim=-1)

        # 缓存 cos 和 sin 值，加速嵌入计算
        self.register_buffer("cos_cached", emb.cos().to(dtype),
persistent=False)
        self.register_buffer("sin_cached", emb.sin().to(dtype),
persistent=False)
```

（5）DeepseekV2DynamicNTKScalingRotaryEmbedding 是 DeepseekV2RotaryEmbedding 的扩展类，参考了 transformers 库中 LlamaDynamicNTKScalingRotaryEmbedding 实现。其主要功能是通过动态 NTK（Neural Tangent Kernel）缩放策略计算旋转位置嵌入，用于处理超出预定义最大位置的长序列输入。该类动态调整频率基数 inv_freq，从而在更长序列上保持模型性能。

```
# Copied from transformers.models.llama.modeling_llama.LlamaDynamicNTKScalin-
  gRotaryEmbedding with Llama->DeepseekV2
class DeepseekV2DynamicNTKScalingRotaryEmbedding(DeepseekV2RotaryEmbedding):
    """DeepseekV2RotaryEmbedding 扩展了动态 NTK 缩放功能。致谢 Reddit 用户 /u/bloc97
和 /u/emozilla"""

    def __init__(
        self,
        dim,
        max_position_embeddings=2048,
        base=10000,
        device=None,
        scaling_factor=1.0,
    ):
        # 动态缩放因子
        self.scaling_factor = scaling_factor
```

```
        # 调用父类初始化
        super().__init__(dim, max_position_embeddings, base, device)

    def _set_cos_sin_cache(self, seq_len, device, dtype):
        """
        设置旋转嵌入的 cos 和 sin 缓存。

        Args:
            seq_len (int): 序列长度。
            device (torch.device): 缓存的设备。
            dtype (torch.dtype): 数据类型。

        计算频率嵌入矩阵并缓存 cos 和 sin 值，以支持动态 NTK 缩放策略。
        """
        # 缓存的最大序列长度
        self.max_seq_len_cached = seq_len

        # 如果序列长度大于最大位置嵌入数，则动态调整频率基数
        if seq_len > self.max_position_embeddings:
            base = self.base * (
                (self.scaling_factor * seq_len / self.max_position_embeddings)
                - (self.scaling_factor - 1)
            ) ** (self.dim / (self.dim - 2))
            inv_freq = 1.0 / (
                base ** (torch.arange(0, self.dim, 2).float().to(device) / self.dim)
            )
            # 缓存动态调整后的频率基数
            self.register_buffer("inv_freq", inv_freq, persistent=False)

        # 计算时间步（位置索引）
        t = torch.arange(
            self.max_seq_len_cached, device=device, dtype=self.inv_freq.dtype
        )

        # 计算频率嵌入
        freqs = torch.outer(t, self.inv_freq)

        # 与论文中不同的方式，通过不同排列得到相同计算结果
        emb = torch.cat((freqs, freqs), dim=-1)

        # 缓存 cos 和 sin 值，加速嵌入计算
        self.register_buffer("cos_cached", emb.cos().to(dtype),
                             persistent=False)
        self.register_buffer("sin_cached", emb.sin().to(dtype),
                             persistent=False)
```

（6）下列代码主要涉及旋转嵌入维度与频率的动态调整与计算功能，提供了基于数学公式计算位置嵌入频率、维度校正范围的工具函数。代码的功能包括根据旋转数量推导嵌入维度、计算维度范围、根据比例获取缩放因子，以及生成线性增长掩码。这些方法可用于优化模型中位置编码策略，以适应不同规模和输入条件。

```
# 根据旋转数量求解嵌入维度的逆公式
def yarn_find_correction_dim(
    num_rotations, dim, base=10000, max_position_embeddings=2048
):
    """
    计算给定旋转数量时的嵌入维度校正值。

    Args:
        num_rotations (int): 旋转次数。
        dim (int): 维度大小。
        base (float, optional): 频率基数。默认为 10000。
        max_position_embeddings (int, optional): 最大位置嵌入数量。默认为 2048。

    Returns:
        float: 校正后的嵌入维度。
    """
    return (dim * math.log(max_position_embeddings / (num_rotations * 2 *
            math.pi))) / (
        2 * math.log(base)
    )

# 根据旋转数量计算嵌入维度范围
def yarn_find_correction_range(
    low_rot, high_rot, dim, base=10000, max_position_embeddings=2048
):
    """
    计算嵌入维度的范围（低、高）。

    Args:
        low_rot (int): 最小旋转数量。
        high_rot (int): 最大旋转数量。
        dim (int): 维度大小。
        base (float, optional): 频率基数。默认为 10000。
        max_position_embeddings (int, optional): 最大位置嵌入数量。默认为 2048。

    Returns:
        tuple: (低范围，高范围)，确保范围在合理区间内。
    """
    low = math.floor(
        yarn_find_correction_dim(low_rot, dim, base, max_position_embeddings)
```

```
    )
    high = math.ceil(
        yarn_find_correction_dim(high_rot, dim, base, max_position_embeddings)
    )
    # 将范围限制在合理的维度范围内
    return max(low, 0), min(high, dim - 1)

# 获取多尺度缩放因子
def yarn_get_mscale(scale=1, mscale=1):
    """
    根据输入比例计算多尺度缩放因子。

    Args:
        scale (float): 缩放比例。
        mscale (float): 缩放权重。

    Returns:
        float: 计算后的缩放因子。
    """
    if scale <= 1:
        return 1.0
    return 0.1 * mscale * math.log(scale) + 1.0

# 生成线性增长掩码
def yarn_linear_ramp_mask(min, max, dim):
    """
    生成从 min 到 max 的线性增长掩码。

    Args:
        min (float): 掩码的最小值。
        max (float): 掩码的最大值。
        dim (int): 掩码的长度。

    Returns:
        torch.Tensor: 线性增长掩码。
    """
    if min == max:
        max += 0.001  # 防止分母为零导致的奇点问题

    # 计算线性函数并限制其在 [0, 1] 范围内
    linear_func = (torch.arange(dim, dtype=torch.float32) - min) / (max - min)
    ramp_func = torch.clamp(linear_func, 0, 1)
    return ramp_func
```

（7）类DeepseekV2YarnRotaryEmbedding是对旋转位置编码机制的进一步扩展，结合Yarn机制

进行频率调整与嵌入优化。它通过引入额外参数（如beta_fast和beta_slow），动态调整频率的计算方式，并采用多尺度缩放 (mscale) 来控制编码的幅度变化，从而优化不同位置和序列长度下的模型性能。这种改进有助于更高效地进行位置嵌入计算，尤其适用于长序列输入的情况。

```
class DeepseekV2YarnRotaryEmbedding(DeepseekV2RotaryEmbedding):

    def __init__(
        self,
        dim,
        max_position_embeddings=2048,
        base=10000,
        device=None,
        scaling_factor=1.0,
        original_max_position_embeddings=4096,
        beta_fast=32,
        beta_slow=1,
        mscale=1,
        mscale_all_dim=0,
    ):
        """
        初始化 Yarn 位置嵌入类，继承自 DeepseekV2RotaryEmbedding，支持动态频率调整和多
        尺度缩放。

        Args:
            dim (int): 位置嵌入维度。
            max_position_embeddings (int, optional): 最大位置嵌入数量。默认为 2048。
            base (float, optional): 频率基数。默认为 10000。
            device (str, optional): 设备名。默认为 None。
            scaling_factor (float, optional): 缩放因子。默认为 1.0。
            original_max_position_embeddings (int, optional): 原始最大位置嵌入
                数量。默认为 4096。
            beta_fast (int, optional): 快速频率校正参数。默认为 32。
            beta_slow (int, optional): 慢速频率校正参数。默认为 1。
            mscale (float, optional): 多尺度缩放系数。默认为 1。
            mscale_all_dim (int, optional): 全维度缩放系数。默认为 0。
        """
        self.scaling_factor = scaling_factor
        self.original_max_position_embeddings = original_max_position_
                                                embeddings
        self.beta_fast = beta_fast
        self.beta_slow = beta_slow
        self.mscale = mscale
        self.mscale_all_dim = mscale_all_dim
        super().__init__(dim, max_position_embeddings, base, device)
```

```
def _set_cos_sin_cache(self, seq_len, device, dtype):
    """
    计算和缓存旋转嵌入的 cos 和 sin 值。

    Args:
        seq_len (int): 序列长度。
        device (str): 设备名。
        dtype (torch.dtype): 数据类型。
    """
    self.max_seq_len_cached = seq_len
    dim = self.dim

    # 计算额外频率和中间频率
    freq_extra = 1.0 / (
        self.base
        ** (torch.arange(0, dim, 2, dtype=torch.float32, device=device) / dim)
    )
    freq_inter = 1.0 / (
        self.scaling_factor
        * self.base
        ** (torch.arange(0, dim, 2, dtype=torch.float32, device=device) / dim)
    )

    # 计算频率校正范围
    low, high = yarn_find_correction_range(
        self.beta_fast,
        self.beta_slow,
        dim,
        self.base,
        self.original_max_position_embeddings,
    )

    # 计算频率掩码并应用频率混合
    inv_freq_mask = 1.0 - yarn_linear_ramp_mask(low, high, dim // 2).to(
        device=device, dtype=torch.float32
    )
    inv_freq = freq_inter * (1 - inv_freq_mask) + freq_extra * inv_freq_mask
    self.register_buffer("inv_freq", inv_freq, persistent=False)

    # 计算频率嵌入
    t = torch.arange(seq_len, device=device, dtype=torch.float32)
    freqs = torch.outer(t, inv_freq)

    # 计算多尺度缩放系数
    _mscale = float(
        yarn_get_mscale(self.scaling_factor, self.mscale)
```

```
            / yarn_get_mscale(self.scaling_factor, self.mscale_all_dim)
        )

        # 将频率嵌入应用到 cos 和 sin 函数并缓存
        emb = torch.cat((freqs, freqs), dim=-1)
        self.register_buffer(
            "cos_cached", (emb.cos() * _mscale).to(dtype), persistent=False
        )
        self.register_buffer(
            "sin_cached", (emb.sin() * _mscale).to(dtype), persistent=False
        )
```

（8）函数rotate_half()用于旋转输入张量的一半维度。它将输入张量的最后一个维度拆分为两部分，然后将后半部分取反与前半部分交换位置并拼接。该操作在位置嵌入的旋转编码中常用于构建旋转不变性特征。

```
def rotate_half(x):
    """
    将输入的隐藏维度的一半进行旋转操作。
    Args:
        x (torch.Tensor): 输入张量，形状为 [..., hidden_size]。
    Returns:
        torch.Tensor: 旋转后的张量。
    """
    # 拆分输入张量的最后一个维度为两部分
    x1 = x[..., : x.shape[-1] // 2]  # 前半部分
    x2 = x[..., x.shape[-1] // 2 :]  # 后半部分
    # 将后半部分取反后与前半部分交换位置并拼接
    return torch.cat((-x2, x1), dim=-1)
```

（9）函数apply_rotary_pos_emb()的功能是将旋转位置嵌入（Rotary Position Embedding）应用于查询（query）和键（key）张量。这种嵌入方式通过结合正弦和余弦函数实现对序列位置的建模，从而使模型能够更好地捕捉序列顺序信息。

```
def apply_rotary_pos_emb(q, k, cos, sin, position_ids, unsqueeze_dim=1):
    """
    将旋转位置嵌入（Rotary Position Embedding）应用于查询（query）和键（key）张量。

    Args:
        q (`torch.Tensor`): 查询张量，形状为 [batch_size, num_heads, seq_len,
                                               head_dim]。
        k (`torch.Tensor`): 键张量，形状与查询张量相同。
        cos (`torch.Tensor`): 旋转嵌入的余弦部分。
        sin (`torch.Tensor`): 旋转嵌入的正弦部分。
        position_ids (`torch.Tensor`):
```

```
            查询和键张量对应的序列位置索引。例如，当使用 KV 缓存时可以传递偏移位置 ID。
        unsqueeze_dim (`int`, *optional*, defaults to 1):
            指定 `cos[position_ids]` 和 `sin[position_ids]` 的 unsqueeze 维度，
            使其能够正确广播到查询和键的维度。
            例如：
            - 当 `cos[position_ids]` 和 `sin[position_ids]` 形状为 [batch_
              size, seq_len, head_dim]，如果 q 和 k 形状为
              [batch_size, num_heads, seq_len, head_dim]，应设置 `unsqueeze_
              dim=1` 以确保维度对齐。
            - 如果 q 和 k 的形状为 [batch_size, seq_len, num_heads, head_dim]，则
              设置 `unsqueeze_dim=2`。

    Returns:
        `tuple(torch.Tensor)`: 旋转嵌入后的查询和键张量。
    """
    # 为 cos 和 sin 添加维度以便广播
    cos = cos[position_ids].unsqueeze(unsqueeze_dim)
    sin = sin[position_ids].unsqueeze(unsqueeze_dim)

    # 对查询张量进行维度变换，将其形状调整为 [..., d // 2, 2] 并交换最后两个维度
    b, h, s, d = q.shape
    q = q.view(b, h, s, d // 2, 2).transpose(4, 3).reshape(b, h, s, d)

    # 对键张量进行同样的处理
    b, h, s, d = k.shape
    k = k.view(b, h, s, d // 2, 2).transpose(4, 3).reshape(b, h, s, d)

    # 应用旋转位置嵌入
    q_embed = (q * cos) + (rotate_half(q) * sin)
    k_embed = (k * cos) + (rotate_half(k) * sin)

    # 返回嵌入后的查询和键张量
    return q_embed, k_embed
```

（10）DeepseekV2MLP是一个自定义的前馈全连接层模块，该模块包括三个线性投影层：gate_proj、up_proj和down_proj，以及一个激活函数。其主要功能是对输入特征进行非线性变换和维度映射，以增强模型的表达能力。

```
class DeepseekV2MLP(nn.Module):
    def __init__(self, config, hidden_size=None, intermediate_size=None):
        super().__init__()
        self.config = config
        self.hidden_size = config.hidden_size if hidden_size is None else
                        hidden_size
        self.intermediate_size = (
```

```
            config.intermediate_size if intermediate_size is None else
                intermediate_size
        )

        self.gate_proj = nn.Linear(self.hidden_size, self.intermediate_size,
                                   bias=False)
        self.up_proj = nn.Linear(self.hidden_size, self.intermediate_size,
                                 bias=False)
        self.down_proj = nn.Linear(self.intermediate_size, self.hidden_size,
                                   bias=False)
        self.act_fn = ACT2FN[config.hidden_act]

    def forward(self, x):
        down_proj = self.down_proj(self.act_fn(self.gate_proj(x)) * self.up_
proj(x))
        return down_proj
```

（11）类MoEGate实现了混合专家网络中的门控模块，用于为输入的特征选择最合适的专家网络。该模块通过计算门控得分（gating scores）并选择Top-K专家来高效处理不同的输入。它支持多种专家选择策略（如贪婪选择和分组选择），并具有辅助损失计算功能，用于优化专家选择过程。

```
class MoEGate(nn.Module):
    """
    混合专家网络（MoE）中的门控模块，用于为输入的特征选择合适的专家网络。

    Args:
        config (`PretrainedConfig`): 包含门控模块的配置参数，例如专家数量、选择策略等。
    """
    def __init__(self, config):
        super().__init__()
        self.config = config
        self.top_k = config.num_experts_per_tok      # 每个 token 选择的专家数量
        self.n_routed_experts = config.n_routed_experts   # 可选择的总专家数量
        self.routed_scaling_factor = config.routed_scaling_factor
                                                                  # 路由缩放系数
        self.scoring_func = config.scoring_func      # 门控得分计算函数类型
        self.alpha = config.aux_loss_alpha           # 辅助损失系数
        self.seq_aux = config.seq_aux                # 是否计算序列辅助损失
        self.topk_method = config.topk_method        # Top-K 选择策略
        self.n_group = config.n_group                # 分组数量
        self.topk_group = config.topk_group          # 选择的分组数量

        # Top-K 选择算法
        self.norm_topk_prob = config.norm_topk_prob # 是否归一化门控权重
        self.gating_dim = config.hidden_size         # 门控维度
```

```
        self.weight = nn.Parameter(
            torch.empty((self.n_routed_experts, self.gating_dim))
        )   # 专家权重
        if self.topk_method == "noaux_tc":
            self.e_score_correction_bias = nn.Parameter(
                torch.empty((self.n_routed_experts))
            )   # 无辅助训练时的偏置
        self.reset_parameters()

    def reset_parameters(self) -> None:
        """
        初始化权重参数。
        """
        import torch.nn.init as init
        init.kaiming_uniform_(self.weight, a=math.sqrt(5))

    def forward(self, hidden_states):
        """
        前向传播函数，计算门控得分并选择专家。

        Args:
            hidden_states (`torch.Tensor`): 输入特征张量，形状为 `(batch_size,
                seq_len, hidden_size)`。

        Returns:
            `tuple`: 包括专家索引、权重和辅助损失。
        """
        bsz, seq_len, h = hidden_states.shape

        # 计算门控得分
        hidden_states = hidden_states.view(-1, h)
        logits = F.linear(hidden_states.type(torch.float32),
                          self.weight.type(torch.float32), None)
        if self.scoring_func == "softmax":
            scores = logits.softmax(dim=-1, dtype=torch.float32)
        elif self.scoring_func == "sigmoid":
            scores = logits.sigmoid()
        else:
            raise NotImplementedError(
                f"不支持的门控评分函数：{self.scoring_func}"
            )

        # 选择 Top-K 专家
        if self.topk_method == "greedy":
            # 贪婪选择策略
            topk_weight, topk_idx = torch.topk(scores, k=self.top_k, dim=-1,
```

```
sorted=False)
        elif self.topk_method == "group_limited_greedy":
            # 分组选择策略
            group_scores = scores.view(bsz * seq_len, self.n_group, -1).
                         max(dim=-1).values
            group_idx = torch.topk(group_scores, k=self.topk_group, dim=-1,
                                   sorted=False)[1]
            group_mask = torch.zeros_like(group_scores)
            group_mask.scatter_(1, group_idx, 1)
            score_mask = group_mask.unsqueeze(-1).expand(bsz * seq_len,
                         self.n_group, self.n_routed_experts // self.n_
                         group).reshape(bsz * seq_len, -1)
            tmp_scores = scores.masked_fill(~score_mask.bool(), 0.0)
            topk_weight, topk_idx = torch.topk(tmp_scores, k=self.top_k,
                                               dim=-1, sorted=False)
        elif self.topk_method == "noaux_tc":
            # 无辅助训练的选择策略
            assert not self.training
            scores_for_choice = scores.view(bsz * seq_len, -1) + self.e_
                              score_correction_bias.unsqueeze(0)
            group_scores = scores_for_choice.view(bsz * seq_len, self.n_
                         group, -1).topk(2, dim=-1)[0].sum(dim=-1)
            group_idx = torch.topk(group_scores, k=self.topk_group, dim=-1,
                        sorted=False)[1]
            group_mask = torch.zeros_like(group_scores)
            group_mask.scatter_(1, group_idx, 1)
            score_mask = group_mask.unsqueeze(-1).expand(bsz * seq_len,
                         self.n_group, self.n_routed_experts // self.n_
                         group).reshape(bsz * seq_len, -1)
            tmp_scores = scores_for_choice.masked_fill(~score_mask.bool(), 0.0)
            _, topk_idx = torch.topk(tmp_scores, k=self.top_k, dim=-1, sorted=False)
            topk_weight = scores.gather(1, topk_idx)

        # 归一化门控权重
        if self.top_k > 1 and self.norm_topk_prob:
            denominator = topk_weight.sum(dim=-1, keepdim=True) + 1e-20
            topk_weight = topk_weight / denominator * self.routed_scaling_factor
        else:
            topk_weight = topk_weight * self.routed_scaling_factor

        # 计算辅助损失
        if self.training and self.alpha > 0.0:
            scores_for_aux = scores
            aux_topk = self.top_k
            topk_idx_for_aux_loss = topk_idx.view(bsz, -1)
            if self.seq_aux:
```

```
            scores_for_seq_aux = scores_for_aux.view(bsz, seq_len, -1)
            ce = torch.zeros(bsz, self.n_routed_experts,
                             device=hidden_states.device)
            ce.scatter_add_(
                1,
                topk_idx_for_aux_loss,
                torch.ones(bsz, seq_len * aux_topk,
                           device=hidden_states.device),
            ).div_(seq_len * aux_topk / self.n_routed_experts)
            aux_loss = (ce * scores_for_seq_aux.mean(dim=1)).sum(dim=1).
                       mean() * self.alpha
        else:
            mask_ce = F.one_hot(topk_idx_for_aux_loss.view(-1),
                                num_classes=self.n_routed_experts)
            ce = mask_ce.float().mean(0)
            Pi = scores_for_aux.mean(0)
            fi = ce * self.n_routed_experts
            aux_loss = (Pi * fi).sum() * self.alpha
    else:
        aux_loss = None

    return topk_idx, topk_weight, aux_loss
```

对上述代码的具体说明如下所示。

- 门控得分计算：使用线性层计算每个输入的专家得分，并通过Softmax或Sigmoid进行归一化。
- Top-K专家选择：支持贪婪选择、分组选择和无辅助训练的选择策略。
- 门控权重归一化：确保选择的专家权重归一化。
- 辅助损失计算：在训练阶段计算辅助损失，以优化专家选择过程。

（12）AddAuxiliaryLoss是一个自定义的torch.autograd.Function类，用于在模型训练时动态添加辅助损失（auxiliary loss）。它通过在前向传播中保存辅助损失信息，并在反向传播阶段计算辅助损失的梯度，确保辅助损失能够影响模型参数的更新，从而优化模型训练效果。

```
class AddAuxiliaryLoss(torch.autograd.Function):
    """
    自定义函数，用于添加辅助（auxiliary）损失，
    在反向传播中包含辅助损失的梯度。
    """

    @staticmethod
    def forward(ctx, x, loss):
        """
        前向传播方法，用于传递输入和辅助损失。

        Args:
```

```
            ctx (`Context`): 上下文对象，用于存储反向传播所需的信息。
            x (`torch.Tensor`): 输入张量。
            loss (`torch.Tensor`): 辅助损失张量，必须是单个标量值。

        Returns:
            `torch.Tensor`: 直接返回输入张量 x。
        """
        assert loss.numel() == 1   # 确保辅助损失是单个标量
        ctx.dtype = loss.dtype     # 记录辅助损失的类型
        ctx.required_aux_loss = loss.requires_grad  # 记录辅助损失是否需要梯度
        return x

    @staticmethod
    def backward(ctx, grad_output):
        """
        反向传播方法，用于计算输入和辅助损失的梯度。

        Args:
            ctx (`Context`): 前向传播保存的上下文信息。
            grad_output (`torch.Tensor`): 上一层传递的梯度。

        Returns:
            `tuple`: 包括输入梯度和辅助损失梯度。
        """
        grad_loss = None
        if ctx.required_aux_loss:
            # 如果辅助损失需要梯度，则返回一个值为 1 的梯度
            grad_loss = torch.ones(1, dtype=ctx.dtype, device=grad_output.
                                   device)
        return grad_output, grad_loss
```

（13）类DeepseekV2MoE是一个可实现MoE机制的深度学习模块，该模块通过门控策略将输入分配给不同的专家网络，并通过辅助损失来优化专家选择和权重分配，从而提升模型的表达能力与计算效率。模型支持多GPU分布式训练环境下的专家网络分布与通信。

```
class DeepseekV2MoE(nn.Module):
    """
    一个包含共享专家的混合专家模块。
    """

    def __init__(self, config):
        super().__init__()
        self.config = config
        self.num_experts_per_tok = config.num_experts_per_tok
```

```
        # 判断是否使用分布式专家网络（专家并行）
        if hasattr(config, "ep_size") and config.ep_size > 1:
            assert config.ep_size == dist.get_world_size()
            self.ep_size = config.ep_size
            self.experts_per_rank = config.n_routed_experts // config.ep_size
            self.ep_rank = dist.get_rank()

            # 定义每个 GPU 上分配的专家网络
            self.experts = nn.ModuleList(
                [
                    (
                        DeepseekV2MLP(
                            config, intermediate_size=config.moe_intermediate_size
                        )
                        if i >= self.ep_rank * self.experts_per_rank
                        and i < (self.ep_rank + 1) * self.experts_per_rank
                        else None
                    )
                    for i in range(config.n_routed_experts)
                ]
            )
        else:
            # 单机单卡专家网络配置
            self.ep_size = 1
            self.experts_per_rank = config.n_routed_experts
            self.ep_rank = 0
            self.experts = nn.ModuleList(
                [
                    DeepseekV2MLP(
                        config, intermediate_size=config.moe_intermediate_size
                    )
                    for i in range(config.n_routed_experts)
                ]
            )

        # 门控机制用于选择专家
        self.gate = MoEGate(config)

        # 定义共享专家模块（如果配置存在）
        if config.n_shared_experts is not None:
            intermediate_size = config.moe_intermediate_size *
                                config.n_shared_experts
            self.shared_experts = DeepseekV2MLP(
                config=config, intermediate_size=intermediate_size
            )
```

```
def forward(self, hidden_states):
    """
    前向传播函数。

    Args:
        hidden_states (`torch.Tensor`): 模型输入特征张量。

    Returns:
        `torch.Tensor`: 模型输出特征。
    """
    identity = hidden_states
    orig_shape = hidden_states.shape

    # 通过门控选择专家并获取辅助损失
    topk_idx, topk_weight, aux_loss = self.gate(hidden_states)

    hidden_states = hidden_states.view(-1, hidden_states.shape[-1])
    flat_topk_idx = topk_idx.view(-1)

    if self.training:
        # 在训练模式下，复制输入以匹配选择的专家数量
        hidden_states = hidden_states.repeat_interleave(
            self.num_experts_per_tok, dim=0
        )
        y = torch.empty_like(hidden_states)

        # 根据门控选择的专家进行计算
        for i, expert in enumerate(self.experts):
            y[flat_topk_idx == i] = expert(hidden_states[flat_topk_idx
                                            == i])

        # 将专家输出按权重加权求和
        y = (y.view(*topk_weight.shape, -1) * topk_weight.unsqueeze(-1)).
            sum(dim=1)
        y = y.to(hidden_states.dtype).view(*orig_shape)
        y = AddAuxiliaryLoss.apply(y, aux_loss)   # 应用辅助损失
    else:
        # 推理模式下使用高效的专家推理方法
        y = self.moe_infer(hidden_states, topk_idx, topk_weight).
                           view(*orig_shape)

    # 加入共享专家的输出
    if self.config.n_shared_experts is not None:
        y = y + self.shared_experts(identity)
    return y
```

```
@torch.no_grad()
def moe_infer(self, x, topk_ids, topk_weight):
    """
    推理模式下的专家选择与计算。

    Args:
        x (`torch.Tensor`): 输入张量。
        topk_ids (`torch.Tensor`): 选择的专家索引。
        topk_weight (`torch.Tensor`): 专家权重。

    Returns:
        `torch.Tensor`: 推理结果。
    """
    cnts = topk_ids.new_zeros((topk_ids.shape[0], len(self.experts)))
    cnts.scatter_(1, topk_ids, 1)
    tokens_per_expert = cnts.sum(dim=0)

    # 对选择的专家进行排序和分配
    idxs = topk_ids.view(-1).argsort()
    sorted_tokens = x[idxs // topk_ids.shape[1]]
    sorted_tokens_shape = sorted_tokens.shape

    # 如果使用分布式环境，进行数据通信
    if self.ep_size > 1:
        tokens_per_ep_rank = tokens_per_expert.view(self.ep_size, -1).
                             sum(dim=1)
        tokens_per_expert_group = tokens_per_expert.new_empty(
            tokens_per_expert.shape[0]
        )
        dist.all_to_all_single(tokens_per_expert_group,
                               tokens_per_expert)

        output_splits = (
            tokens_per_expert_group.view(self.ep_size, -1)
            .sum(1)
            .cpu()
            .numpy()
            .tolist()
        )

        gathered_tokens = sorted_tokens.new_empty(
            tokens_per_expert_group.sum(dim=0).cpu().item(),
            sorted_tokens.shape[1]
        )
        input_split_sizes = tokens_per_ep_rank.cpu().numpy().tolist()
```

```
        # 数据传输
        dist.all_to_all(
            list(gathered_tokens.split(output_splits)),
            list(sorted_tokens.split(input_split_sizes)),
        )

        tokens_per_expert_post_gather = tokens_per_expert_group.view(
            self.ep_size, self.experts_per_rank
        ).sum(dim=0)

        gatherd_idxs = np.zeros(shape=(gathered_tokens.shape[0],),
                                dtype=np.int32)
        s = 0
        for i, k in enumerate(tokens_per_expert_group.cpu().numpy()):
            gatherd_idxs[s : s + k] = i % self.experts_per_rank
            s += k
        gatherd_idxs = gatherd_idxs.argsort()
        sorted_tokens = gathered_tokens[gatherd_idxs]
        tokens_per_expert = tokens_per_expert_post_gather

    tokens_per_expert = tokens_per_expert.cpu().numpy()

    outputs = []
    start_idx = 0

    # 计算每个专家的输出
    for i, num_tokens in enumerate(tokens_per_expert):
        end_idx = start_idx + num_tokens
        if num_tokens == 0:
            continue
        expert = self.experts[i + self.ep_rank * self.experts_per_rank]
        tokens_for_this_expert = sorted_tokens[start_idx:end_idx]
        expert_out = expert(tokens_for_this_expert)
        outputs.append(expert_out)
        start_idx = end_idx

    outs = torch.cat(outputs, dim=0) if len(outputs) else sorted_tokens.
        new_empty(0)

    # 数据通信以重构输出
    if self.ep_size > 1:
        new_x = torch.empty_like(outs)
        new_x[gatherd_idxs] = outs
        gathered_tokens = new_x.new_empty(*sorted_tokens_shape)
        dist.all_to_all(
            list(gathered_tokens.split(input_split_sizes)),
```

```
                list(new_x.split(output_splits)),
            )
            outs = gathered_tokens

        new_x = torch.empty_like(outs)
        new_x[idxs] = outs
        final_out = (
            new_x.view(*topk_ids.shape, -1)
            .type(topk_weight.dtype)
            .mul_(topk_weight.unsqueeze(dim=-1))
            .sum(dim=1)
            .type(new_x.dtype)
        )
        return final_out
```

（14）函数repeat_kv()用于在注意力机制中通过重复n_rep次扩展键值张量。它接受一个形状为 (batch, num_key_value_heads, seqlen, head_dim) 的张量，并返回一个形状为 (batch, num_attention_heads, seqlen, head_dim) 的张量，其中注意力头的数量会被乘以n_rep。

```
def repeat_kv(hidden_states: torch.Tensor, n_rep: int) -> torch.Tensor:
    """
    这是等效于 torch.repeat_interleave(x, dim=1, repeats=n_rep) 的操作。隐状态从
    形状 (batch,
    num_key_value_heads, seqlen, head_dim) 转变为 (batch, num_attention_
    heads, seqlen, head_dim)
    """
    batch, num_key_value_heads, slen, head_dim = hidden_states.shape
    if n_rep == 1:
        return hidden_states
    hidden_states = hidden_states[:, :, None, :, :].expand(
        batch, num_key_value_heads, n_rep, slen, head_dim
    )
    return hidden_states.reshape(batch, num_key_value_heads * n_rep, slen,
head_dim)
```

（15）类DeepseekV2Attention实现了多头自注意力机制，设计理念来源于*Attention Is All You Need*论文的结构，并进行了深度定制。类DeepseekV2Attention支持多种参数配置，如LoRA（低秩适应）和RoPE（旋转位置编码）。该类包含了对查询、键值对的处理、位置编码、注意力权重计算和最终输出的多个阶段。它还支持缓存机制（用于自回归推理）和不同的注意力缩放策略，如线性、动态和yarn缩放等。

```
class DeepseekV2Attention(nn.Module):
    """来自 'Attention Is All You Need' 论文的多头注意力 """

    def __init__(self, config: DeepseekV2Config, layer_idx: Optional[int] =
```

```
None):
        super().__init__()
        self.config = config
        self.layer_idx = layer_idx
        if layer_idx is None:
            logger.warning_once(
                f"没有传递 `layer_idx` 来实例化 {self.__class__.__name__} 是不
                  推荐的，如果使用缓存，"
                "这会导致前向调用时出错。请确保在创建该类时提供 `layer_idx`。"
            )

        self.attention_dropout = config.attention_dropout
        self.hidden_size = config.hidden_size
        self.num_heads = config.num_attention_heads

        self.max_position_embeddings = config.max_position_embeddings
        self.rope_theta = config.rope_theta
        self.q_lora_rank = config.q_lora_rank
        self.qk_rope_head_dim = config.qk_rope_head_dim
        self.kv_lora_rank = config.kv_lora_rank
        self.v_head_dim = config.v_head_dim
        self.qk_nope_head_dim = config.qk_nope_head_dim
        self.q_head_dim = config.qk_nope_head_dim + config.qk_rope_head_dim

        self.is_causal = True

        if self.q_lora_rank is None:
            self.q_proj = nn.Linear(
                self.hidden_size, self.num_heads * self.q_head_dim, bias=False
            )
        else:
            self.q_a_proj = nn.Linear(
                self.hidden_size, config.q_lora_rank, bias=config.attention_bias
            )
            self.q_a_layernorm = DeepseekV2RMSNorm(config.q_lora_rank)
            self.q_b_proj = nn.Linear(
                config.q_lora_rank, self.num_heads * self.q_head_dim,
                bias=False
            )

        self.kv_a_proj_with_mqa = nn.Linear(
            self.hidden_size,
            config.kv_lora_rank + config.qk_rope_head_dim,
            bias=config.attention_bias,
        )
        self.kv_a_layernorm = DeepseekV2RMSNorm(config.kv_lora_rank)
```

```
        self.kv_b_proj = nn.Linear(
            config.kv_lora_rank,
            self.num_heads
            * (self.q_head_dim - self.qk_rope_head_dim + self.v_head_dim),
            bias=False,
        )

        self.o_proj = nn.Linear(
            self.num_heads * self.v_head_dim,
            self.hidden_size,
            bias=config.attention_bias,
        )
        self._init_rope()

        self.softmax_scale = self.q_head_dim ** (-0.5)
        if self.config.rope_scaling is not None:
            mscale_all_dim = self.config.rope_scaling.get("mscale_all_dim", 0)
            scaling_factor = self.config.rope_scaling["factor"]
            if mscale_all_dim:
                mscale = yarn_get_mscale(scaling_factor, mscale_all_dim)
                self.softmax_scale = self.softmax_scale * mscale * mscale

    def _init_rope(self):
        if self.config.rope_scaling is None:
            self.rotary_emb = DeepseekV2RotaryEmbedding(
                self.qk_rope_head_dim,
                max_position_embeddings=self.max_position_embeddings,
                base=self.rope_theta,
            )
        else:
            scaling_type = self.config.rope_scaling["type"]
            scaling_factor = self.config.rope_scaling["factor"]
            if scaling_type == "linear":
                self.rotary_emb = DeepseekV2LinearScalingRotaryEmbedding(
                    self.qk_rope_head_dim,
                    max_position_embeddings=self.max_position_embeddings,
                    scaling_factor=scaling_factor,
                    base=self.rope_theta,
                )
            elif scaling_type == "dynamic":
                self.rotary_emb = DeepseekV2DynamicNTKScalingRotaryEmbedding(
                    self.qk_rope_head_dim,
                    max_position_embeddings=self.max_position_embeddings,
                    scaling_factor=scaling_factor,
                    base=self.rope_theta,
                )
```

```
        elif scaling_type == "yarn":
            kwargs = {
                key: self.config.rope_scaling[key]
                for key in [
                    "original_max_position_embeddings",
                    "beta_fast",
                    "beta_slow",
                    "mscale",
                    "mscale_all_dim",
                ]
                if key in self.config.rope_scaling
            }
            self.rotary_emb = DeepseekV2YarnRotaryEmbedding(
                self.qk_rope_head_dim,
                max_position_embeddings=self.max_position_embeddings,
                scaling_factor=scaling_factor,
                base=self.rope_theta,
                **kwargs,
            )
        else:
            raise ValueError(f"未知的 RoPE 缩放类型 {scaling_type}")

def _shape(self, tensor: torch.Tensor, seq_len: int, bsz: int):
    return (
        tensor.view(bsz, seq_len, self.num_heads, self.v_head_dim)
        .transpose(1, 2)
        .contiguous()
    )

def forward(
    self,
    hidden_states: torch.Tensor,
    attention_mask: Optional[torch.Tensor] = None,
    position_ids: Optional[torch.LongTensor] = None,
    past_key_value: Optional[Cache] = None,
    output_attentions: bool = False,
    use_cache: bool = False,
    **kwargs,
) -> Tuple[torch.Tensor, Optional[torch.Tensor], Optional[Tuple[torch.
        Tensor]]]:
    if "padding_mask" in kwargs:
        warnings.warn(
            "传递 `padding_mask` 已被弃用，将在 v4.37 中删除，请确保使用
             `attention_mask`。"
        )
    bsz, q_len, _ = hidden_states.size()
```

```
if self.q_lora_rank is None:
    q = self.q_proj(hidden_states)
else:
    q = self.q_b_proj(self.q_a_layernorm(self.q_a_proj(hidden_states)))
q = q.view(bsz, q_len, self.num_heads, self.q_head_dim).transpose(1, 2)
q_nope, q_pe = torch.split(
    q, [self.qk_nope_head_dim, self.qk_rope_head_dim], dim=-1
)

compressed_kv = self.kv_a_proj_with_mqa(hidden_states)
compressed_kv, k_pe = torch.split(
    compressed_kv, [self.kv_lora_rank, self.qk_rope_head_dim], dim=-1
)
compressed_kv = self.kv_a_layernorm(compressed_kv)
k_pe = k_pe.view(bsz, q_len, 1, self.qk_rope_head_dim).transpose(1, 2)

kv_seq_len = k_pe.shape[-2]
if past_key_value is not None:
    if self.layer_idx is None:
        raise ValueError(
            f"自 v4.36 版本以来，缓存结构已发生变化。如果你在使用 {self.__
             class__.__name__} "
            "进行自回归推理并使用 k/v 缓存，请确保用层索引初始化注意力类。"
        )
    kv_seq_len += past_key_value.get_usable_length(kv_seq_len,
                                                    self.layer_idx)

cos, sin = self.rotary_emb(q_pe, seq_len=kv_seq_len)
q_pe, k_pe = apply_rotary_pos_emb(q_pe, k_pe, cos, sin, position_ids)

if past_key_value is not None:
    cache_kwargs = {"sin": sin, "cos": cos}  # 特定于 RoPE 模型
    compressed_kv = compressed_kv.unsqueeze(1)
    k_pe, compressed_kv = past_key_value.update(k_pe, compressed_kv,
                        self.layer_idx, cache_kwargs)
    compressed_kv = compressed_kv.squeeze(1)

kv_b_proj = self.kv_b_proj.weight.view(self.num_heads, -1, self.kv_
                                       lora_rank)
q_absorb = kv_b_proj[:, :self.qk_nope_head_dim, :]
out_absorb = kv_b_proj[:, self.qk_nope_head_dim:, :]

q_nope = torch.matmul(q_nope, q_absorb)
attn_weights = (torch.matmul(q_pe, k_pe.mT) +
                torch.matmul(q_nope, compressed_kv.unsqueeze(-3).
```

```
                                mT)) * self.softmax_scale
        if attn_weights.size() != (bsz, self.num_heads, q_len, kv_seq_len):
            raise ValueError(
                f"注意力权重的尺寸应为 {(bsz, self.num_heads, q_len, kv_seq_
                  len)}, 但却是 "
                f" {attn_weights.size()}"
            )
        assert attention_mask is not None
        if attention_mask is not None:
            if attention_mask.size() != (bsz, 1, q_len, kv_seq_len):
                raise ValueError(
                    f"注意力掩码的尺寸应为 {(bsz, 1, q_len, kv_seq_len)},
                      但却是 {attention_mask.size()}"
                )
            attn_weights = attn_weights + attention_mask

        # 将注意力权重转为 fp32
        attn_weights = nn.functional.softmax(
            attn_weights, dim=-1, dtype=torch.float32
        ).to(q_pe.dtype)
        attn_weights = nn.functional.dropout(
            attn_weights, p=self.attention_dropout, training=self.training
        )
        attn_output = torch.einsum('bhql,blc->bhqc', attn_weights,
                                   compressed_kv)

        attn_output = torch.matmul(attn_output, out_absorb.mT)

        if attn_output.size() != (bsz, self.num_heads, q_len, self.v_head_dim):
            raise ValueError(
                f"`attn_output` 的尺寸应为 {(bsz, self.num_heads, q_len, self.
                  v_head_dim)}, 但却是 "
                f"{attn_output.size()}"
            )
        attn_output = self._shape(attn_output, q_len, bsz)

        output = self.o_proj(attn_output)

        if not output_attentions:
            attn_weights = None

        return attn_output, attn_weights, past_key_value
```

（16）类DeepseekV2FlashAttention2 是一个用于深度学习模型的注意力机制模块，继承自DeepseekV2Attention。DeepseekV2FlashAttention2集成了Flash Attention的优化，在处理序列数据时

能显著提高性能，尤其是在大型模型和复杂计算中。该模块通过对查询（Q）、键（K）和值（V）向量进行旋转位置编码（RoPE），增强了对序列顺序的理解，同时使用了Flash Attention来高效计算注意力得分。此外，还处理了padding tokens和缓存（cache）机制，以提高计算效率。此模块特别适用于处理包含padding的输入序列，并且支持低精度训练与推理的优化。

```
import torch
import torch.nn as nn
import torch.nn.functional as F
from transformers.activations import ACT2FN
from transformers.utils import logging

logger = logging.get_logger(__name__)

class DeepseekV2FlashAttention2(DeepseekV2Attention):
    """
    DeepseekV2 flash attention模块。该模块继承自 `DeepseekV2Attention`,权重保持不变。
    唯一需要更改的是前向传播，正确调用Flash Attention的公共API，并处理输入中的填充部分。
    """

    def __init__(self, *args, **kwargs):
        super().__init__(*args, **kwargs)

        # TODO: 一旦Flash Attention for RoCm升级到 2.1，应移除此代码
        # flash_attn<2.1 生成左上角对齐的因果掩码，而这里需要的是右下角对齐，这在
          flash_attn>=2.1 中已成为默认值
        # 此属性用于处理这种差异。参考：https://github.com/Dao-AILab/flash-
          attention/releases/tag/v2.1.0
        # 注意：在flash_attn<2.1 中，使用q_seqlen != k_seqlen（除了q_seqlen ==
          1 的情况）会产生错误的掩码（左上角）
        self._flash_attn_uses_top_left_mask = not is_flash_attn_greater_or_
                                               equal_2_10()

    def forward(
        self,
        hidden_states: torch.Tensor,
        attention_mask: Optional[torch.LongTensor] = None,
        position_ids: Optional[torch.LongTensor] = None,
        past_key_value: Optional[Cache] = None,
        output_attentions: bool = False,
        use_cache: bool = False,
        **kwargs,
    ) -> Tuple[torch.Tensor, Optional[torch.Tensor], Optional[Tuple[torch.
              Tensor]]]:
        # DeepseekV2FlashAttention2 不支持output_attentions
        if "padding_mask" in kwargs:
```

```
        warnings.warn(
            "传递 `padding_mask` 已被弃用，将在 v4.37 中移除。请确保使用
             `attention_mask` 替代。"
        )

        # 覆盖 attention_mask 为 padding_mask
        attention_mask = kwargs.pop("padding_mask")

    output_attentions = False

    bsz, q_len, _ = hidden_states.size()

    if self.q_lora_rank is None:
        q = self.q_proj(hidden_states)
    else:
        q = self.q_b_proj(self.q_a_layernorm(self.q_a_proj(hidden_states)))
    q = q.view(bsz, q_len, self.num_heads, self.q_head_dim).transpose(1, 2)
    q_nope, q_pe = torch.split(
        q, [self.qk_nope_head_dim, self.qk_rope_head_dim], dim=-1
    )

    # Flash attention要求输入的形状为batch_size x seq_length x head_dim x hidden_dim
    # 因此我们只需保持原始形状
    compressed_kv = self.kv_a_proj_with_mqa(hidden_states)
    compressed_kv, k_pe = torch.split(
        compressed_kv, [self.kv_lora_rank, self.qk_rope_head_dim], dim=-1
    )
    k_pe = k_pe.view(bsz, q_len, 1, self.qk_rope_head_dim).transpose(1, 2)
    kv = (
        self.kv_b_proj(self.kv_a_layernorm(compressed_kv))
        .view(bsz, q_len, self.num_heads,
              self.qk_nope_head_dim + self.v_head_dim)
        .transpose(1, 2)
    )

    k_nope, value_states = torch.split(
        kv, [self.qk_nope_head_dim, self.v_head_dim], dim=-1
    )
    kv_seq_len = value_states.shape[-2]

    if past_key_value is not None:
        kv_seq_len += past_key_value.get_usable_length(kv_seq_len, self.
                                                        layer_idx)

    cos, sin = self.rotary_emb(value_states, seq_len=kv_seq_len)
    q_pe, k_pe = apply_rotary_pos_emb(q_pe, k_pe, cos, sin, position_ids)
```

```
query_states = k_pe.new_empty(bsz, self.num_heads, q_len,
                              self.q_head_dim)
query_states[:, :, :, : self.qk_nope_head_dim] = q_nope
query_states[:, :, :, self.qk_nope_head_dim :] = q_pe

key_states = k_pe.new_empty(bsz, self.num_heads, q_len, self.q_head_dim)
key_states[:, :, :, : self.qk_nope_head_dim] = k_nope
key_states[:, :, :, self.qk_nope_head_dim :] = k_pe

if self.q_head_dim != self.v_head_dim:
    value_states = F.pad(value_states,
                         [0, self.q_head_dim - self.v_head_dim])

# TODO: 支持压缩的 kv 缓存（而不是 key_states 和 value_states）在 flash_
  attention 版本中
if past_key_value is not None:
    cache_kwargs = {"sin": sin, "cos": cos}  # 特定于 RoPE 模型
    key_states, value_states = past_key_value.update(
        key_states, value_states, self.layer_idx, cache_kwargs
    )

# TODO: 这些转置操作效率较低，但 Flash Attention 要求布局为 [batch_size,
  sequence_length, num_heads, head_dim]
# 我们需要重构 KV 缓存，以避免这些转置 / 重塑 / 视图操作
query_states = query_states.transpose(1, 2)
key_states = key_states.transpose(1, 2)
value_states = value_states.transpose(1, 2)

dropout_rate = self.attention_dropout if self.training else 0.0

# 在 PEFT 中，通常我们将层归一化（LayerNorm）转换为 float32，以确保训练稳定性
# 因此，输入的隐藏状态会默默地转换为 float32
# 我们需要将输入重新转换为正确的数据类型，以确保一切按预期工作
# 这可能会减慢训练和推理的速度，因此建议不要将 LayerNorm 转换为 fp32
# （DeepseekV2RMSNorm 正确处理了这一点）

input_dtype = query_states.dtype
if input_dtype == torch.float32:
    # 处理模型被量化的情况
    if hasattr(self.config, "_pre_quantization_dtype"):
        target_dtype = self.config._pre_quantization_dtype
    elif torch.is_autocast_enabled():
        target_dtype = torch.get_autocast_gpu_dtype()
    else:
        target_dtype = (
```

```
                    self.q_proj.weight.dtype
                    if self.q_lora_rank is None
                    else self.q_a_proj.weight.dtype
                )

            logger.warning_once(
                f"输入的隐藏状态似乎被默默地转换为 float32，这可能与您将嵌入或层归一化
                  层转换为 float32 有关。"
                f"我们将把输入重新转换为 {target_dtype}。"
            )

            query_states = query_states.to(target_dtype)
            key_states = key_states.to(target_dtype)
            value_states = value_states.to(target_dtype)

        attn_output = self._flash_attention_forward(
            query_states,
            key_states,
            value_states,
            attention_mask,
            q_len,
            dropout=dropout_rate,
            softmax_scale=self.softmax_scale,
        )
        if self.q_head_dim != self.v_head_dim:
            attn_output = attn_output[:, :, :, : self.v_head_dim]

        attn_output = attn_output.reshape(
            bsz, q_len, self.num_heads * self.v_head_dim
        ).contiguous()
        attn_output = self.o_proj(attn_output)

        if not output_attentions:
            attn_weights = None

        return attn_output, attn_weights, past_key_value

    def _flash_attention_forward(
        self,
        query_states,
        key_states,
        value_states,
        attention_mask,
        query_length,
        dropout=0.0,
        softmax_scale=None,
    ):
```

```
"""
调用 Flash Attention 的前向方法。如果输入的隐藏状态中至少包含一个填充（padding）
标记，首先去除输入的填充部分，然后计算注意力分数，最后对最终的注意力分数进行填充。

参数：
    query_states (`torch.Tensor`):
        要传递给 Flash Attention API 的输入查询状态。
    key_states (`torch.Tensor`):
        要传递给 Flash Attention API 的输入键状态。
    value_states (`torch.Tensor`):
        要传递给 Flash Attention API 的输入值状态。
    attention_mask (`torch.Tensor`):
        填充掩码 - 对应于大小为 `(batch_size, seq_len)` 的张量，其中 0 表
                示填充标记的位置，1 表示非填充标记的位置。
    dropout (`int`, 可选):
        注意力 dropout。
    softmax_scale (`float`, 可选):
        在应用 softmax 之前对 QK^T 的缩放比例。默认为 1 / sqrt(head_dim)。
"""
if not self._flash_attn_uses_top_left_mask:
    causal = self.is_causal
else:
    # TODO: 一旦 Flash Attention for RoCm 升级到 2.1，移除 `query_length
      != 1` 检查。
    # 详情请参见 DeepseekV2FlashAttention2 __init__ 中的注释。
    causal = self.is_causal and query_length != 1

# 如果序列中至少包含一个填充标记
if attention_mask is not None:
    batch_size = query_states.shape[0]
    (
        query_states,
        key_states,
        value_states,
        indices_q,
        cu_seq_lens,
        max_seq_lens,
    ) = self._upad_input(
        query_states, key_states, value_states, attention_mask, query_length
    )

    cu_seqlens_q, cu_seqlens_k = cu_seq_lens
    max_seqlen_in_batch_q, max_seqlen_in_batch_k = max_seq_lens

    attn_output_unpad = flash_attn_varlen_func(
        query_states,
```

```
                key_states,
                value_states,
                cu_seqlens_q=cu_seqlens_q,
                cu_seqlens_k=cu_seqlens_k,
                max_seqlen_q=max_seqlen_in_batch_q,
                max_seqlen_k=max_seqlen_in_batch_k,
                dropout_p=dropout,
                softmax_scale=softmax_scale,
                causal=causal,
            )

            attn_output = pad_input(
                attn_output_unpad, indices_q, batch_size, query_length
            )
        else:
            attn_output = flash_attn_func(
                query_states,
                key_states,
                value_states,
                dropout,
                softmax_scale=softmax_scale,
                causal=causal,
            )

        return attn_output

    def _upad_input(
        self, query_layer, key_layer, value_layer, attention_mask, query_length
    ):
        """
        处理输入的填充部分，去除填充部分并返回相关数据。

        参数：
            query_layer (`torch.Tensor`): 查询层。
            key_layer (`torch.Tensor`): 键层。
            value_layer (`torch.Tensor`): 值层。
            attention_mask (`torch.Tensor`): 填充掩码。
            query_length (`int`): 查询序列的长度。
        """
        indices_k, cu_seqlens_k,
        max_seqlen_in_batch_k = _get_unpad_data(attention_mask)
        batch_size, kv_seq_len, num_key_value_heads, head_dim = key_layer.shape

        key_layer = index_first_axis(
            key_layer.reshape(batch_size * kv_seq_len, num_key_value_heads, head_dim),
            indices_k,
```

```
        )
        value_layer = index_first_axis(
            value_layer.reshape(batch_size * kv_seq_len,
                                num_key_value_heads, head_dim),
            indices_k,
        )
        if query_length == kv_seq_len:
            query_layer = index_first_axis(
                query_layer.reshape(batch_size * kv_seq_len,
                                    self.num_heads, head_dim),
                indices_k,
            )
            cu_seqlens_q = cu_seqlens_k
            max_seqlen_in_batch_q = max_seqlen_in_batch_k
            indices_q = indices_k
        elif query_length == 1:
            max_seqlen_in_batch_q = 1
            cu_seqlens_q = torch.arange(
                batch_size + 1, dtype=torch.int32, device=query_layer.device
            )  # 这里有一个内存拷贝，非常不好。
            indices_q = cu_seqlens_q[:-1]
            query_layer = query_layer.squeeze(1)
        else:
            # -q_len: 切片假设左填充。
            attention_mask = attention_mask[:, -query_length:]
            query_layer, indices_q, cu_seqlens_q,
            max_seqlen_in_batch_q = unpad_input(
                query_layer, attention_mask
            )

        return (
            query_layer,
            key_layer,
            value_layer,
            indices_q,
            (cu_seqlens_q, cu_seqlens_k),
            (max_seqlen_in_batch_q, max_seqlen_in_batch_k),
        )
```

（17）下列代码定义了字典ATTENTION_CLASSES，用于映射不同的注意力机制类型到具体的实现类。它支持多种注意力机制，包括普通注意力（eager）、Flash Attention 2（flash_attention_2），以及针对多语言模型（MLA）和多头注意力（MHA）的变体。通过这种方式，用户可以根据需要选择不同的注意力机制，而无须修改代码逻辑。

```
ATTENTION_CLASSES = {
```

```
    "eager": DeepseekV2Attention,        # 普通的 DeepseekV2 注意力机制
    "flash_attention_2": DeepseekV2FlashAttention2,
                                        # 使用 Flash Attention 2 的 DeepseekV2 注意力机制

    "mla_eager": DeepseekV2Attention,   # 多语言模型（MLA）的普通注意力机制
    "mla_flash_attention_2": DeepseekV2FlashAttention2,
                                        # 多语言模型（MLA）的 Flash Attention 2 注意力机制

    "mha_eager": LlamaAttention,         # 多头注意力（MHA）的普通注意力机制
    "mha_flash_attention_2": LlamaFlashAttention2
                                        # 多头注意力（MHA）的 Flash Attention 2 注意力机制
}
```

（18）DeepseekV2DecoderLayer是一个基于PyTorch的Transformer解码器层，用于构建深度学习模型中的解码器部分。它包含以下主要组件：

- 输入归一化：使用DeepseekV2RMSNorm对输入进行归一化。
- 自注意力机制：根据配置选择不同的注意力实现（如普通注意力、Flash Attention 2 等）。
- 前馈网络（MLP）：根据配置选择普通MLP或MoE模块。
- 残差连接：在自注意力和MLP模块后应用残差连接，以增强模型的训练稳定性。
- 输出：返回解码器层的输出，以及可选的注意力权重和缓存的键值对。

```
class DeepseekV2DecoderLayer(nn.Module):
    def __init__(self, config: DeepseekV2Config, layer_idx: int):
        """
        初始化 DeepseekV2DecoderLayer。

        参数：
            config (DeepseekV2Config): 模型配置。
            layer_idx (int): 当前层的索引。
        """
        super().__init__()
        self.hidden_size = config.hidden_size   # 隐藏层维度

        # 根据配置选择注意力机制的实现
        if config.use_mla:
            attn_implementation = "mla_" + config._attn_implementation
        else:
            attn_implementation = "mha_" + config._attn_implementation

        self.self_attn = ATTENTION_CLASSES[attn_implementation](
            config=config, layer_idx=layer_idx
        )  # 自注意力模块

        # 根据配置选择 MLP 或 MoE 模块
```

```
        self.mlp = (
            DeepseekV2MoE(config)
            if (
                config.n_routed_experts is not None
                and layer_idx >= config.first_k_dense_replace
                and layer_idx % config.moe_layer_freq == 0
            )
            else DeepseekV2MLP(config)
        )
        self.input_layernorm = DeepseekV2RMSNorm(
            config.hidden_size, eps=config.rms_norm_eps
        )  # 输入归一化
        self.post_attention_layernorm = DeepseekV2RMSNorm(
            config.hidden_size, eps=config.rms_norm_eps
        )  # 注意力后归一化

    def forward(
        self,
        hidden_states: torch.Tensor,
        attention_mask: Optional[torch.Tensor] = None,
        position_ids: Optional[torch.LongTensor] = None,
        past_key_value: Optional[Tuple[torch.Tensor]] = None,
        output_attentions: Optional[bool] = False,
        use_cache: Optional[bool] = False,
        **kwargs,
    ) -> Tuple[torch.FloatTensor, Optional[Tuple[torch.FloatTensor, torch.
              FloatTensor]]]:
        """
        前向传播。

        参数:
            hidden_states (`torch.FloatTensor`): 输入张量，形状为 `(batch, seq_
                          len, embed_dim)`。
            attention_mask (`torch.FloatTensor`, 可选):
                注意力掩码，形状为 `(batch_size, sequence_length)`
                (如果使用 Flash Attention)或
                `(batch_size, 1, query_sequence_length, key_sequence_
                length)`(如果使用默认注意力)。
            output_attentions (`bool`, 可选):
                是否返回所有注意力层的注意力张量。详情请参见返回张量中的 `attentions`。
            use_cache (`bool`, 可选):
                如果设置为 `True`，将返回 `past_key_values` 键值状态，
                可用于加速解码(参见 `past_key_values`)。
            past_key_value (`Tuple(torch.FloatTensor)`, 可选): 缓存的键和值投影状态。
        """
        if "padding_mask" in kwargs:
```

```
        warnings.warn(
            "传递 `padding_mask` 已被弃用，将在 v4.37 中移除。
             请确保使用 `attention_mask` 替代。"
        )

    residual = hidden_states  # 保存残差连接

    # 输入归一化
    hidden_states = self.input_layernorm(hidden_states)

    # 自注意力模块
    hidden_states, self_attn_weights, present_key_value = self.self_attn(
        hidden_states=hidden_states,
        attention_mask=attention_mask,
        position_ids=position_ids,
        past_key_value=past_key_value,
        output_attentions=output_attentions,
        use_cache=use_cache,
        **kwargs,
    )
    hidden_states = residual + hidden_states  # 残差连接

    # 前馈网络（MLP 或 MoE）
    residual = hidden_states
    hidden_states = self.post_attention_layernorm(hidden_states)
    hidden_states = self.mlp(hidden_states)
    hidden_states = residual + hidden_states  # 残差连接

    outputs = (hidden_states,)

    if output_attentions:
        outputs += (self_attn_weights,)

    if use_cache:
        outputs += (present_key_value,
```

（19）DeepseekV2PreTrainedModel是一个基于Hugging Face PreTrainedModel的基类，用于定义DeepseekV2模型的预训练模型结构。其提供了以下功能。

- 配置类：指定模型的配置类为DeepseekV2Config。
- 模型前缀：定义模型的基本前缀为"model"，用于保存和加载模型时的模块路径。
- 支持梯度检查点（gradient checkpointing）：支持梯度检查点，以减少内存占用。
- 不可分割模块：指定DeepseekV2DecoderLayer为不可分割模块，这在模型并行化时有用。
- 设备放置跳过键：在放置模型到设备时跳过past_key_values，这在使用缓存时有用。
- 支持Flash Attention 2和缓存类：明确支持Flash Attention 2和缓存机制。

● 权重初始化：在 _init_weights方法中，对线性层和嵌入层的权重进行初始化，使用正态分布（均值为 0，标准差为config.initializer_range），并对偏置项进行零初始化。

```
class DeepseekV2PreTrainedModel(PreTrainedModel):
    """
    DeepseekV2 预训练模型基类。
    """
    config_class = DeepseekV2Config                        # 指定模型的配置类
    base_model_prefix = "model"                            # 模型的基本前缀
    supports_gradient_checkpointing = True                 # 支持梯度检查点
    _no_split_modules = ["DeepseekV2DecoderLayer"]         # 不可分割模块
    _skip_keys_device_placement = "past_key_values"        # 在放置设备时跳过的键
    _supports_flash_attn_2 = True                          # 支持Flash Attention 2
    _supports_cache_class = True                           # 支持缓存类

    def _init_weights(self, module):
        """
        初始化权重。

        参数：
            module: 要初始化的模块。
        """
        std = self.config.initializer_range   # 获取初始化范围
        if isinstance(module, nn.Linear):
            # 如果是线性层，使用正态分布初始化权重
            module.weight.data.normal_(mean=0.0, std=std)
            if module.bias is not None:
                # 如果有偏置项，进行零初始化
                module.bias.data.zero_()
        elif isinstance(module, nn.Embedding):
            # 如果是嵌入层，使用正态分布初始化权重
            module.weight.data.normal_(mean=0.0, std=std)
            if module.padding_idx is not None:
                # 如果有填充索引，将对应权重置为零
                module.weight.data[module.padding_idx].zero_()
```

（20）下列代码定义了DeepseekV2_INPUTS_DOCSTRING，这是一个多行字符串文档，详细描述了DeepSeek V2 模型的输入参数，包括input_ids, attention_mask, position_ids, past_key_values, inputs_embeds等。它用于为模型的输入提供清晰的说明，帮助开发者理解各参数的作用及使用方法。

```
DeepseekV2_INPUTS_DOCSTRING = r"""
    参数说明：
        input_ids (`torch.LongTensor`, 形状为 `(batch_size, sequence_length)`):
            输入序列中词汇表中各个标记的索引。如果提供了填充标记，默认会忽略它们。
```

```
    可以使用 [`AutoTokenizer`] 获取索引。详见 [`PreTrainedTokenizer.
    encode`] 和 [`PreTrainedTokenizer.__call__`]。

    [什么是输入 ID？](../glossary#input-ids)
attention_mask (`torch.Tensor`, 形状为 `(batch_size, sequence_
               length)`, *可选*):
    掩码，用于避免对填充标记执行注意力计算。掩码值范围 `[0, 1]`:

    - 1 表示标记 **未被掩码**,
    - 0 表示标记 **已被掩码**。

    [什么是注意力掩码？](../glossary#attention-mask)

    可以使用 [`AutoTokenizer`] 获取索引。详见 [`PreTrainedTokenizer.
    encode`] 和 [`PreTrainedTokenizer.__call__`]。

    如果使用了 `past_key_values`，可以选择仅输入最后的 `input_ids`（见
    `past_key_values`）。

    如果希望改变填充行为，可以阅读 [`modeling_opt._prepare_decoder_
    attention_mask`] 并按需修改。
    参考 [论文中的图1](https://arxiv.org/abs/1910.13461) 以了解默认策略。

    - 1 表示头部 **未被掩码**,
    - 0 表示头部 **已被掩码**。
position_ids (`torch.LongTensor`, 形状为 `(batch_size, sequence_
             length)`, *可选*):
    每个输入序列标记在位置嵌入中的索引。选择范围 `[0, config.n_positions - 1]`。

    [什么是位置 ID？](../glossary#position-ids)
past_key_values (`Cache` 或 `tuple(tuple(torch.FloatTensor))`,*可选*):
    预计算的隐藏状态（包括自注意力块和交叉注意力块中的键和值）用于加速序列解码。这通
    常由模型在前一阶段返回，当 `use_cache=True` 或 `config.use_cache=True` 时。

    允许两种格式：
    - [`~cache_utils.Cache`] 实例;
    - 长度为 `config.n_layers` 的 `tuple(torch.FloatTensor)` 元组，每个
      元组包含两个张量，形状为
      `(batch_size, num_heads, sequence_length, embed_size_per_
      head)`。这也被称为传统缓存格式。

    模型会输出与输入一致的缓存格式。如果没有传入 `past_key_values`，将返回传统
    缓存格式。

    如果使用 `past_key_values`，用户可以选择仅输入最后的 `input_ids`（没有
    提供其先前键值状态的部分），
```

```
        其形状为 `(batch_size, 1)`，而非所有 `input_ids` 的形状 `(batch_
        size, sequence_length)`。
    inputs_embeds (`torch.FloatTensor`, 形状为 `(batch_size, sequence_
                length, hidden_size)`, *可选*):
        可选项：可以直接传入嵌入表示，而不是通过 `input_ids`。如果希望对 `input_
        ids` 索引如何转换为关联向量进行更灵活的控制，可以使用该选项
```

（21）定义模型类DeepseekV2Model，是一个基于Transformer结构的解码器模型。DeepseekV2Model继承自DeepseekV2PreTrainedModel，包括嵌入层、多个解码器层、RMS归一化层。其主要功能是接收输入序列的token、注意力掩码和位置编码，并通过多层解码器进行处理，输出模型的隐藏状态及相关信息。该模型支持缓存机制来优化解码过程，同时能够选择输出注意力权重和隐藏状态。

```
@add_start_docstrings(
    "裸 DeepseekV2 模型，不带任何特定的头部输出原始隐藏状态。",
    DeepseekV2_START_DOCSTRING,
)
class DeepseekV2Model(DeepseekV2PreTrainedModel):
    """
    Transformer 解码器，由 `config.num_hidden_layers` 层组成。
    每层都是 [`DeepseekV2DecoderLayer`]。

    参数：
        config: DeepseekV2Config
    """

    def __init__(self, config: DeepseekV2Config):
        super().__init__(config)
        self.padding_idx = config.pad_token_id
        self.vocab_size = config.vocab_size

        # 嵌入层，用于将词 ID 转换为向量表示
        self.embed_tokens = nn.Embedding(
            config.vocab_size, config.hidden_size, self.padding_idx
        )
        # 定义多个解码器层
        self.layers = nn.ModuleList(
            [
                DeepseekV2DecoderLayer(config, layer_idx)
                for layer_idx in range(config.num_hidden_layers)
            ]
        )
        self._use_flash_attention_2 = config._attn_implementation ==
                                        "flash_attention_2"
```

```
        # RMS 归一化层
        self.norm = DeepseekV2RMSNorm(config.hidden_size,
                                      eps=config.rms_norm_eps)

        self.gradient_checkpointing = False
        # 初始化权重并应用最终处理
        self.post_init()

    def get_input_embeddings(self):
        return self.embed_tokens

    def set_input_embeddings(self, value):
        self.embed_tokens = value

    @add_start_docstrings_to_model_forward(DeepseekV2_INPUTS_DOCSTRING)
    def forward(
        self,
        input_ids: torch.LongTensor = None,
        attention_mask: Optional[torch.Tensor] = None,
        position_ids: Optional[torch.LongTensor] = None,
        past_key_values: Optional[List[torch.FloatTensor]] = None,
        inputs_embeds: Optional[torch.FloatTensor] = None,
        use_cache: Optional[bool] = None,
        output_attentions: Optional[bool] = None,
        output_hidden_states: Optional[bool] = None,
        return_dict: Optional[bool] = None,
        cache_position: Optional[torch.LongTensor] = None
    ) -> Union[Tuple, BaseModelOutputWithPast]:
        output_attentions = (
            output_attentions
            if output_attentions is not None
            else self.config.output_attentions
        )
        output_hidden_states = (
            output_hidden_states
            if output_hidden_states is not None
            else self.config.output_hidden_states
        )
        use_cache = use_cache if use_cache is not None else self.config.use_cache

        return_dict = (
            return_dict if return_dict is not None else self.config.use_return_dict
        )

        # 检索 input_ids 和 inputs_embeds
        if input_ids is not None and inputs_embeds is not None:
```

```
        raise ValueError(
            "input_ids 和 inputs_embeds 不能同时指定。"
        )
    elif input_ids is not None:
        batch_size, seq_length = input_ids.shape[:2]
    elif inputs_embeds is not None:
        batch_size, seq_length = inputs_embeds.shape[:2]
    else:
        raise ValueError("必须指定 input_ids 或 inputs_embeds 之一。")

    if self.gradient_checkpointing and self.training:
        if use_cache:
            logger.warning_once(
                "`use_cache=True` 与梯度检查点功能不兼容，将 `use_cache` 设为 False。"
            )
            use_cache = False

    past_key_values_length = 0
    if use_cache:
        # 判断是否使用旧版缓存
        use_legacy_cache = not isinstance(past_key_values, Cache)
        if use_legacy_cache:
            past_key_values = DynamicCache.from_legacy_cache(past_key_values)
        past_key_values_length = past_key_values.get_usable_length(seq_length)

    # 如果未指定位置 ID，则自动生成
    if position_ids is None:
        device = input_ids.device if input_ids is not None else inputs_
                embeds.device
        position_ids = torch.arange(
            past_key_values_length,
            seq_length + past_key_values_length,
            dtype=torch.long,
            device=device,
        )
        position_ids = position_ids.unsqueeze(0)

    # 如果未提供嵌入层，则通过词嵌入层获取
    if inputs_embeds is None:
        inputs_embeds = self.embed_tokens(input_ids)

    # 根据是否使用 flash attention 构造不同的注意力掩码
    if self._use_flash_attention_2:
        attention_mask = (
            attention_mask
            if (attention_mask is not None and 0 in attention_mask)
```

```
            else None
        )
    else:
        attention_mask = _prepare_4d_causal_attention_mask(
            attention_mask,
            (batch_size, seq_length),
            inputs_embeds,
            past_key_values_length,
        )

    # 嵌入位置编码
    hidden_states = inputs_embeds

    # 处理解码器层
    all_hidden_states = () if output_hidden_states else None
    all_self_attns = () if output_attentions else None
    next_decoder_cache = None

    for decoder_layer in self.layers:
        if output_hidden_states:
            all_hidden_states += (hidden_states,)

        # 梯度检查点功能
        if self.gradient_checkpointing and self.training:
            layer_outputs = self._gradient_checkpointing_func(
                decoder_layer.__call__,
                hidden_states,
                attention_mask,
                position_ids,
                past_key_values,
                output_attentions,
                use_cache,
            )
        else:
            layer_outputs = decoder_layer(
                hidden_states,
                attention_mask=attention_mask,
                position_ids=position_ids,
                past_key_value=past_key_values,
                output_attentions=output_attentions,
                use_cache=use_cache,
            )

        hidden_states = layer_outputs[0]

        if use_cache:
```

```
                next_decoder_cache = layer_outputs[2 if output_attentions else 1]

            if output_attentions:
                all_self_attns += (layer_outputs[1],)

        # 对隐藏状态进行归一化
        hidden_states = self.norm(hidden_states)

        # 将最后一层的隐藏状态添加到输出中
        if output_hidden_states:
            all_hidden_states += (hidden_states,)

        next_cache = None
        if use_cache:
            next_cache = (
                next_decoder_cache.to_legacy_cache()
                if use_legacy_cache
                else next_decoder_cache
            )
        if not return_dict:
            return tuple(
                v
                for v in [hidden_states, next_cache, all_hidden_states,
                          all_self_attns]
                if v is not None
            )
        return BaseModelOutputWithPast(
            last_hidden_state=hidden_states,
            past_key_values=next_cache,
            hidden_states=all_hidden_states,
            attentions=all_self_attns,
        )
```

（22）下列代码定义了一个名为DeepseekV2ForCausalLM的模型类，继承自DeepseekV2PreTrainedModel。DeepseekV2ForCausalLM用于自回归语言建模任务（Causal Language Modeling），支持条件文本生成、语言建模任务及模型微调。主要功能包括：

- 定义了模型的基本组件，包括编码器、语言模型头等模块。
- 实现了模型的前向传播逻辑forward，计算输出logits并支持损失计算。
- 提供了生成相关的方法prepare_inputs_for_generation和缓存重排序 _reorder_cache。
- 支持在生成任务中通过缓存past key values提升计算效率。
- 注重位置编码与缓存优化，适配不同的缓存机制。

```
class DeepseekV2ForCausalLM(DeepseekV2PreTrainedModel):
    _tied_weights_keys = ["lm_head.weight"]
```

```
    def __init__(self, config):
        super().__init__(config)
        self.model = DeepseekV2Model(config)
        self.vocab_size = config.vocab_size
        self.lm_head = nn.Linear(config.hidden_size, config.vocab_size,
                                 bias=False)

        # 初始化权重并进行最终处理
        self.post_init()

    def get_input_embeddings(self):
        return self.model.embed_tokens

    def set_input_embeddings(self, value):
        self.model.embed_tokens = value

    def get_output_embeddings(self):
        return self.lm_head

    def set_output_embeddings(self, new_embeddings):
        self.lm_head = new_embeddings

    def set_decoder(self, decoder):
        self.model = decoder

    def get_decoder(self):
        return self.model

    @add_start_docstrings_to_model_forward(DeepseekV2_INPUTS_DOCSTRING)
    @replace_return_docstrings(
        output_type=CausalLMOutputWithPast, config_class=_CONFIG_FOR_DOC
    )
    def forward(
            self,
            input_ids: torch.LongTensor = None,
            attention_mask: Optional[torch.Tensor] = None,
            position_ids: Optional[torch.LongTensor] = None,
            past_key_values: Optional[List[torch.FloatTensor]] = None,
            inputs_embeds: Optional[torch.FloatTensor] = None,
            labels: Optional[torch.LongTensor] = None,
            use_cache: Optional[bool] = None,
            output_attentions: Optional[bool] = None,
            output_hidden_states: Optional[bool] = None,
            return_dict: Optional[bool] = None,
            cache_position: Optional[torch.LongTensor] = None
```

```
) -> Union[Tuple, CausalLMOutputWithPast]:
    """
    参数：
        labels (`torch.LongTensor`, 形状为 `(batch_size, sequence_
                length)`, 可选)：
            用于计算掩码语言建模损失的标签。索引应在 `[0, transformers.config.
            vocab_size]` 或 `-100` 之间。设置为 `-100` 的标记将被忽略（掩码），
            仅计算在 `[0, config.vocab_size]` 内的标记损失。
    """
    output_attentions = (
        output_attentions
        if output_attentions is not None
        else self.config.output_attentions
    )
    output_hidden_states = (
        output_hidden_states
        if output_hidden_states is not None
        else self.config.output_hidden_states
    )
    return_dict = (
        return_dict if return_dict is not None else self.config.use_
            return_dict
    )

    # 解码器输出包括 (dec_features, layer_state, dec_hidden, dec_attn)
    outputs = self.model(
        input_ids=input_ids,
        attention_mask=attention_mask,
        position_ids=position_ids,
        past_key_values=past_key_values,
        inputs_embeds=inputs_embeds,
        use_cache=use_cache,
        output_attentions=output_attentions,
        output_hidden_states=output_hidden_states,
        return_dict=return_dict,
        cache_position=cache_position
    )

    hidden_states = outputs[0]
    logits = self.lm_head(hidden_states)
    logits = logits.float()

    loss = None
    if labels is not None:
        # 将 logits 和标签进行位移，使每个位置预测下一个 token
```

```
            shift_logits = logits[..., :-1, :].contiguous()
            shift_labels = labels[..., 1:].contiguous()
            # 将 token 展平
            loss_fct = CrossEntropyLoss()
            shift_logits = shift_logits.view(-1, self.config.vocab_size)
            shift_labels = shift_labels.view(-1)
            # 启用模型并行
            shift_labels = shift_labels.to(shift_logits.device)
            loss = loss_fct(shift_logits, shift_labels)

        if not return_dict:
            output = (logits,) + outputs[1:]
            return (loss,) + output if loss is not None else output

        return CausalLMOutputWithPast(
            loss=loss,
            logits=logits,
            past_key_values=outputs.past_key_values,
            hidden_states=outputs.hidden_states,
            attentions=outputs.attentions,
        )

    def prepare_inputs_for_generation(
            self,
            input_ids,
            past_key_values=None,
            attention_mask=None,
            inputs_embeds=None,
            **kwargs,
    ):
        past_length = 0
        if past_key_values is not None:
            if isinstance(past_key_values, Cache):
                cache_length = past_key_values.get_seq_length()
                past_length = past_key_values.seen_tokens
                max_cache_length = past_key_values.get_max_length()
            else:
                cache_length = past_length = past_key_values[0][0].shape[2]
                max_cache_length = None

            # 仅保留未处理的 token:
            if attention_mask is not None and attention_mask.shape[1] >
                input_ids.shape[1]:
                input_ids = input_ids[:, -(attention_mask.shape[1] - past_
                                       length):]
            elif past_length < input_ids.shape[1]:
```

```
        input_ids = input_ids[:, past_length:]

    # 如果即将超出最大缓存长度，则裁剪输入的attention mask
    if (
            max_cache_length is not None
            and attention_mask is not None
            and cache_length + input_ids.shape[1] > max_cache_length
    ):
        attention_mask = attention_mask[:, -max_cache_length:]

position_ids = kwargs.get("position_ids", None)
if attention_mask is not None and position_ids is None:
    # 动态创建position_ids
    position_ids = attention_mask.long().cumsum(-1) - 1
    position_ids.masked_fill_(attention_mask == 0, 1)
    if past_key_values:
        position_ids = position_ids[:, -input_ids.shape[1]:]

if self.generation_config.cache_implementation == "static":
    # 静态缓存生成
    cache_position = kwargs.get("cache_position", None)
    if cache_position is None:
        past_length = 0
    else:
        past_length = cache_position[-1] + 1
    input_ids = input_ids[:, past_length:]
    position_ids = position_ids[:, past_length:]

cache_position = torch.arange(past_length,
                    past_length + position_ids.shape[-1],
                    device=position_ids.device)

if inputs_embeds is not None and past_key_values is None:
    model_inputs = {"inputs_embeds": inputs_embeds}
else:
    model_inputs = {"input_ids": input_ids.contiguous()}

model_inputs.update(
    {
        "position_ids": position_ids.contiguous(),
        "cache_position": cache_position,
        "past_key_values": past_key_values,
        "use_cache": kwargs.get("use_cache"),
        "attention_mask": attention_mask,
    }
)
```

```
        return model_inputs

    @staticmethod
    def _reorder_cache(past_key_values, beam_idx):
        reordered_past = ()
        for layer_past in past_key_values:
            reordered_past += (
                tuple(
                    past_state.index_select(0, beam_idx.to(past_state.device))
                    for past_state in layer_past
                ),
            )
        return reordered_past
```

（23）下列代码定义了DeepseekV2ForSequenceClassification模型，这是一个基于DeepseekV2 模型结构的序列分类模型。该模型使用序列中的最后一个token来进行分类预测（类似GPT-2 等因果模型的设计）。模型需要准确定位最后一个token的位置。如果配置中定义了pad_token_id，模型就会跳过填充token；如果没有定义，则默认使用序列中的最后一个值。当传入的是inputs_embeds而非input_ids时，模型也只能直接使用每行最后一个token。

```
@add_start_docstrings(
    """
    DeepseekV2 模型的一个变体，在其顶部加入了一个序列分类头（线性层）。

    [`DeepseekV2ForSequenceClassification`] 使用序列中的最后一个 token 来进行分类预
    测，与其他因果模型（如 GPT-2）类似。

    由于基于最后一个 token 进行分类，因此模型需要知道该 token 的位置。
    - 如果配置中定义了 `pad_token_id`，模型会在每一行中找到最后一个不是填充的 token。
    - 如果没有定义 `pad_token_id`，则模型直接取每一行的最后一个值作为分类输入。

    在传入 `inputs_embeds` 而不是 `input_ids` 时，由于模型无法猜测填充 token 的位置，
    也只能采用上述相同的方式（直接取每一行的最后一个值）。
    """,
    DeepseekV2_START_DOCSTRING,
)
```

（24）下列代码定义了模型类DeepseekV2ForSequenceClassification，其继承自DeepseekV2PreTrainedModel。该模型主要用于序列分类任务，核心功能如下所示。

● 模型初始化：加载DeepseekV2Model并添加线性分类头。

● 前向传播：通过调用forward方法，支持输入input_ids、注意力掩码、位置编码等参数，返回分类logits和其他模型输出。

● 分类处理：在最后一个有效token上进行分类计算，支持单标签分类、多标签分类和回归任务。

- 损失计算：根据标签和配置自动选择损失函数（MSE、交叉熵或带有logits的二元交叉熵）。
- 兼容性处理：考虑填充token的处理逻辑，确保对批量输入有效。

```
class DeepseekV2ForSequenceClassification(DeepseekV2PreTrainedModel):
    def __init__(self, config):
        super().__init__(config)
        self.num_labels = config.num_labels
        self.model = DeepseekV2Model(config)
        self.score = nn.Linear(config.hidden_size, self.num_labels, bias=False)

        # 初始化权重并执行最终处理
        self.post_init()

    def get_input_embeddings(self):
        return self.model.embed_tokens

    def set_input_embeddings(self, value):
        self.model.embed_tokens = value

    @add_start_docstrings_to_model_forward(DeepseekV2_INPUTS_DOCSTRING)
    def forward(
            self,
            input_ids: torch.LongTensor = None,
            attention_mask: Optional[torch.Tensor] = None,
            position_ids: Optional[torch.LongTensor] = None,
            past_key_values: Optional[List[torch.FloatTensor]] = None,
            inputs_embeds: Optional[torch.FloatTensor] = None,
            labels: Optional[torch.LongTensor] = None,
            use_cache: Optional[bool] = None,
            output_attentions: Optional[bool] = None,
            output_hidden_states: Optional[bool] = None,
            return_dict: Optional[bool] = None,
    ) -> Union[Tuple, SequenceClassifierOutputWithPast]:
        r"""
        labels (`torch.LongTensor` of shape `(batch_size,)`, *可选*):
            用于计算序列分类/回归损失的标签。索引值应在 `[0, config.num_labels - 1]` 之间。
            如果 `config.num_labels == 1`，计算回归损失（均方误差）；如果 `config.
            num_labels > 1`，则计算分类损失（交叉熵）。
        """
        return_dict = (
            return_dict if return_dict is not None else self.config.use_return_dict
        )

        # 获取 Transformer 模型的输出
        transformer_outputs = self.model(
```

```
        input_ids,
        attention_mask=attention_mask,
        position_ids=position_ids,
        past_key_values=past_key_values,
        inputs_embeds=inputs_embeds,
        use_cache=use_cache,
        output_attentions=output_attentions,
        output_hidden_states=output_hidden_states,
        return_dict=return_dict,
    )
    hidden_states = transformer_outputs[0]
    logits = self.score(hidden_states)

    # 计算批量大小
    if input_ids is not None:
        batch_size = input_ids.shape[0]
    else:
        batch_size = inputs_embeds.shape[0]

    # 检查是否定义填充 token，避免多样本输入处理错误
    if self.config.pad_token_id is None and batch_size != 1:
        raise ValueError(
            "如果未定义填充 token，则无法处理 batch size 大于 1 的情况。"
        )
    if self.config.pad_token_id is None:
        sequence_lengths = -1
    else:
        # 找到最后一个非填充 token 的位置
        if input_ids is not None:
            sequence_lengths = (
                    torch.eq(input_ids, self.config.pad_token_id).int().
                            argmax(-1) - 1
            ).to(logits.device)
        else:
            sequence_lengths = -1

    # 获取最后一个有效 token 的分类结果
    pooled_logits = logits[
        torch.arange(batch_size, device=logits.device), sequence_lengths
    ]

    # 损失计算
    loss = None
    if labels is not None:
        labels = labels.to(logits.device)
        # 根据标签类型选择问题类型
```

```
        if self.config.problem_type is None:
            if self.num_labels == 1:
                self.config.problem_type = "regression"
            elif self.num_labels > 1 and (
                    labels.dtype == torch.long or labels.dtype == torch.int
            ):
                self.config.problem_type = "single_label_classification"
            else:
                self.config.problem_type = "multi_label_classification"

        # 回归损失计算
        if self.config.problem_type == "regression":
            loss_fct = MSELoss()
            if self.num_labels == 1:
                loss = loss_fct(pooled_logits.squeeze(), labels.squeeze())
            else:
                loss = loss_fct(pooled_logits, labels)
        # 单标签分类损失计算
        elif self.config.problem_type == "single_label_classification":
            loss_fct = CrossEntropyLoss()
            loss = loss_fct(
                pooled_logits.view(-1, self.num_labels), labels.view(-1)
            )
        # 多标签分类损失计算
        elif self.config.problem_type == "multi_label_classification":
            loss_fct = BCEWithLogitsLoss()
            loss = loss_fct(pooled_logits, labels)

    # 返回模型输出
    if not return_dict:
        output = (pooled_logits,) + transformer_outputs[1:]
        return ((loss,) + output) if loss is not None else output

    return SequenceClassifierOutputWithPast(
        loss=loss,
        logits=pooled_logits,
        past_key_values=transformer_outputs.past_key_values,
        hidden_states=transformer_outputs.hidden_states,
        attentions=transformer_outputs.attentions,
    )
```

注意

文件modeling_deepseek.py**与**modeling_deepseek_vl_v2.py**的区别。**

（1）主要功能范围不同。

- modeling_deepseek.py主要侧重于DeepSeek模型的基本架构设计，包括解码层和复杂的注意力机制。
- modeling_deepseek_vl_v2.py则关注多模态模型的实现，特别是结合视觉和语言信息的模型结构。

（2）多模态支持。

- modeling_deepseek.py并未涉及图像处理模块或多模态信息处理。
- modeling_deepseek_vl_v2.py定义了视觉编码器和投影模块，用于处理图像特征并结合语言模型。

（3）注意力机制的实现。

- modeling_deepseek.py提供了多种注意力机制的实现，如Rotary Position Embedding和Flash Attention。
- modeling_deepseek_vl_v2.py中注意力机制更聚焦在多模态输入的处理上。

（4）模型结构。

- modeling_deepseek.py定义了基本的DeepSeek解码层和MLP模块。
- modeling_deepseek_vl_v2.py包括视觉、语言以及投影模块的集成，适用于多模态任务。

总之，这两个文件分别专注于单模态与多模态DeepSeek模型的实现，并在架构、注意力机制、输入处理上有显著差异。

6.5.5 Vision Transformer（ViT）的视觉模型

在本项目中，文件siglip_vit.py实现了一种基于ViT的视觉模型结构及其相关组件，这些组件包括Attention模块、LayerScale模块、Transformer Block模块等。同时还定义了权重初始化策略和模型创建方法。代码中详细介绍了SigLIP Vision模型的参数配置，支持根据系统配置创建Vision Transformer模型及进行相关层的加载和计算。模型适用于视觉相关任务，例如图像分类和特征提取。

文件siglip_vit.py的核心要点如下所示。

- Attention机制的实现：使用标准的Scaled Dot-Product Attention，具备查询、键和值的计算与归一化操作。如果支持Flash Attention加速，则优先使用以提升计算效率。
- Transformer Block设计：采用标准的Transformer块设计，包含注意力机制、MLP层及其归一化操作。支持LayerScale进行可学习权重缩放。
- 模型权重初始化：定义了init_weights_vit_timm等权重初始化方法，通过截断正态分布进行权重初始化，确保模型收敛性能。
- Positional Embedding实现：使用可重采样的绝对位置编码resample_abs_pos_embed，支持图像动态尺寸的适应性。
- 模型创建与加载功能：定义了create_siglip_vit方法，可以根据预设的模型配置创建ViT模型，并支持从模型检查点加载权重。
- 模型兼容性：采用与huggingface和timm库兼容的模块设计，支持分布式训练、梯度检查点和头部自由定制等功能。
- SigLIP模型配置与选择：提供了不同规模的模型配置，如siglip_so400m_patch14_384和

siglip_large_patch16_384，满足不同任务需求。

这个文件展示了如何基于Transformer构建一个高性能的视觉模型，同时通过多种优化策略（如LayerScale、权重初始化、Flash Attention）来提升模型的计算效率与性能。

文件siglip_vit.py的具体实现流程如下。

（1）下列代码用于检测flash_attn库版本是否可用，并在可用时导入flash_attn_qkvpacked_func函数。 flash_attn是一种优化的注意力计算库，可以显著加速Transformer模型中的注意力计算过程，通过更高效的内存管理和计算路径提升性能。如果Flash Attention 2 可用，该功能将被引入以替代传统的注意力计算，从而优化模型的计算速度与内存使用。

```
if is_flash_attn_2_available():
    from flash_attn import flash_attn_qkvpacked_func
```

（2）下列代码实现了一个无梯度截断正态分布初始化函数 _no_grad_trunc_normal_()，用于在给定范围 [a, b] 内初始化张量的权重值。其实现基于截断正态分布，通过截断后的均值、标准差及正态分布的累积分布函数（CDF）进行反变换采样。该方法有效避免了传统正态分布在极端值区间采样时的数值不稳定问题，使模型参数更加符合截断范围要求。

```
def _no_grad_trunc_normal_(tensor, mean, std, a, b):
    # 从 PyTorch 官方代码中复制而来，直到正式发布版本中包含此功能
    # 该方法基于 https://people.sc.fsu.edu/~jburkardt/presentations/truncated_
      normal.pdf

    def norm_cdf(x):
        # 计算标准正态分布的累积分布函数 (CDF)
        return (1.0 + math.erf(x / math.sqrt(2.0))) / 2.0

    # 检查给定的均值是否离 [a, b] 超过了 2 倍的标准差范围
    if (mean < a - 2 * std) or (mean > b + 2 * std):
        warnings.warn(
            "均值超过了 [a, b] 的两倍标准差范围，可能导致初始化分布不正确。",
            stacklevel=2,
        )

    # 关闭梯度计算以避免不必要的图追踪
    with torch.no_grad():
        # 使用截断后的均匀分布生成值，并通过正态分布的逆 CDF 进行变换
        # 计算上下限对应的 cdf 值
        l = norm_cdf((a - mean) / std)  # noqa: E741
        u = norm_cdf((b - mean) / std)

        # 用 [l, u] 范围的均匀分布填充张量
        # 然后将其平移到 [2l-1, 2u-1] 区间
        tensor.uniform_(2 * l - 1, 2 * u - 1)
```

```
    # 通过正态分布的逆 CDF 变换生成截断标准正态分布值
    tensor.erfinv_()

    # 变换为指定的均值和标准差
    tensor.mul_(std * math.sqrt(2.0))
    tensor.add_(mean)

    # 截断数值以确保在指定范围内
    tensor.clamp_(min=a, max=b)

    return tensor
```

（3）函数trunc_normal_()用于将张量中的元素通过截断正态分布进行初始化。相比标准的torch.nn.init.trunc_normal_，该实现额外处理了bfloat16 数据类型的兼容性问题。具体做法是将输入张量临时转换为float32 类型，在其上应用截断正态分布初始化，再将结果还原为原始数据类型。

```
def trunc_normal_(tensor, mean=0.0, std=1.0, a=-2.0, b=2.0):
    # type: (torch.Tensor, float, float, float, float) -> torch.Tensor
    r"""
    timm.models.layers.weight_init.trunc_normal_ 的原始版本尚不支持 bfloat16 数
    据类型。
    在该实现中，我们首先将张量转换为 float32 类型，应用截断正态分布初始化，然后再将结果转
    换回原始数据类型。

    使用截断正态分布填充输入张量。值实际上来自正态分布 :math:`\mathcal{N}(\text
    {mean}, \text{std}^2)`，但任何位于区间 :math:`[a, b]` 之外的值都会重新生成，直
    到符合范围要求。
    当满足 :math:`a \leq \text{mean} \leq b` 条件时，该方法效果最佳。

    Args:
        tensor (torch.Tensor): n 维张量。
        mean (float): 正态分布的均值。
        std (float): 正态分布的标准差。
        a (float): 最小截断值。
        b (float): 最大截断值。

    示例:
        >>> w = torch.empty(3, 5)
        >>> nn.init.trunc_normal_(w)
    """

    with torch.no_grad():
        # 保存原始数据类型
        dtype = tensor.dtype
```

```
    # 将张量临时转换为 float32，以确保截断正态初始化的准确性
    tensor_fp32 = tensor.float()

    # 调用无梯度截断正态初始化
    tensor_fp32 = _no_grad_trunc_normal_(tensor_fp32, mean, std, a, b)

    # 将结果转换回原始数据类型
    tensor_dtype = tensor_fp32.to(dtype=dtype)

    # 将转换后的张量内容复制回原张量
    tensor.copy_(tensor_dtype)
```

（4）函数 init_weights() 用于为模型中位置嵌入 (pos_embed) 和潜在变量 (latent) 张量进行权重初始化操作。它使用截断正态分布初始化权重，并依据维度大小调整标准差，以确保权重初始化的数值在合理范围内，从而避免训练过程中的梯度爆炸或消失。

```
def init_weights(self):
    if self.pos_embed is not None:
        trunc_normal_(self.pos_embed, std=self.pos_embed.shape[1] ** -0.5)
    trunc_normal_(self.latent, std=self.latent_dim ** -0.5)
```

（5）函数 init_weights_vit_timm() 用于初始化 ViT 模型的权重。它支持线性层 (nn.Linear) 的标准初始化，以及模块自身定义的权重初始化方法。此方法基于 timm 库实现，用于保证权重初始化过程的可复现性。

```
def init_weights_vit_timm(module: nn.Module, name: str = '') -> None:
    """
    ViT 模型权重初始化，基于 timm 库的实现（为了保证结果的可复现性）。

    Args:
        module (nn.Module): 需要初始化权重的模型模块。
        name (str, optional): 模块名称，用于调试与追踪（默认值为空字符串）。

    功能说明：
    - 对于线性层 `nn.Linear`:
      - 使用截断正态分布初始化权重，标准差为 0.02。
      - 若存在偏置项，则将其初始化为零。
    - 如果模块包含 `init_weights` 方法，则调用该方法进行初始化。
    """

    # 初始化线性层的权重与偏置
    if isinstance(module, nn.Linear):
        trunc_normal_(module.weight, std=.02)
        if module.bias is not None:
```

```
            nn.init.zeros_(module.bias)

    # 调用模块自身的初始化方法
    elif hasattr(module, 'init_weights'):
        module.init_weights()
```

（6）类Attention实现了一个多头自注意力模块，支持标准和高效注意力计算方式。Attention提供了对查询（query）、键（key）、值（value）的线性变换和注意力计算，支持标准、FlashAttention 2和高效内存注意力计算等多种实现方式。此外，类Attention支持查询和键的归一化操作，从而提升注意力计算的数值稳定性。

```
class Attention(nn.Module):
    fused_attn: Final[bool]

    def __init__(
            self,
            dim: int,
            num_heads: int = 8,
            qkv_bias: bool = False,
            qk_norm: bool = False,
            attn_drop: float = 0.,
            proj_drop: float = 0.,
            norm_layer: nn.Module = nn.LayerNorm,
            deterministic: bool = False,
    ) -> None:
        """
        多头自注意力模块（支持 FlashAttention 与高效内存注意力计算）。

        Args:
            dim (int): 输入特征维度。
            num_heads (int): 注意力头的数量。
            qkv_bias (bool): 是否为 Q、K、V 矩阵添加偏置项。
            qk_norm (bool): 是否对查询与键向量进行归一化。
            attn_drop (float): 注意力概率的丢弃率。
            proj_drop (float): 投影层的丢弃率。
            norm_layer (nn.Module): 归一化层的类型。
            deterministic (bool): 是否在注意力计算中保持确定性。
        """
        super().__init__()
        assert dim % num_heads == 0, 'dim 应当能被 num_heads 整除'

        self.num_heads = num_heads
        self.head_dim = dim // num_heads
        self.scale = self.head_dim ** -0.5
        self.qk_norm = qk_norm
```

```
        self.fused_attn = True
        self.deterministic = deterministic

        # 定义查询、键、值的线性变换
        self.qkv = nn.Linear(dim, dim * 3, bias=qkv_bias)

        # 定义查询和键的归一化层
        self.q_norm = norm_layer(self.head_dim) if qk_norm else nn.Identity()
        self.k_norm = norm_layer(self.head_dim) if qk_norm else nn.Identity()

        self.attn_drop = nn.Dropout(attn_drop)
        self.proj = nn.Linear(dim, dim)
        self.proj_drop = nn.Dropout(proj_drop) if proj_drop > 0. else nn.Identity()

    def forward(self, x: torch.Tensor) -> torch.Tensor:
        """
        前向计算注意力输出。

        Args:
            x (torch.Tensor): 输入特征张量，形状为 [B, N, C]，其中 B 是批量大小，N 是
                              序列长度，C 是特征维度。

        Returns:
            torch.Tensor: 注意力输出，形状与输入相同。
        """
        B, N, C = x.shape
        qkv = self.qkv(x).reshape(B, N, 3, self.num_heads, self.head_dim)

        # 如果不使用查询和键的归一化
        if not self.qk_norm:
            # 使用 FlashAttention 进行高效计算
            if self.head_dim % 32 == 0 and is_flash_attn_2_available():
                # FlashAttention 要求 head_dim 是 32 的倍数
                x = flash_attn_qkvpacked_func(
                    qkv, dropout_p=self.attn_drop.p if self.training else 0.,
                    deterministic=self.deterministic)
            else:
                # 使用标准的高效注意力计算
                q, k, v = qkv.unbind(2)
                x = memory_efficient_attention(q, k, v, p=self.attn_drop.p
                                               if self.training else 0.)

            x = x.reshape(B, N, C)
            x = self.proj(x)
            x = self.proj_drop(x)
            return x
```

```
        # 如果需要归一化，将 qkv 矩阵调整维度
        qkv = qkv.permute(2, 0, 3, 1, 4)
        q, k, v = qkv.unbind(0)
        q, k = self.q_norm(q), self.k_norm(k)

        # 如果启用融合的注意力计算
        if self.fused_attn:
            with torch.backends.cuda.sdp_kernel(enable_math=False, enable_
                                                mem_efficient=False):
                # 使用上下文管理启用 Scale Dot Product Attention
                x = F.scaled_dot_product_attention(
                    q, k, v,
                    dropout_p=self.attn_drop.p if self.training else 0.,
                )
        else:
            # 手动计算点积注意力
            q = q * self.scale
            attn = q @ k.transpose(-2, -1)
            attn = attn.softmax(dim=-1)
            attn = self.attn_drop(attn)
            x = attn @ v

        # 还原形状
        x = x.transpose(1, 2).reshape(B, N, C)
        x = self.proj(x)
        x = self.proj_drop(x)
        return x
```

（7）类LayerScale实现了一种层缩放机制，通过一个可学习参数gamma对输入张量进行缩放操作。这种机制在Transformer模型中可以用于增强模型的稳定性，并通过调节每一层的输出幅度改善模型的收敛性能。

```
class LayerScale(nn.Module):
    def __init__(
            self,
            dim: int,
            init_values: float = 1e-5,
            inplace: bool = False,
    ) -> None:
        super().__init__()
        self.inplace = inplace
        self.gamma = nn.Parameter(init_values * torch.ones(dim))

    def forward(self, x: torch.Tensor) -> torch.Tensor:
```

```
        return x.mul_(self.gamma) if self.inplace else x * self.gamma
```

（8）类Block实现了一个基础的Transformer编码块，包括多头注意力（Attention）、前馈网络（MLP）、残差连接（Residual Connection）、层缩放（LayerScale）及丢弃路径（DropPath）等组件。这种模块是Transformer和ViT模型的核心构建单元，用于对输入特征进行编码。

```
class Block(nn.Module):
    """
    Transformer 编码块，实现多头注意力、前馈网络、残差连接、层缩放和丢弃路径等机制。

    Args:
        dim (int): 输入特征维度。
        num_heads (int): 注意力头的数量。
        mlp_ratio (float): 前馈层隐藏层维度与输入维度的比例，默认为 4。
        qkv_bias (bool): 是否为 qkv 线性层添加偏置，默认为 False。
        qk_norm (bool): 是否为 qk 添加归一化层，默认为 False。
        proj_drop (float): 输出投影的 dropout 概率，默认为 0。
        attn_drop (float): 注意力 dropout 概率，默认为 0。
        init_values (Optional[float]): 层缩放参数的初始值。如果为None,则不使用层缩放。
        drop_path (float): 丢弃路径的概率，默认为 0。
        act_layer (nn.Module): 激活函数模块，默认为 nn.GELU。
        norm_layer (nn.Module): 归一化层模块，默认为 nn.LayerNorm。
        mlp_layer (nn.Module): MLP 层模块，默认为 Mlp。
        deterministic (bool): 是否在训练中启用确定性计算。

    """

    def __init__(
            self,
            dim: int,
            num_heads: int,
            mlp_ratio: float = 4.,
            qkv_bias: bool = False,
            qk_norm: bool = False,
            proj_drop: float = 0.,
            attn_drop: float = 0.,
            init_values: Optional[float] = None,
            drop_path: float = 0.,
            act_layer: nn.Module = nn.GELU,
            norm_layer: nn.Module = nn.LayerNorm,
            mlp_layer: nn.Module = Mlp,
            deterministic: bool = False,
    ) -> None:
        super().__init__()
```

```
        # 第一部分：多头注意力层
        self.norm1 = norm_layer(dim)  # 归一化层
        self.attn = Attention(
            dim,
            num_heads=num_heads,
            qkv_bias=qkv_bias,
            qk_norm=qk_norm,
            attn_drop=attn_drop,
            proj_drop=proj_drop,
            norm_layer=norm_layer,
            deterministic=deterministic,
        )
        # 层缩放与丢弃路径
        self.ls1 = LayerScale(dim, init_values=init_values) if init_values
                   else nn.Identity()
        self.drop_path1 = DropPath(drop_path) if drop_path > 0. else
                          nn.Identity()

        # 第二部分：前馈网络层
        self.norm2 = norm_layer(dim)  # 归一化层
        self.mlp = mlp_layer(
            in_features=dim,
            hidden_features=int(dim * mlp_ratio),
            act_layer=act_layer,
            drop=proj_drop,
        )
        self.ls2 = LayerScale(dim, init_values=init_values) if init_values
                   else nn.Identity()
        self.drop_path2 = DropPath(drop_path) if drop_path > 0. else
                          nn.Identity()

    def forward(self, x: torch.Tensor) -> torch.Tensor:
        """
        前向传播过程。

        Args:
            x (torch.Tensor): 输入特征张量，形状为 [batch_size, seq_len, dim]。

        Returns:
            torch.Tensor: 经过编码块处理后的特征张量。
        """
        # 多头注意力模块 + 残差连接
        x = x + self.drop_path1(self.ls1(self.attn(self.norm1(x))))
        # 前馈网络模块 + 残差连接
        x = x + self.drop_path2(self.ls2(self.mlp(self.norm2(x))))
        return x
```

（9）类 VisionTransformer 实现了一个基于 ViT 的图像分类模型，此模型的核心思想是将图像划分为若干个小块（patches），然后通过 Transformer 架构对这些小块进行处理和学习。该实现基于论文 *An Image is Worth 16x16 Words: Transformers for Image Recognition at Scale*。VisionTransformer 模型的主要组件包括以下内容。

- Patch Embedding：图像被划分成若干个固定大小的块，每个块被映射到一个向量空间中。这些向量随后输入 Transformer 网络中。
- Transformer Blocks：由多个 Transformer 层堆叠而成，每个 Transformer 层由自注意力机制（Self-Attention）和前馈神经网络（Feed-Forward Network）组成。
- 位置编码：为每个图像块添加位置编码，使得 Transformer 能够感知各个图像块在图像中的位置。
- 分类头：经过 Transformer 处理后的特征会被传递到分类头，进行最终的类别预测。可以选择不同的全局池化策略，如使用 class_token 或 avg 池化。
- 模型参数：模型有多个可调参数，包括图像尺寸、patch 大小、Transformer 深度、嵌入维度、注意力头数等。
- 训练与初始化：支持不同的权重初始化方式，并且可以通过 reset_classifier 函数调整分类头。
- 动态图片大小：支持动态输入尺寸的图像，通过调整输入图片的尺寸来适应不同的应用场景。

总的来说，该代码实现了一个灵活的 Vision Transformer 架构，支持多种配置选项，可以在不同任务和数据集上进行图像分类等任务。

类 VisionTransformer 的实现流程如下所示：

- 方法 __init__ 是 VisionTransformer 类的初始化方法，负责设置和初始化该模型的各个组件和参数。该方法接受多个参数，用于控制模型的不同方面，例如输入图像的大小、补丁大小、嵌入维度、注意力头的数量、Transformer 层数等。它会初始化图像补丁嵌入层、位置嵌入、Transformer 层、归一化层、激活层、分类头等，并根据给定的配置选项调整相应的超参数。此外，__init__ 方法还处理一些模型行为的控制，如是否使用类 token、是否启用动态图像大小、是否启用梯度检查点等。

```
class VisionTransformer(nn.Module):
    """ Vision Transformer

    一个 PyTorch 实现，参考论文：`An Image is Worth 16x16 Words: Transformers for
    Image Recognition at Scale`
        - https://arxiv.org/abs/2010.11929
    """
    dynamic_img_size: Final[bool]

    def __init__(
            self,
            img_size: Union[int, Tuple[int, int]] = 224,
            patch_size: Union[int, Tuple[int, int]] = 16,
            in_chans: int = 3,
```

```
        num_classes: int = 1000,
        global_pool: Literal['', 'avg', 'token', 'map'] = 'token',
        embed_dim: int = 768,
        depth: int = 12,
        num_heads: int = 12,
        mlp_ratio: float = 4.,
        qkv_bias: bool = True,
        qk_norm: bool = False,
        init_values: Optional[float] = None,
        class_token: bool = True,
        no_embed_class: bool = False,
        reg_tokens: int = 0,
        pre_norm: bool = False,
        fc_norm: Optional[bool] = None,
        dynamic_img_size: bool = False,
        dynamic_img_pad: bool = False,
        drop_rate: float = 0.,
        pos_drop_rate: float = 0.,
        patch_drop_rate: float = 0.,
        proj_drop_rate: float = 0.,
        attn_drop_rate: float = 0.,
        drop_path_rate: float = 0.,
        weight_init: Literal['skip', 'jax', 'jax_nlhb', 'moco', ''] = '',
        embed_layer: Callable = PatchEmbed,
        norm_layer: Optional[LayerType] = None,
        act_layer: Optional[LayerType] = None,
        block_fn: Type[nn.Module] = Block,
        mlp_layer: Type[nn.Module] = Mlp,
        ignore_head: bool = False,
        deterministic: bool = False,
        num_recomputing_layers: int = 0
) -> None:
    """
    参数:
        img_size: 输入图像的尺寸。
        patch_size: 图像划分块的尺寸。
        in_chans: 输入图像的通道数。
        num_classes: 分类头的类别数。
        global_pool: 最终序列的全局池化类型（默认: 'token'）。
        embed_dim: Transformer 的嵌入维度。
        depth: Transformer 的深度。
        num_heads: 注意力头的数量。
        mlp_ratio: MLP 隐藏层的维度与嵌入维度的比例。
        qkv_bias: 如果为 True，则启用 qkv 投影的偏置。
        init_values: 层级初始化值（如果不为 None，则启用层级缩放）。
        class_token: 是否使用类标记。
```

```
        no_embed_class: 是否不包含类标记(或回归)位置的嵌入。
        reg_tokens: 回归标记的数量。
        fc_norm: 是否在池化后对分类头进行归一化(如果为 None，则在 global_pool
                 为 'avg' 时启用)。
        drop_rate: 分类头的丢弃率。
        pos_drop_rate: 位置编码的丢弃率。
        attn_drop_rate: 注意力丢弃率。
        drop_path_rate: 随机路径丢弃率。
        weight_init: 权重初始化方案。
        embed_layer: 图像块嵌入层。
        norm_layer: 归一化层。
        act_layer: MLP 激活层。
        block_fn: Transformer 块层。
    """
    super().__init__()
    assert global_pool in ('', 'avg', 'token', 'map')
    assert class_token or global_pool != 'token'
    use_fc_norm = global_pool == 'avg' if fc_norm is None else fc_norm
    norm_layer = partial(nn.LayerNorm, eps=1e-6)
    act_layer = partial(nn.GELU, approximate='tanh')

    self.num_classes = num_classes
    self.global_pool = global_pool
    self.num_features = self.embed_dim = embed_dim
                                    # 为了和其他模型一致，命名为 num_features
    self.num_prefix_tokens = 1 if class_token else 0
    self.num_prefix_tokens += reg_tokens
    self.num_reg_tokens = reg_tokens
    self.has_class_token = class_token
    self.no_embed_class = no_embed_class  # 不为类标记(或回归标记)嵌入位置
    self.dynamic_img_size = dynamic_img_size
    self.grad_checkpointing = False
    self.ignore_head = ignore_head
    self.num_recomputing_layers = num_recomputing_layers

    embed_args = {}
    if dynamic_img_size:
        # 在位置嵌入之后延迟展开
        embed_args.update(dict(strict_img_size=False, output_fmt='NHWC'))
    self.patch_embed = embed_layer(
        img_size=img_size,
        patch_size=patch_size,
        in_chans=in_chans,
        embed_dim=embed_dim,
        bias=not pre_norm,  # 如果使用预归一化，则禁用偏置(例如，CLIP)
        dynamic_img_pad=dynamic_img_pad,
```

```
        **embed_args,
    )
    num_patches = self.patch_embed.num_patches

    self.cls_token = nn.Parameter(torch.zeros(1, 1, embed_dim)) if
                     class_token else None
    self.reg_token = nn.Parameter(torch.zeros(1, reg_tokens, embed_dim))
                     if reg_tokens else None
    embed_len = num_patches if no_embed_class else num_patches +
                self.num_prefix_tokens
    self.pos_embed = nn.Parameter(torch.randn(1, embed_len, embed_dim) * .02)
    self.pos_drop = nn.Dropout(p=pos_drop_rate)
    if patch_drop_rate > 0:
        self.patch_drop = PatchDropout(
            patch_drop_rate,
            num_prefix_tokens=self.num_prefix_tokens,
        )
    else:
        self.patch_drop = nn.Identity()
    self.norm_pre = norm_layer(embed_dim) if pre_norm else nn.Identity()

    dpr = [x.item() for x in torch.linspace(0, drop_path_rate, depth)]
                                                          # 随机深度衰减规则
    self.blocks = nn.Sequential(*[
        block_fn(
            dim=embed_dim,
            num_heads=num_heads,
            mlp_ratio=mlp_ratio,
            qkv_bias=qkv_bias,
            qk_norm=qk_norm,
            init_values=init_values,
            proj_drop=proj_drop_rate,
            attn_drop=attn_drop_rate,
            drop_path=dpr[i],
            norm_layer=norm_layer,
            act_layer=act_layer,
            mlp_layer=mlp_layer,
            deterministic=deterministic,
        )
        for i in range(depth)])
    self.norm = norm_layer(embed_dim) if not use_fc_norm else nn.Identity()

    # 分类头
    if global_pool == 'map':
        AttentionPoolLatent.init_weights = init_weights
        self.attn_pool = AttentionPoolLatent(
```

```
                self.embed_dim,
                num_heads=num_heads,
                mlp_ratio=mlp_ratio,
                norm_layer=norm_layer,
            )
        else:
            self.attn_pool = None
        self.fc_norm = norm_layer(embed_dim) if use_fc_norm else nn.Identity()
        self.head_drop = nn.Dropout(drop_rate)
        self.head = nn.Linear(self.embed_dim, num_classes) if num_classes >
                0 else nn.Identity()

        if weight_init != 'skip':
            self.init_weights(weight_init)
```

- 方法init_weights()负责初始化模型的权重。它首先检查传入的初始化模式是否有效（'jax'，'jax_nlhb'，'moco'或空字符串）。其次根据指定的模式为类头（head）设置偏置值，并使用截断正态分布初始化位置嵌入。对于类token（如果存在），它会用一个非常小的标准差进行正态初始化。最后通过named_apply函数调用init_weights_vit_timm，进一步初始化Vision Transformer模型的其他权重。

```
def init_weights(self, mode: Literal['jax', 'jax_nlhb', 'moco', ''] = '') ->
None:
    # 确保初始化模式在有效范围内
    assert mode in ('jax', 'jax_nlhb', 'moco', '')

    # 如果在初始化模式中包含 'nlhb'，则将头部偏置设为类别数（num_classes）的负对数；否
      则头部偏置为 0
    head_bias = -math.log(self.num_classes) if 'nlhb' in mode else 0.

    # 使用标准差为 0.02 的截断正态分布初始化位置嵌入
    trunc_normal_(self.pos_embed, std=.02)

    # 如果类 token 存在，使用标准差为 1e-6 的正态分布初始化
    if self.cls_token is not None:
        nn.init.normal_(self.cls_token, std=1e-6)

    # 调用 named_apply 函数初始化其他权重
    named_apply(init_weights_vit_timm, self)
```

- 方法no_weight_decay()用来指定哪些参数在训练过程中不参与权重衰减。该方法返回一个包含不需要进行权重衰减的参数名称的集合（pos_embed, cls_token, dist_token）。这些参数通常是模型中的嵌入、类token或其他特殊token，在一些情况下，它们可能不需要通过正则化来更新。

```
    @torch.jit.ignore
```

```
    def no_weight_decay(self) -> Set:
        return {'pos_embed', 'cls_token', 'dist_token'}
```

- 方法group_matcher()用于匹配不同部分的模型参数组。它根据正则表达式匹配模型中的各个部分，返回一个字典，其中包含了模型的不同子模块的参数组。通过这个方法，能够方便地对模型的不同层或部分应用不同的策略（例如学习率调度或权重衰减）。

```
    @torch.jit.ignore
    def group_matcher(self, coarse: bool = False) -> Dict:
        return dict(
            stem=r'^cls_token|pos_embed|patch_embed',  # stem and embed
            blocks=[(r'^blocks\.(\d+)', None), (r'^norm', (99999,))]
        )
```

- 方法set_grad_checkpointing()用于启用或禁用梯度检查点（gradient checkpointing），这一策略可以在训练过程中减少内存消耗，通过在反向传播时重新计算某些中间结果。get_classifier方法返回模型的分类器部分（最后的全连接层self.head），通常用于模型的最终预测。

```
    @torch.jit.ignore
    def set_grad_checkpointing(self, enable: bool = True) -> None:
        self.grad_checkpointing = enable

    @torch.jit.ignore
    def get_classifier(self) -> nn.Module:
        return self.head
```

- 方法reset_classifier()用于重置模型的分类头。它接受两个参数：num_classes（分类的类别数量）和global_pool（全局池化策略）。根据global_pool的值，方法会选择是否启用全局池化，并调整分类头的结构。如果num_classes大于0，则会创建一个新的线性分类器；否则，将使用一个身份层。

```
def reset_classifier(self, num_classes: int, global_pool=None) -> None:
    # 重置分类头
    self.num_classes = num_classes
    if global_pool is not None:
        # 确保全局池化方式有效
        assert global_pool in ('', 'avg', 'token', 'map')
        if global_pool == 'map' and self.attn_pool is None:
            assert False, "当前无法在reset_classifier()中添加注意力池化
                           (attention pooling)。"
        elif global_pool != 'map' and self.attn_pool is not None:
            self.attn_pool = None  # 移除注意力池化
        self.global_pool = global_pool
    self.head = nn.Linear(self.embed_dim, num_classes) if num_classes > 0
else nn.Identity()
```

● 方法_pos_embed()用于计算并应用位置嵌入功能。根据dynamic_img_size的标志，决定是使用静态位置嵌入还是动态调整位置嵌入。在动态模式下，会根据输入图像的大小重新调整位置嵌入的尺寸。如果有类别令牌 (cls_token) 或回归令牌 (reg_token)，将被添加到输入张量x中，并与位置嵌入一起使用。最后，返回添加了位置嵌入和可能的类别令牌的张量，并应用丢弃层。

```
def _pos_embed(self, x: torch.Tensor) -> torch.Tensor:
    if self.dynamic_img_size:
        # 如果启用了动态图像大小
        B, H, W, C = x.shape
        pos_embed = resample_abs_pos_embed(
            self.pos_embed,
            (H, W),
            num_prefix_tokens=0 if self.no_embed_class else self.num_prefix_
                                tokens,
        )
        x = x.view(B, -1, C)
    else:
        pos_embed = self.pos_embed

    to_cat = []
    if self.cls_token is not None:
        # 如果存在类别令牌（cls_token），则将其添加到 to_cat 中
        to_cat.append(self.cls_token.expand(x.shape[0], -1, -1))
    if self.reg_token is not None:
        # 如果存在回归令牌（reg_token），则将其添加到 to_cat 中
        to_cat.append(self.reg_token.expand(x.shape[0], -1, -1))

    if self.no_embed_class:
        # 对于不使用类别嵌入的情况（如 deit-3，更新后的 JAX）
        # 位置嵌入与类别令牌不重叠，先添加再拼接
        x = x + pos_embed
        if to_cat:
            x = torch.cat(to_cat + [x], dim=1)
    else:
        # 对于原始的 timm、JAX 和 deit vit 实现
        # 位置嵌入包含类别令牌，先拼接再添加
        if to_cat:
            x = torch.cat(to_cat + [x], dim=1)
        x = x + pos_embed

    return self.pos_drop(x)
```

● 方法_intermediate_layers()的主要功能是从一个深度神经网络（如Transformer架构）中提取指定数量或指定索引的中间层输出。

```
def _intermediate_layers(
        self,
        x: torch.Tensor,
        n: Union[int, Sequence] = 1,
) -> List[torch.Tensor]:
    outputs, num_blocks = [], len(self.blocks)
    take_indices = set(range(num_blocks - n, num_blocks) if isinstance(n,
                   int) else n)

    # forward pass
    x = self.patch_embed(x)
    x = self._pos_embed(x)
    x = self.patch_drop(x)
    x = self.norm_pre(x)
    for i, blk in enumerate(self.blocks):
        x = blk(x)
        if i in take_indices:
            outputs.append(x)

    return outputs
```

- 方法_intermediate_layers()用于提取Transformer模型中的中间层输出。首先，通过patch_embed()和 _pos_embed()方法处理输入数据。其次，在通过各个Transformer block进行前向传播时，根据指定的层索引 (n) 收集并返回特定的中间层输出。n可以是一个整数或一个索引序列，表示需要提取的块的数量或特定块的索引。最后，返回一个包含所选中间层输出的列表。

```
def get_intermediate_layers(
        self,
        x: torch.Tensor,
        n: Union[int, Sequence] = 1,
        reshape: bool = False,
        return_prefix_tokens: bool = False,
        norm: bool = False,
) -> Tuple[Union[torch.Tensor, Tuple[torch.Tensor]]]:
    outputs = self._intermediate_layers(x, n)
    if norm:
        outputs = [self.norm(out) for out in outputs]
    prefix_tokens = [out[:, 0:self.num_prefix_tokens] for out in outputs]
    outputs = [out[:, self.num_prefix_tokens:] for out in outputs]

    if reshape:
        grid_size = self.patch_embed.grid_size
        outputs = [
            out.reshape(x.shape[0], grid_size[0],
            grid_size[1], -1).permute(0, 3, 1, 2).contiguous()
```

```
                for out in outputs
            ]

        if return_prefix_tokens:
            return tuple(zip(outputs, prefix_tokens))
        return tuple(outputs)
```

● 方法forward_features()用于处理输入张量x并通过网络的不同层进行前向传播。首先，如果is_first_stage属性为True，则对输入进行patch嵌入、位置嵌入、patch丢弃和预归一化处理。其次，如果启用了梯度检查点，并且脚本化未启用，则使用checkpoint_seq对Transformer block层进行梯度检查点处理，否则直接通过所有block进行前向传播。最后，如果is_last_stage为True，则对输出进行归一化。该方法返回最后的输出特征张量。

```
    def forward_features(self, x: torch.Tensor) -> torch.Tensor:
        if getattr(self, "is_first_stage", True):
            x = self.patch_embed(x)
            x = self._pos_embed(x)
            x = self.patch_drop(x)
            x = self.norm_pre(x)
        if self.grad_checkpointing and not torch.jit.is_scripting():
            skip_last = max(1, len(self.blocks) -
                        self.num_recomputing_layers)
            x = checkpoint_seq(self.blocks, x, skip_last=skip_last)
        else:
            x = self.blocks(x)
        if getattr(self, "is_last_stage", True):
            x = self.norm(x)
        return x
```

● 方法forward_head()负责处理网络的头部部分。首先，根据is_last_stage属性判断是否进行头部处理，如果是最后阶段并且attn_pool存在，则应用该池化方法。否则，根据global_pool设置的类型决定如何进行池化：如果是 'avg'，则对输入的后缀部分（不包括前缀tokens）进行均值池化；如果是其他值，则使用类标记class token。其次，进行归一化处理和丢弃操作。最后，根据pre_logits参数的值决定是否输出头部计算前的特征向量或通过头部线性层计算输出。

```
    def forward_head(self, x: torch.Tensor, pre_logits: bool = False) ->
        torch.Tensor:
        if not getattr(self, "is_last_stage", True):
            return x
        if self.attn_pool is not None:
            x = self.attn_pool(x)
        elif self.global_pool == 'avg':
            x = x[:, self.num_prefix_tokens:].mean(dim=1)
        elif self.global_pool:
```

```
            x = x[:, 0]  # class token
        x = self.fc_norm(x)
        x = self.head_drop(x)
        return x if pre_logits else self.head(x)
```

● 方法forward()负责调用forward_features和forward_head完成网络的前向传播。如果ignore_head为False，则会进一步处理头部部分。

```
    def forward(self, x: torch.Tensor) -> torch.Tensor:
        x = self.forward_features(x)
        if not self.ignore_head:
            x = self.forward_head(x)
        return x
```

● 方法to_pipeline()用于在分布式训练中将模型的各个部分分配到不同的计算节点上。它根据pp_rank（当前设备的排名）和pp_size（总设备数量）决定模型是否属于管道的第一阶段或最后阶段，并根据这些信息对模型进行裁剪，删除不需要的模块。

```
    def to_pipeline(self, pp_size, pp_rank, pp_splits: Optional[List[int]] = None):
        self.is_first_stage = pp_rank == 0
        self.is_last_stage = pp_rank == pp_size - 1
        if not self.is_first_stage and hasattr(self, "patch_embed"):
            del self.patch_embed, self.cls_token, self.reg_token, self.pos_
                embed, self.pos_drop, self.patch_drop, self.norm_pre
        if not self.is_last_stage and hasattr(self, "norm"):
            del self.norm, self.attn_pool, self.fc_norm, self.head_drop, self.head
        if pp_splits is not None:
            assert len(self.blocks) == sum(pp_splits)
            splits = np.cumsum([0] + pp_splits)
            self.blocks = self.blocks[splits[pp_rank]:splits[pp_rank + 1]]
        return self
```

对上述代码的具体说明如下所示：

● 如果当前设备是第一阶段（pp_rank == 0），则保留与输入处理相关的模块（如patch_embed, cls_token, reg_token等）。

● 如果当前设备不是最后阶段（pp_rank != pp_size - 1），则删除与输出处理相关的模块（如norm, attn_pool, head_drop, head等）。

● 如果提供了pp_splits，则根据该列表将模型的块（blocks）划分为多个部分，每个部分对应不同的计算节点。

最终，返回修改后的模型。

（10）SigLIPVisionCfg是一个数据类，用于配置SigLIP模型的视觉部分的超参数。类SigLIPVisionCfg包含了模型的各类设置，比如网络宽度（width）、层数（layers）、注意力头数（heads）、图像大小（image_size）、全局池化方式（global_pool）等。这些配置可以帮助构建一个视觉模型的架构，并且可以根据需要进行调整。通过这个类，可以轻松管理和传递超参数设置。

```
@dataclass
class SigLIPVisionCfg:
    width: int = 1152
    layers: Union[Tuple[int, int, int, int], int] = 27
    heads: int = 16
    patch_size: int = 14
    image_size: Union[Tuple[int, int], int] = 336
    global_pool: str = "map"
    mlp_ratio: float = 3.7362
    class_token: bool = False
    num_classes: int = 0
    use_checkpoint: bool = False
```

（11）下列代码中的SigLIP_MODEL_CONFIG是一个字典，包含了多个SigLIP模型配置的超参数设置。每个配置项代表一个不同版本的SigLIP模型，键值对应模型的名称，值是一个字典，包含该模型的图像大小、图像块大小、网络宽度、层数、注意力头数、MLP比例（mlp_ratio）、全局池化方式及是否使用检查点（use_checkpoint）等超参数。这些配置可以用于构建和调整不同版本的SigLIP模型。

```
SigLIP_MODEL_CONFIG = {
    "siglip_so400m_patch14_384": {
        "image_size": 384,
        "patch_size": 14,
        "width": 1152,
        "layers": 27,
        "heads": 16,
        "mlp_ratio": 3.7362,
        "global_pool": "map",
        "use_checkpoint": False
    },

    "siglip_so400m_patch14_224": {
        "image_size": 224,
        "patch_size": 14,
        "width": 1152,
        "layers": 27,
        "heads": 16,
        "mlp_ratio": 3.7362,
        "global_pool": "map",
        "use_checkpoint": False
    },

    "siglip_large_patch16_384": {
```

```
        "image_size": 384,
        "patch_size": 16,
        "width": 1024,
        "layers": 24,
        "heads": 16,
        "mlp_ratio": 4,
        "global_pool": "map",
        "use_checkpoint": False
    }
}
```

（12）下列代码定义了函数create_siglip_vit()，用于创建并返回一个SigLIP版本的ViT模型。该函数接受多个参数，用于根据给定的模型配置（例如模型名称、图像大小、选择的层数等）来初始化一个ViT模型。

```
def create_siglip_vit(
        model_name: str = "siglip_so400m_patch14_384",
        image_size: int = 384,
        select_layer: int = -1,
        ckpt_path: str = "",
        **kwargs
):
    assert model_name in SigLIP_MODEL_CONFIG.keys(), f"model name should be
        in {SigLIP_MODEL_CONFIG.keys()}"

    vision_cfg = SigLIPVisionCfg(**SigLIP_MODEL_CONFIG[model_name])

    if select_layer <= 0:
        layers = min(vision_cfg.layers, vision_cfg.layers + select_layer + 1)
    else:
        layers = min(vision_cfg.layers, select_layer)

    model = VisionTransformer(
        img_size=image_size,
        patch_size=vision_cfg.patch_size,
        embed_dim=vision_cfg.width,
        depth=layers,
        num_heads=vision_cfg.heads,
        mlp_ratio=vision_cfg.mlp_ratio,
        class_token=vision_cfg.class_token,
        global_pool=vision_cfg.global_pool,
        ignore_head=kwargs.get("ignore_head", True),
        weight_init=kwargs.get("weight_init", "skip"),
        num_classes=0,
        deterministic=kwargs.get("deterministic", False),
```

```
        num_recomputing_layers=kwargs.get("num_recomputing_layers", 0)
    )

    if ckpt_path:
        state_dict = torch.load(ckpt_path, map_location="cpu")

        incompatible_keys = model.load_state_dict(state_dict, strict=False)
        print(f"SigLIP-ViT restores from {ckpt_path},\n"
              f"\tincompatible_keys:', {incompatible_keys}.")

    return model
```

6.5.6 对话模板和历史记录管理

文件conversation.py实现了一个用于管理对话模板和历史记录的类Conversation，并通过不同的SeparatorStyle样式来构造对话内容。此文件的核心功能包括：创建和维护不同的对话模板、管理消息历史、生成用于模型训练的对话提示（prompt）、将对话转换为不同格式（如Gradio或OpenAI API格式），以及注册和获取预设的对话模板。通过不同的分隔符样式（如DeepSeek、DeepSeekV2、PLAIN和ALIGNMENT），用户可以定制对话内容的格式化方式。该代码还提供了一个全局注册表，允许根据模板名称注册和获取对话模板。最后，示例代码演示了如何使用这些模板进行对话和生成提示。

（1）定义枚举类型SeparatorStyle，用于表示不同的分隔符样式。DeepSeek、DeepSeekV2、PLAIN和ALIGNMENT分别代表四种不同的分隔符样式，它们可以用于格式化对话或文本的分隔方式，以满足不同的处理需求。

```
class SeparatorStyle(IntEnum):
    """分隔符样式。"""

    DeepSeek = auto()      # DeepSeek样式
    DeepSeekV2 = auto()    # DeepSeekV2样式
    PLAIN = auto()         # 简单样式
    ALIGNMENT = auto()     # 对齐样式
```

（2）定义类Conversation，用于管理对话的历史记录和生成提示模板。类Conversation中包含了多个功能，如设置和更新系统消息、追加新消息、重置消息、转换为不同格式的对话数据（如Gradio或OpenAI API格式），以及复制或字典化对话内容。

● 下列代码定义了类Conversation，用于管理对话历史和生成提示模板。包含了多个属性来存储对话的不同元素，如系统提示、角色名称、消息列表、分隔符样式等。该类提供了方法来设置系统消息、追加新消息、更新最后一条消息、重置消息，以及将对话转换为不同格式（如Gradio或OpenAI API格式）。此外，还支持自定义分隔符样式，并允许控制生成的停止条件。

```
@dataclasses.dataclass
class Conversation:
    """一个管理提示模板并保持所有对话历史的类。"""

    # 这个模板的名称
    name: str
    # 系统提示的模板
    system_template: str = "{system_message}"
    # 系统消息
    system_message: str = ""
    # 两个角色的名称
    roles: List[str] = (("USER", "ASSISTANT"),)
    # 所有消息。每个项是 (role, message) 的组合
    messages: List[List[str]] = ()
    # 少样本示例的数量
    offset: int = 0
    # 分隔符样式和配置
    sep_style: SeparatorStyle = SeparatorStyle.DeepSeek
    sep: str = "\n"
    sep2: str = None
    # 停止标准(默认是 EOS 标记)
    stop_str: str = None
    # 如果遇到列表中的任何令牌,则停止生成
    stop_token_ids: List[int] = None
```

● 方法get_prompt(self)用于生成对话提示字符串，依据不同的分隔符样式（如DeepSeek、DeepSeekV2、PLAIN、ALIGNMENT）构建系统消息和角色消息的提示内容，返回一个格式化的字符串用于生成对话。

```
    def get_prompt(self) -> str:
        """获取生成的提示。"""
        system_prompt = self.system_template.format(system_message=
                                                    self.system_message)
        if self.sep_style == SeparatorStyle.DeepSeek:
            seps = [self.sep, self.sep2]
            if system_prompt == "" or system_prompt is None:
                ret = ""
            else:
                ret = system_prompt + seps[0]
            for i, (role, message) in enumerate(self.messages):
                if message:
                    ret += role + ": " + message + seps[i % 2]
                else:
                    ret += role + ":"
            return ret
```

```
        elif self.sep_style == SeparatorStyle.DeepSeekV2:
            seps = [self.sep, self.sep2]
            if system_prompt == "" or system_prompt is None:
                ret = ""
            else:
                ret = system_prompt + seps[0]
            for i, (role, message) in enumerate(self.messages):
                if message:
                    if role == "User":
                        ret += "<｜sft▁begin｜>\n" + message + self.sep
                               #<｜sft▁begin｜>User Input<｜sft▁end｜
                               >\nResponse<｜end▁of▁sentence｜>
                    else:
                        ret += message + self.sep2
                else:
                    ret = ret
            return ret

        elif self.sep_style == SeparatorStyle.PLAIN:
            seps = [self.sep, self.sep2]
            ret = ""
            for i, (role, message) in enumerate(self.messages):
                if message:
                    if type(message) is tuple:
                        message, _, _ = message
                    if i % 2 == 0:
                        ret += message + seps[i % 2]
                    else:
                        ret += message + seps[i % 2]
                else:
                    ret += ""
            return ret
        elif self.sep_style == SeparatorStyle.ALIGNMENT:
            seps = [self.sep, self.sep2]
            ret = ""
            for i, (role, message) in enumerate(self.messages):
                if message:
                    if type(message) is tuple:
                        message, _, _ = message
                    if i % 2 == 0:
                        ret += '<image>\n' + seps[i % 2]
                    else:
                        ret += message + seps[i % 2]
                else:
                    ret += ""
            return ret
```

```
        else:
            raise ValueError(f"无效的样式 : {self.sep_style}")
```

- 方法set_system_message(self, system_message: str)用于设置系统消息、更新system_message属性，从而在生成提示时使用新的系统消息内容。

```
    def set_system_message(self, system_message: str):
        """设置系统消息。"""
  self.system_message = system_message
```

- 方法append_message(self, role: str, message: str)用于向对话中添加一条新消息，并指定消息的角色（如USER或ASSISTANT），将其追加到messages列表中。

```
    def append_message(self, role: str, message: str):
        """追加一条新消息。"""
        self.messages.append([role, message])
```

- 方法update_last_message(self, message: str)用于更新最后一条消息的内容，通常用于更新模型响应后，修改messages列表中的最后一条消息。

```
    def update_last_message(self, message: str):
        """更新最后一条输出消息。

        通常在构建提示时，最后一条消息会被设置为None，
        所以在从模型获得响应后，我们需要就地更新它。
        """
        self.messages[-1][1] = message
```

- 方法reset_message(self)用于清空当前所有的消息，将messages列表重置为空列表，便于开始新的对话。

```
    def reset_message(self):
        """重置一条新消息。"""
        self.messages = []
```

- 方法to_gradio_chatbot(self)的功能是将对话格式化为适用于Gradio聊天机器人的格式，将消息按角色分配为［用户消息,助手消息］格式的对话记录列表。

```
    def to_gradio_chatbot(self):
        """将对话转换为Gradio聊天机器人格式。"""
        ret = []
        for i, (role, msg) in enumerate(self.messages[self.offset :]):
            if i % 2 == 0:
                ret.append([msg, None])
            else:
                ret[-1][-1] = msg
```

```
        return ret
```

● 方法to_openai_api_messages(self)的功能是将对话转换为适合OpenAI API的聊天完成格式，生成一个包含系统消息和按角色排列的用户及助手消息的字典列表。

```
    def to_openai_api_messages(self):
        """将对话转换为 OpenAI 聊天完成格式。"""
        system_prompt = self.system_template.format(system_message=self.
                                                system_message)
        ret = [{"role": "system", "content": system_prompt}]

        for i, (_, msg) in enumerate(self.messages[self.offset :]):
            if i % 2 == 0:
                ret.append({"role": "user", "content": msg})
            else:
                if msg is not None:
                    ret.append({"role": "assistant", "content": msg})
        return ret
```

● 方法copy(self)的功能是创建并返回当前Conversation对象的一个副本，所有属性都被复制到新的实例中，方便保留对话的原始状态。

```
    def copy(self):
        return Conversation(
            name=self.name,
            system_template=self.system_template,
            system_message=self.system_message,
            roles=self.roles,
            messages=[[x, y] for x, y in self.messages],
            offset=self.offset,
            sep_style=self.sep_style,
            sep=self.sep,
            sep2=self.sep2,
            stop_str=self.stop_str,
            stop_token_ids=self.stop_token_ids,
        )
```

● 方法dict(self)的功能是将Conversation对象的主要属性（如name、system_message、roles、messages等）转换为字典格式，便于序列化或进一步处理。

```
    def dict(self):
        return {
            "template_name": self.name,
            "system_message": self.system_message,
            "roles": self.roles,
            "messages": self.messages,
```

```
            "offset": self.offset,
        }
```

（3）下列代码的功能是定义一个全局的对话模板注册表conv_templates，并提供一个函数register_conv_template()用于注册新的对话模板。该函数可以选择性地覆盖已有的模板，如果override参数为False，则会检查模板名是否已存在，若存在则抛出异常，若模板名不存在则将新的模板添加到注册表中。

```
# 一个全局注册表，保存所有对话模板
conv_templates: Dict[str, Conversation] = {}

def register_conv_template(template: Conversation, override: bool = False):
    """ 注册一个新的对话模板。"""
    if not override:
        assert template.name not in conv_templates, f"{template.name}
        已经被注册过了。"
    conv_templates[template.name] = template
```

（4）定义函数get_conv_template()，用于根据给定的模板名称从全局注册表conv_templates中获取对应的对话模板，并返回该模板的副本。

```
def get_conv_template(name: str) -> Conversation:
    """ 获取一个对话模板。"""
    return conv_templates[name].copy()
```

（5）下列代码的功能是注册一个新的对话模板，模板的名称为 "deepseek"，并定义了相关的系统消息、角色、消息分隔符等配置。此模板用于构建对话，并配置了停止标记和停止符号的设定。

```
register_conv_template(
    Conversation(
        name="deepseek",
        system_template="{system_message}",
        # system_message=" 你是一个有作用的助手，请如实回答并一步步写出你的思考过程，确
                            保你得到正确的答案。",
        system_message="",
        roles=("<|User|>", "<|Assistant|>"),
        messages=(),
        offset=0,
        sep_style=SeparatorStyle.DeepSeek,
        sep="\n\n",
        sep2="< | end_of_sentence | >",
        stop_token_ids=[100001],
        stop_str=["User:", "< | end_of_sentence | >"]
    )
)
```

（6）下列代码的功能是注册一个新的对话模板，名为 "deepseekv2"，用于构建对话。该模板定义了系统消息、角色及消息分隔符。此模版采用了 "DeepSeekV2" 的分隔符样式，并配置了停止标记和停止符号。

```
register_conv_template(
    Conversation(
        name="deepseekv2",
        system_template="{system_message}",
        system_message="",
        roles=("|<User>|", "|<Assistant>|"),
        messages=(),
        offset=0,
        sep_style=SeparatorStyle.DeepSeekV2,
        sep="\n< | sft_end | >",
        sep2="< | end_of_sentence | >",
        stop_token_ids=[100001],
        stop_str=["User:", "< | end_of_sentence | >"]
    )
)
```

（7）下列代码的功能是注册一个新的对话模板，名为 "plain"，用于构建简单对话。该模板没有系统消息和角色名称，采用 "PLAIN" 的分隔符样式，且配置了停止标记和停止符号。

```
register_conv_template(
    Conversation(
        name="plain",
        system_template="",
        system_message="",
        roles=("", ""),
        messages=(),
        offset=0,
        sep_style=SeparatorStyle.PLAIN,
        sep="",
        sep2="",
        stop_token_ids=[100001],
        stop_str=['</s>'],
    )
)
```

（8）下列代码的功能是注册一个新的对话模板，名为 "alignment"，用于构建具有对齐样式的对话。该模板没有系统消息和角色名称，采用 "ALIGNMENT" 的分隔符样式，且配置了停止标记和停止符号。

```
register_conv_template(
    Conversation(
```

```
        name="alignment",
        system_template="",
        system_message="",
        roles=("", ""),
        messages=(),
        offset=0,
        sep_style=SeparatorStyle.ALIGNMENT,
        sep="",
        sep2="",
        stop_token_ids=[100001],
        stop_str=['</s>'],
    )
)
```

（9）下列代码演示了使用不同的对话模板（"deepseek" 和 "deepseekv2"）来创建并生成对话提示的用法。通过获取这些模板的信息，追加对话消息，并打印输出生成的提示信息。

```
if __name__ == "__main__":
    print("deepseek template:")
    conv = get_conv_template("deepseek")
    conv.append_message(conv.roles[0], "Hello!")
    conv.append_message(conv.roles[1], "Hi! This is Tony.")
    conv.append_message(conv.roles[0], "Who are you?")
    conv.append_message(conv.roles[1], "I am a helpful assistant.")
    conv.append_message(conv.roles[0], "How are you?")
    conv.append_message(conv.roles[1], None)
    print(conv.get_prompt())

    print("deepseekv2 template:")
    conv = get_conv_template("deepseekv2")
    conv.append_message(conv.roles[0], "Hello!")
    conv.append_message(conv.roles[1], "Hi! This is Tony.")
    conv.append_message(conv.roles[0], "Who are you?")
    conv.append_message(conv.roles[1], "I am a helpful assistant.")
    conv.append_message(conv.roles[0], "How are you?")
    conv.append_message(conv.roles[1], None)
    print(conv.get_prompt())
```

上述代码展示了如何使用已注册的对话模板生成实际的对话历史，并且根据不同模板的样式生成适当输出的用法。

6.5.7 DeepSeek-VL2 模型总结

DeepSeek-VL2模型通过模块化设计，将模型的各个部分分解为配置管理、文本对话处理、视觉编码、语言生成、混合专家模块、数据预处理及权重初始化等组件，构成了一个多模态

Transformer框架。其中文本和语言部分采用标准Transformer解码器结构，并集成了多头注意力、层缩放、丢弃路径及辅助损失等技术；视觉部分则利用动态平铺策略和SigLIP Vision Transformer对高分辨率图像进行高效特征提取。此外，混合专家模块通过门控机制实现了输入特征的专家选择和分布式计算，进一步提高了模型的表达能力和推理效率。整个架构紧密结合了Hugging Face Transformers和timm的实现理念，通过精细的配置和多种优化策略，确保模型在多模态理解任务上具有高性能、可扩展性和稳定性。

在下面的内容中，将详细总结“models”目录的技术架构和原理，概括每个模块的主要功能、核心实现及关键技术。

1. configuration_deepseek.py

（1）主要功能：定义DeepSeek模型的各类配置类，为模型各部分（如语言模型、视觉编码器、MLP投影器等）提供超参数设置和初始化参数。

（2）核心实现：

- DeepseekV2Config包含语言模型的配置（如隐藏层维度、层数、注意力头数量、MoE专家设置等）。
- VisionEncoderConfig定义视觉编码器参数（如图像尺寸、补丁大小、Transformer层数、注意力头数等）。
- MlpProjectorConfig用于配置MLP投影器的参数，如输入维度、嵌入维度、层数及下采样比例等。

（3）技术原理：所有配置类均继承自Hugging Face的PretrainedConfig，实现与Transformers框架的无缝对接，并支持从预训练模型加载参数。

2. conversation.py

（1）主要功能：实现对话逻辑和会话模板，支持多轮对话交互。

（2）核心实现：

- 提供获取对话模板（如SFT模板）的接口。
- 实现对话消息的追加、格式化和生成完整对话提示。

（3）技术原理：通过将用户和助手的对话按照预定义模板格式化，生成符合模型输入要求的文本串，使多模态输入中的文本部分能与视觉信息对齐。

3. modeling_deepseek.py

（1）主要功能：实现DeepSeek模型的基本架构，兼容DeepSeekV2与DeepSeekV3。

（2）核心实现：

- 借鉴GPT-NeoX和OPT的Transformer架构，实现解码器、注意力层、前馈网络等核心组件。
- 包含权重初始化、位置编码（如旋转位置编码）及部分注意力机制的实现。

（3）技术原理：基于Transformer的解码器结构，采用残差连接、层归一化和截断正态初始化等技术，确保文本生成与语言建模任务的高效性和稳定性。

4. modeling_deepseek_vl_v2.py

（1）主要功能：扩展基本模型，实现视觉与语言的联合建模，构建DeepSeek-VL2多模态模型。

（2）核心实现：

- 集成视觉编码器（如基于SigLIP的VisionTransformer）以提取高分辨率图像特征。
- 引入视觉语言适配器，将视觉特征映射到语言模型嵌入空间。
- 使用MoE语言模型，结合MLA机制，提高推理效率和模型表达能力。

（3）技术原理：通过动态平铺策略和图像token格式化，将图像信息与文本数据有效融合，并利用稀疏计算技术实现专家网络的高效选择和融合。

5. processing_deepseek_vl_v2.py

（1）主要功能：实现多模态数据的预处理和后处理，包括图像分辨率选择、图像转换、对话格式化和批处理操作。

（2）核心实现：

- 定义函数select_best_resolution，用于根据输入图像的尺寸和候选分辨率选择最佳裁剪方案。
- 实现ImageTransform类，用于将PIL图像转换为张量，并进行归一化处理。
- 提供DeepseekVLV2Processor类，负责将对话与图像数据转换成模型所需的token序列、图像嵌入及对应掩码，并支持批量化处理。

（3）技术原理：通过结合torchvision变换和自定义对话模板，确保输入数据在转换过程中保持信息一致性和格式正确性，从而使多模态模型能够准确接收和处理输入。

6. siglip_vit.py

（1）主要功能：实现SigLIP Vision Transformer模型，用于高效提取图像特征。

（2）核心实现：

- 基于Vision Transformer架构，利用patch embedding、位置编码和Transformer块提取图像特征。
- 支持多种权重初始化方法（如截断正态分布初始化）和动态调整位置编码（如使用重采样位置编码）。
- 提供了不同规模模型配置（如siglip_so400m_patch14_384、siglip_large_patch16_384等），以适应不同图像尺寸和任务需求。

（3）技术原理：融合timm库的ViT实现和Hugging Face的Transformers框架，通过对图像进行分割、嵌入和自注意力编码，生成丰富的图像特征表示，为多模态模型提供高质量视觉输入。

7. 模型注册与集成

（1）主要功能：在目录末尾，通过AutoConfig.register和AutoModelForCausalLM.register等函数，将自定义的配置和模型类注册到Hugging Face Transformers框架中。

（2）技术原理：利用Transformers的自动化API，实现模型的便捷加载、配置和推理，使DeepSeek-VL2能够与现有生态系统无缝对接。

总之，“models”目录采用模块化设计，将模型的各个组成部分（配置、对话逻辑、视觉编码、语言生成、混合专家及数据预处理等）分离出来，每个模块都专注于特定任务，这样既保证了整体架构的灵活性和可扩展性，又通过引入先进的技术（如动态平铺、Flash Attention、MoE、LayerScale、旋转位置编码等）提升了模型的性能与训练稳定性。通过这些模块的协同工作，DeepSeek-VL2 构建了一个高效、灵活、面向多模态任务的技术框架。

6.6 模型部署和在线服务

在DeepSeek-VL2项目的“serve”目录中主要保存了模型部署相关的源码，包含了用于在线服务和Web Demo的代码，比如Gradio工具的封装、前端静态资源（如JS、CSS、图标、字体）及推理接口，实现了模型的加载、交互和部署，方便用户快速进行多模态任务的演示和测试。

6.6.1 设置部署参数

文件presets.py定义了部署DeepSeek-VL2 模型时的默认界面配置和参数设置，这些配置包括页面标题、顶部描述、并发请求数、事件最大数、图像尺寸限制，以及一组用于标记边框颜色的预设颜色映射。此外，还定义了一个标记用于表示已由解析器转换的内容，并且设置了一个名为“small_and_beautiful_theme”的Gradio主题，涵盖了按钮背景、边框、文本颜色、输入框背景等UI风格参数，旨在为模型的Web Demo提供一致且美观的前端展示。

```
title = """<h1 align="left" style="min-width:200px; margin-top:0;">Chat with
    DeepSeek-VL2 </h1>"""
description_top = """Special Tokens: `<image>`,     Visual
    Grounding: `<|ref|>{query}</|ref|>`,    Grounding Conversation:
    `<|grounding|>{question}`"""
description = """"""
CONCURRENT_COUNT = 1
MAX_EVENTS = 10
MAX_IMAGE_SIZE = 800
MIN_IMAGE_SIZE = 400

BOX2COLOR = {
    0: (255, 0, 0),          # 红色
    1: (0, 255, 0),          # 绿色
    2: (0, 0, 255),          # 蓝色
    3: (0, 255, 255),        # 青色
    4: (255, 255, 0),        # 黄色
    5: (255, 0, 255),        # 品红色
    6: (127, 127, 127),      # 灰色
    7: (255, 255, 127),      # 淡黄色
```

```
    8: (255, 127, 255),     # 淡品红色
    9: (127, 255, 255),     # 淡青色
    10: (127, 127, 255),    # 淡蓝色
    11: (127, 255, 127),    # 淡绿色
    12: (255, 127, 127),    # 淡红色
}


ALREADY_CONVERTED_MARK = "<!-- ALREADY CONVERTED BY PARSER. -->"

small_and_beautiful_theme = gr.themes.Soft(
    primary_hue=gr.themes.Color(
        c50="#EBFAF2",
        c100="#CFF3E1",
        c200="#A8EAC8",
        c300="#77DEA9",
        c400="#3FD086",
        c500="#02C160",
        c600="#06AE56",
        c700="#05974E",
        c800="#057F45",
        c900="#04673D",
        c950="#2E5541",
        name="small_and_beautiful",
    ),
    secondary_hue=gr.themes.Color(
        c50="#576b95",
        c100="#576b95",
        c200="#576b95",
        c300="#576b95",
        c400="#576b95",
        c500="#576b95",
        c600="#576b95",
        c700="#576b95",
        c800="#576b95",
        c900="#576b95",
        c950="#576b95",
    ),
    neutral_hue=gr.themes.Color(
        name="gray",
        c50="#f6f7f8",
        # c100="#f3f4f6",  # （原注释：旧版备用颜色）
        c100="#F2F2F2",
        c200="#e5e7eb",
        c300="#d1d5db",
        c400="#B2B2B2",
```

```
        c500="#808080",
        c600="#636363",
        c700="#515151",
        c800="#393939",
        # c900="#272727",  # （原注释：旧版备用颜色）
        c900="#2B2B2B",
        c950="#171717",
    ),
    radius_size=gr.themes.sizes.radius_sm,
).set(
    # 以下注释的配置项为深色模式下的按钮和界面风格设置
    # button_primary_background_fill="*primary_500",
                                        # 主按钮背景填充色（默认，已注释）
    button_primary_background_fill_dark="*primary_600",
                                        # 主按钮深色模式背景填充色
    # button_primary_background_fill_hover="*primary 400",
                                        # 主按钮悬停时背景填充色（默认，已注释）
    # button_primary_border_color="*primary_500",
                                        # 主按钮边框颜色（默认，已注释）
    button_primary_border_color_dark="*primary_600",  # 主按钮深色模式边框颜色
    button_primary_text_color="white",          # 主按钮文本颜色
    button_primary_text_color_dark="white",   # 主按钮深色模式文本颜色
    button_secondary_background_fill="*neutral_100",  # 次级按钮背景填充色
    button_secondary_background_fill_hover="*neutral_50",
                                        # 次级按钮悬停时背景填充色
    button_secondary_background_fill_dark="*neutral_900",
                                        # 次级按钮深色模式背景填充色
    button_secondary_text_color="*neutral_800",# 次级按钮文本颜色
    button_secondary_text_color_dark="white",  # 次级按钮深色模式文本颜色
    # background_fill_primary="#F7F7F7",        # 主背景填充色（默认，已注释）
    # background_fill_primary_dark="#1F1F1F",   # 主背景深色模式填充色(默认,已注释)
    # block_title_text_color="*primary_500",    # 模块标题文本颜色（默认，已注释）
    block_title_background_fill_dark="*primary_900",
                                        # 模块标题深色模式背景填充色
    block_label_background_fill_dark="*primary_900",  # 模块标签深色模式背景填充色
    input_background_fill="#F6F6F6",# 输入框背景填充色
    # chatbot_code_background_color_dark="*neutral_950",
                                # 聊天机器人代码块深色模式背景颜色(默认,已注释)
)
```

对上述代码的具体说明如下所示：

（1）页面元素与描述。

- title：定义页面标题的HTML代码，设置为“Chat with DeepSeek-VL2”。
- description_top：定义顶部描述信息，列出了特殊token及其用途，如 <image>、用于视觉定位的 <|ref|>{query}<|/ref|>，以及用于对话中的 <|grounding|>{question}。

（2）并发与图像参数：CONCURRENT_COUNT、MAX_EVENTS、MAX_IMAGE_SIZE和MIN_IMAGE_SIZE分别定义了并发数量、最大事件数及图像尺寸的上限和下限。

（3）边框颜色映射：BOX2COLOR定义了一组数字到RGB颜色的映射，便于在可视化过程中为不同类别的边框赋予不同颜色。

（4）转换标记：ALREADY_CONVERTED_MARK是一个标记，用于指示内容已经被解析器转换过。

（5）Gradio主题设置：small_and_beautiful_theme定义了一个Gradio主题，包含了主色调（primary_hue）、次级色调（secondary_hue）和中性色调（neutral_hue）的具体颜色配置，以及按钮、背景、输入框等组件在不同模式下的风格设置。

总之，文件presets.py用于为DeepSeek-VL2 的Web服务界面提供预设的文本和界面风格配置，以确保前端展示美观且统一，同时也为后续部署和交互提供必要的参数支持。

6.6.2 工具函数

文件utils.py提供了一系列工具函数，用于配置日志记录、文本处理、Markdown格式转换、图像与HTML的转换，以及一些辅助函数（如检测停用词、转换代码中的语言标签、解析视觉定位的边界框等）。这些工具主要用于DeepSeek-VL2 项目的前端部署部分，帮助格式化输出、转换和显示多模态内容，为Web Demo和在线服务提供支持。

（1）函数configure_logger() 的功能是配置并返回日志记录器（logger），该记录器名为 "gradio_logger"，设置了DEBUG级别，并将日志同时输出到控制台和文件。日志文件存放在“deepseek_vl2/serve/logs”目录下，并以当前时间戳命名，确保记录的日志格式统一且便于追踪调试信息。

```
# 日志配置函数
def configure_logger():
    """
    配置并返回日志记录器，用于记录 Gradio 相关日志。
    """
    logger = logging.getLogger("gradio_logger")
    logger.setLevel(logging.DEBUG)

    timestr = time.strftime("%Y%m%d-%H%M%S")
    os.makedirs("deepseek_vl2/serve/logs", exist_ok=True)
    file_handler = logging.FileHandler(
        f"deepseek_vl2/serve/logs/{timestr}_gradio_log.log"
    )
    console_handler = logging.StreamHandler()

    formatter = logging.Formatter(
        "%(asctime)s - %(name)s - %(levelname)s - %(message)s"
    )
    console_handler.setFormatter(formatter)
```

```
    file_handler.setFormatter(formatter)

    console_handler.setLevel(logging.INFO)
    file_handler.setLevel(logging.INFO)

    logger.addHandler(console_handler)
    logger.addHandler(file_handler)

    return logger
```

（2）函数strip_stop_words(x, stop_words) 的功能是检查输入字符串x中是否包含停用词列表stop_words中的任意一个单词，如果存在，则返回停用词首次出现位置之前的子字符串，并去除首尾空白；否则直接返回去除首尾空白后的整个字符串。

```
# 去除停用词（或停止词）的函数
def strip_stop_words(x, stop_words):
    """
    去除字符串中首次出现停用词后的部分。

    Args:
        x (str): 输入字符串。
        stop_words (list): 停用词列表。

    Returns:
        str: 截取停用词前的字符串，去除首尾空白。
    """
    for w in stop_words:
        if w in x:
            return x[: x.index(w)].strip()
    return x.strip()
```

（3）函数format_output(history, text, x) 的功能是将当前对话中的新问答对［text, x］添加到已有的对话历史history中，并将每个回复内容转换为Markdown格式。函数返回格式化后的对话输出和更新后的对话历史，便于后续展示和进一步处理。

```
# 格式化输出，将历史对话和当前文本合并，并转换为 Markdown 格式
def format_output(history, text, x):
    """
    格式化输出，将历史对话与当前文本合并，并转换为 Markdown 格式。

    Args:
        history: 历史对话记录。
        text (str): 当前文本。
        x (str): 当前回复文本。
```

```
    Returns:
        tuple: （格式化后的 Markdown 输出，更新后的历史记录）
    """
    updated_history = history + [[text, x]]
    a = [[y[0], convert_to_markdown(y[1])] for y in updated_history]
    return a, updated_history
```

（4）函数markdown_to_html_with_syntax_highlight(md_str) 目前已经被废弃，功能是将Markdown格式的字符串md_str转换为HTML，并对代码块应用语法高亮。它通过正则表达式提取代码块，使用Pygments库对代码进行高亮处理，最后将转换后的文本通过Markdown转换器生成HTML。

```
# 将 Markdown 转换为带有语法高亮的 HTML（已废弃）
def markdown_to_html_with_syntax_highlight(md_str):  # deprecated
    """
    将 Markdown 字符串转换为带有语法高亮的 HTML。（已废弃）

    Args:
        md_str (str): Markdown 字符串。

    Returns:
        str: 转换后的 HTML 字符串。
    """
    def replacer(match):
        lang = match.group(1) or "text"
        code = match.group(2)

        try:
            lexer = get_lexer_by_name(lang, stripall=True)
        except ValueError:
            lexer = get_lexer_by_name("text", stripall=True)

        formatter = HtmlFormatter()
        highlighted_code = highlight(code, lexer, formatter)

        return f'<pre><code class="{lang}">{highlighted_code}</code></pre>'

    code_block_pattern = r"```(\w+)?\n([\s\S]+?)\n```"
    md_str = re.sub(code_block_pattern, replacer, md_str, flags=re.MULTILINE)

    html_str = markdown(md_str)
    return html_str
```

（5）函数normalize_markdown(md_text) 目前已经被废弃，功能是对输入的Markdown文本md_text进行归一化处理，主要针对列表项的格式调整，确保列表前后空行合理，使得Markdown格式更加统一和规范。

```
# 归一化 Markdown 文本（已废弃）
def normalize_markdown(md_text: str) -> str:  # deprecated
    """
    对 Markdown 文本进行归一化处理，调整列表格式的间距。（已废弃）

    Args:
        md_text (str): 输入的 Markdown 文本。

    Returns:
        str: 归一化后的 Markdown 文本。
    """
    lines = md_text.split("\n")
    normalized_lines = []
    inside_list = False

    for i, line in enumerate(lines):
        if re.match(r"^(\d+\.|-|\*|\+)\s", line.strip()):
            if not inside_list and i > 0 and lines[i - 1].strip() != "":
                normalized_lines.append("")
            inside_list = True
            normalized_lines.append(line)
        elif inside_list and line.strip() == "":
            if i < len(lines) - 1 and not re.match(r"^(\d+\.|-|\*|\+)\s",
                lines[i + 1].strip()):
                normalized_lines.append(line)
            continue
        else:
            inside_list = False
            normalized_lines.append(line)

    return "\n".join(normalized_lines)
```

（6）函数 convert_mdtext(md_text) 的功能是将 Markdown 文本 md_text 转换为 HTML 格式，同时处理代码块和内联代码，调用归一化和高亮函数对文本进行格式化，并在最终结果后附加转换标记，表明文本已被转换。

```
# 将 Markdown 文本转换为 HTML 格式，同时添加转换标记
def convert_mdtext(md_text):
    """
    将 Markdown 文本转换为 HTML 格式，并附加转换标记。

    Args:
        md_text (str): 输入的 Markdown 文本。
```

```
    Returns:
        str: 转换后的 HTML 字符串，并附加转换标记。
    """
    code_block_pattern = re.compile(r"```(.*?)(?:```|$)", re.DOTALL)
    inline_code_pattern = re.compile(r"`(.*?)`", re.DOTALL)
    code_blocks = code_block_pattern.findall(md_text)
    non_code_parts = code_block_pattern.split(md_text)[::2]

    result = []
    for non_code, code in zip(non_code_parts, code_blocks + [""]):
        if non_code.strip():
            non_code = normalize_markdown(non_code)
            if inline_code_pattern.search(non_code):
                result.append(markdown(non_code, extensions=["tables"]))
            else:
                result.append(mdtex2html.convert(non_code,
                                    extensions=["tables"]))
        if code.strip():
            code = f"\n```{code}\n\n```"
            code = markdown_to_html_with_syntax_highlight(code)
            result.append(code)
    result = "".join(result)
    result += ALREADY_CONVERTED_MARK
    return result
```

（7）函数convert_asis(userinput) 的功能是将用户输入的文本直接转换为HTML格式的段落（使用 <p> 标签，设置white-space: pre-wrap），同时对文本进行HTML转换，并在末尾附加转换标记，确保原样显示用户输入内容。

```
# 直接将输入原样转换为 HTML 格式显示
def convert_asis(userinput):
    """
    将用户输入的文本直接转换为 HTML 格式，并附加转换标记。

    Args:
        userinput (str): 用户输入文本。

    Returns:
        str: HTML 格式的文本。
    """
    return f'<p style="white-space:pre-wrap;">{html.escape(userinput)}</
            p>{ALREADY_CONVERTED_MARK}'
```

（8）函数is_stop_word_or_prefix(s, stop_words) 的功能是判断字符串s是否以停用词列表stop_words中的任一停用词结尾，若是则返回True，否则返回False，用于辅助判断文本是否应被截断或停止处理。

```
# 判断字符串是否以停用词或停用词前缀结束
def is_stop_word_or_prefix(s: str, stop_words: list) -> bool:
    """
    判断字符串是否以停用词或停用词前缀结尾。

    Args:
        s (str): 输入字符串。
        stop_words (list): 停用词列表。

    Returns:
        bool: 如果字符串以任一停用词结尾，则返回 True，否则返回 False。
    """
    return any(s.endswith(stop_word) for stop_word in stop_words)
```

（9）函数 detect_converted_mark(userinput) 的功能是检查用户输入的文本是否以预定义的转换标记（ALREADY_CONVERTED_MARK）结尾，从而判断文本是否已经格式转换，若是则返回 True，否则返回 False。

```
# 检测用户输入是否已经包含转换标记
def detect_converted_mark(userinput):
    """
    检测用户输入是否以预定义的转换标记结尾。

    Args:
        userinput (str): 用户输入文本。

    Returns:
        bool: 如果包含转换标记，则返回 True，否则返回 False。
    """
    return bool(userinput.endswith(ALREADY_CONVERTED_MARK))
```

（10）函数 detect_language(code) 的功能是从给定的代码字符串中提取第一行，判断该行内容是否为语言提示（通常代表代码语言），然后返回检测到的语言（以小写形式）和去除语言提示后的代码文本。

```
# 检测代码的语言标签
def detect_language(code):
    """
    检测代码的语言标签，并返回第一行文本作为语言及剥离后的代码内容。

    Args:
        code (str): 包含代码的字符串。
```

```
    Returns:
        tuple: (语言标签，剥离语言标签后的代码)
    """
    first_line = "" if code.startswith("\n") else code.strip().split("\n", 1)[0]
    language = first_line.lower() if first_line else ""
    code_without_language = code[len(first_line):].lstrip() if first_line
else code
    return language, code_without_language
```

（11）函数convert_to_markdown(text) 的功能是将输入文本text转换为Markdown格式。它先替换特殊字符（如将 $ 转义为HTML实体），处理换行符，并对每行的前导空格和制表符进行转换，然后对部分Markdown语法（如标题标记）进行转义，生成规范的Markdown文本。

```
# 将文本转换为 Markdown 格式
def convert_to_markdown(text):
    """
    将文本转换为 Markdown 格式。

    Args:
        text (str): 输入文本。

    Returns:
        str: 转换后的 Markdown 文本，其中特殊字符被转义以防止 HTML 注入。
    """
    text = text.replace("$", "&#36;")
    text = text.replace("\r\n", "\n")

    def replace_leading_tabs_and_spaces(line):
        new_line = []

        for char in line:
            if char == "\t":
                new_line.append("&#9;")
            elif char == " ":
                new_line.append(" ")
            else:
                break
        return "".join(new_line) + line[len(new_line):]

    markdown_text = ""
    lines = text.split("\n")
    in_code_block = False

    for line in lines:
        if in_code_block is False and line.startswith("```"):
```

```
            in_code_block = True
            markdown_text += f"{line}\n"
        elif in_code_block is True and line.startswith("```"):
            in_code_block = False
            markdown_text += f"{line}\n"
        elif in_code_block:
            markdown_text += f"{line}\n"
        else:
            line = replace_leading_tabs_and_spaces(line)
            line = re.sub(r"^(#)", r"\\\1", line)
            markdown_text += f"{line}  \n"

    return markdown_text
```

（12）函数add_language_tag(text) 的功能是为输入文本中的代码块自动添加语言标签。通过正则表达式匹配代码块，并利用Pygments库猜测代码语言，将相应的语言标识添加到代码块的Markdown语法中，从而增强语法高亮效果。

```
# 为代码块添加语言标签
def add_language_tag(text):
    """
    为代码块添加适当的语言标签。

    Args:
        text (str): 输入文本，可能包含代码块。

    Returns:
        str: 处理后带有语言标签的文本。
    """
    def detect_language(code_block):
        try:
            lexer = guess_lexer(code_block)
            return lexer.name.lower()
        except ClassNotFound:
            return ""

    code_block_pattern = re.compile(r"(```)(\w*\n[^`]+```)", re.MULTILINE)

    def replacement(match):
        code_block = match.group(2)
        if match.group(2).startswith("\n"):
            language = detect_language(code_block)
            return (
                f"```{language}{code_block}```" if language else f"```\
                  n{code_block}```"
```

```
        )
    else:
        return match.group(1) + code_block + "```"

text2 = code_block_pattern.sub(replacement, text)
return text2
```

（13）函数is_variable_assigned(var_name) 的功能是检查当前局部变量作用域中是否存在名称为var_name的变量，若存在则返回True，否则返回False，主要用于调试或动态检测变量状态。

```
# 检测变量是否被赋值（此函数可能无实际作用，仅用于调试）
def is_variable_assigned(var_name: str) -> bool:
    """
    检查本地变量中是否存在指定变量名称。

    Args:
        var_name (str): 变量名称。

    Returns:
        bool: 如果变量存在则返回 True，否则返回 False。
    """
    return var_name in locals()
```

（14）函数pil_to_base64()的功能是将PIL图像转换为Base64 编码的HTML <img> 标签字符串。它可根据参数选择是否调整图像大小，然后将图像保存到内存缓冲区，进行Base64 编码，并生成包含指定替代文本（alt）的HTML图像标签。

```
# 将 PIL 图像转换为 Base64 编码的 HTML <img> 标签字符串
def pil_to_base64(
    image: Image.Image,
    alt: str = "user upload image",
    resize: bool = True,
    max_size: int = MAX_IMAGE_SIZE,
    min_size: int = MIN_IMAGE_SIZE,
    format: str = "JPEG",
    quality: int = 95
) -> str:
    """
    将 PIL 图像转换为 Base64 编码的 HTML <img> 标签字符串。

    Args:
        image (PIL.Image.Image): 输入图像。
        alt (str): 图片的替代文本。默认为 "user upload image"。
        resize (bool): 是否调整图像大小。默认为 True。
        max_size (int): 图像的最大尺寸。默认为 MAX_IMAGE_SIZE。
```

```
        min_size (int): 图像的最小尺寸。默认为 MIN_IMAGE_SIZE。
        format (str): 保存图像的格式，默认为 "JPEG"。
        quality (int): 图像保存质量，默认为 95。

    Returns:
        str: 包含 Base64 编码图像数据的 HTML <img> 标签字符串。
    """
    if resize:
        max_hw, min_hw = max(image.size), min(image.size)
        aspect_ratio = max_hw / min_hw
        shortest_edge = int(min(max_size / aspect_ratio, min_size, min_hw))
        longest_edge = int(shortest_edge * aspect_ratio)
        W, H = image.size
        if H > W:
            H, W = longest_edge, shortest_edge
        else:
            H, W = shortest_edge, longest_edge
        image = image.resize((W, H))

    buffered = io.BytesIO()
    image.save(buffered, format=format, quality=quality)
    img_b64_str = base64.b64encode(buffered.getvalue()).decode()
    img_str = f'<img src="data:image/png;base64,{img_b64_str}" alt="{alt}" />'

    return img_str
```

（15）函数 parse_ref_bbox(response, image) 的功能是解析模型响应 response 中的视觉定位标记（如 <|ref|> 与 <|det|>），提取边界框坐标和对应的标签，并在传入的图像上绘制边界框和标签的图像；如果解析失败，则返回 None。

```
# 解析视觉定位响应中的边界框
def parse_ref_bbox(response, image: Image.Image):
    """
    从模型的响应中解析视觉定位的边界框信息，并在图像上绘制对应的矩形和标签。

    Args:
        response (str): 模型的响应文本，包含 <|ref|> 和 <|det|> 标签。
        image (PIL.Image.Image): 输入图像。

    Returns:
        PIL.Image.Image or None: 绘制边界框和标签后的图像，如果解析失败则返回 None。
    """
    try:
        image = image.copy()
        image_h, image_w = image.size
```

```
        draw = ImageDraw.Draw(image)

        ref = re.findall(r'<\|ref\|>.*?<\|/ref\|>', response)
        bbox = re.findall(r'<\|det\|>.*?<\|/det\|>', response)
        assert len(ref) == len(bbox)

        if len(ref) == 0:
            return None

        boxes, labels = [], []
        for box, label in zip(bbox, ref):
            box = box.replace('<|det|>', '').replace('<|/det|>', '')
            label = label.replace('<|ref|>', '').replace('<|/ref|>', '')
            box = box[1:-1]
            for onebox in re.findall(r'\[.*?\]', box):
                boxes.append(eval(onebox))
                labels.append(label)

        for indice, (box, label) in enumerate(zip(boxes, labels)):
            box = (
                int(box[0] / 999 * image_h),
                int(box[1] / 999 * image_w),
                int(box[2] / 999 * image_h),
                int(box[3] / 999 * image_w),
            )

            box_color = BOX2COLOR[indice % len(BOX2COLOR.keys())]
            box_width = 3
            draw.rectangle(box, outline=box_color, width=box_width)

            text_x = box[0]
            text_y = box[1] - 20
            text_color = box_color
            font = ImageFont.truetype("deepseek_vl2/serve/assets/simsun.
                                      ttc", size=20)
            draw.text((text_x, text_y), label, font=font, fill=text_color)

        # 返回绘制了边界框和标签的图像
        return image
    except:
        return None
```

（16）函数display_example(image_list) 的功能是接收图片路径列表image_list，依次打开每张图片，将图片转换为Base64 编码的HTML <img> 标签，并将所有标签组合成flex布局的HTML字符串，用于在前端界面中展示示例图片。

```
# 显示示例图片，将图片转换为 HTML 格式展示
def display_example(image_list):
    """
    将多个示例图片转换为 HTML 格式，以便在前端展示。

    Args:
        image_list (list): 图片路径列表。

    Returns:
        str: 包含 HTML <img> 标签的字符串，用于展示图片。
    """
    images_html = ""
    for i, img_path in enumerate(image_list):
        image = Image.open(img_path)
        buffered = io.BytesIO()
        image.save(buffered, format="PNG", quality=100)
        img_b64_str = base64.b64encode(buffered.getvalue()).decode()
        img_str = f'<img src="data:image/png;base64,{img_b64_str}" alt="{img_
                  path}" style="height:80px; margin-right: 10px;" />'
        images_html += img_str

    result_html = f"""
    <div style="display: flex; align-items: center; margin-bottom: 10px;">
        <div style="flex: 1; margin-right: 10px;">{images_html}</div>
    </div>
    """

    return result_html
```

上述函数共同构成了DeepSeek-VL2项目中前端展示和数据预处理的辅助工具，涵盖日志记录、文本和Markdown格式转换、图像处理与转换，以及视觉定位结果的解析与可视化。

6.6.3 Gradio 工具

文件gradio_utils.py实现了Gradio应用程序中的辅助功能与状态管理工具，包括生成函数包装、对话历史的管理与重置、文本框内容的重置，以及输出中断的状态控制。其中，State类用于控制全局中断状态。通过这些功能，能够有效地简化交互界面中各组件的逻辑与操作。

```
from functools import wraps
import gradio as gr

def wrap_gen_fn(gen_fn):
    """
    包装生成函数（generation function），捕获异常并将其封装为 Gradio 的错误提示。
```

```
    Args:
        gen_fn (function): 文本生成函数。

    Returns:
        wrapped_gen_fn (function): 包装后的生成函数。
    """
    @wraps(gen_fn)
    def wrapped_gen_fn(prompt, *args, **kwargs):
        try:
            # 直接调用原始生成函数
            yield from gen_fn(prompt, *args, **kwargs)
        except gr.Error as g_err:
            # 捕获 Gradio 自定义错误
            raise g_err
        except Exception as e:
            # 捕获一般异常并将其包装为 Gradio 错误
            raise gr.Error(f"Failed to generate text: {e}") from e

    return wrapped_gen_fn


def delete_last_conversation(chatbot, history):
    """
    删除对话历史中的最后一条会话记录。

    Args:
        chatbot (list): Chatbot 的显示内容列表。
        history (list): 对话历史记录。

    Returns:
        tuple: 更新后的 Chatbot 内容、对话历史和删除状态消息。
    """
    if len(history) % 2 != 0:
        gr.Error("history length is not even")
        return (
            chatbot,
            history,
            "Delete Done",
        )

    if len(chatbot) > 0:
        chatbot.pop()
    if len(history) > 0 and len(history) % 2 == 0:
        history.pop()
        history.pop()
    return (
```

```
        chatbot,
        history,
        "Delete Done",
    )


def reset_state():
    """
    重置对话历史和状态。

    Returns:
        tuple: 清空后的 chatbot 内容、历史记录及重置完成消息。
    """
    return [], [], None, "Reset Done"


def reset_textbox():
    """
    重置文本框的内容。

    Returns:
        tuple: 重置后的文本框更新对象和空字符串。
    """
    return gr.update(value=""), ""


def cancel_outputing():
    """
    取消正在输出的文本生成过程。

    Returns:
        str: 取消完成的状态消息。
    """
    return "Stop Done"


class State:
    """
    用于管理输出中断的全局状态类。
    """
    interrupted = False

    def interrupt(self):
        """
        将中断状态设置为 True，标识需要停止输出。
        """
```

```
        self.interrupted = True

    def recover(self):
        """
        恢复中断状态为 False，标识输出可以继续。
        """
        self.interrupted = False

# 共享的状态对象，用于控制输出中断
shared_state = State()
```

通过上述代码封装了与Gradio相关的工具函数，简化了前端组件的调用和配置，帮助快速构建交互式界面。

6.6.4 模板覆盖与扩展

文件overwrites.py的主要功能是对DeepSeek-VL2 项目中的文本处理和界面输出进行自定义扩展，同时提供了JavaScript加载与模板覆盖功能，以优化用户交互体验。其核心功能如下所示。

- 文本块压缩（compact_text_chunks）：将输入文本块进行合并与重新分割，以适应模型处理需求。
- 后处理（postprocess）：对模型生成的对话进行后处理，将用户和模型生成的Markdown格式消息转换为HTML。
- JavaScript文件加载（reload_javascript）：动态加载自定义的JavaScript文件并插入Gradio模板中。
- 覆盖Gradio模型响应（GradioTemplateResponseOriginal）：以重新定义模板响应方法来支持自定义JavaScript。

```
def compact_text_chunks(self, prompt, text_chunks: List[str]) -> List[str]:
    """
    将文本块进行压缩和重新分割。

    Args:
        prompt (str): 输入的提示文本。
        text_chunks (List[str]): 待处理的文本块列表。

    Returns:
        List[str]: 重新分割后的文本块列表。
    """
    logging.debug("Compacting text chunks...🚀🚀🚀")
    # 去除空白内容的块
    combined_str = [c.strip() for c in text_chunks if c.strip()]
    # 为每个块添加编号
    combined_str = [f"[{index+1}] {c}" for index, c in enumerate(combined_str)]
```

```
    # 使用两个换行符连接所有块
    combined_str = "\n\n".join(combined_str)
    # 根据 self.max_chunk_overlap 重新切分
    text_splitter = self.get_text_splitter_given_prompt(prompt, 1, padding=1)
    return text_splitter.split_text(combined_str)


def postprocess(
    self, y: List[Tuple[str | None, str | None]]
) -> List[Tuple[str | None, str | None]]:
    """
    对模型生成的对话进行后处理，将 Markdown 转换为 HTML。

    Args:
        y (List[Tuple[str | None, str | None]]): 用户和模型消息的元组列表，每个消
                                                  息可能为 Markdown 格式。

    Returns:
        List[Tuple[str | None, str | None]]: 用户和模型消息的元组列表，每个消息为
                                             HTML 格式。
    """
    if y is None or y == []:
        return []
    temp = []
    for x in y:
        user, bot = x
        # 检查并转换用户消息
        if not detect_converted_mark(user):
            user = convert_asis(user)
        # 检查并转换模型消息
        if not detect_converted_mark(bot):
            bot = convert_mdtext(bot)
        temp.append((user, bot))
    return temp


with open("deepseek_vl2/serve/assets/custom.js", "r", encoding="utf-8") as f, open(
    "deepseek_vl2/serve/assets/Kelpy-Codos.js", "r", encoding="utf-8"
) as f2:
    # 读取自定义 JavaScript 文件
    customJS = f.read()
    kelpyCodos = f2.read()


def reload_javascript():
    """
```

```
    重新加载并插入自定义 JavaScript 代码。
    """
    print("Reloading javascript...")
    js = f"<script>{customJS}</script><script>{kelpyCodos}</script>"

    def template_response(*args, **kwargs):
        """
        在 Gradio 模板响应中插入自定义 JavaScript。
        """
        res = GradioTemplateResponseOriginal(*args, **kwargs)
        # 插入自定义脚本到 HTML 结构中
        res.body = res.body.replace(b"</html>", f"{js}</html>".encode("utf8"))
        res.init_headers()
        return res

    # 覆盖 Gradio 模板响应函数
    gr.routes.templates.TemplateResponse = template_response


# 保存原始的 Gradio 模板响应类
GradioTemplateResponseOriginal = gr.routes.templates.TemplateResponse
```

6.6.5 Web 前端

“serve/assets/”目录主要用于存放DeepSeek-VL2 项目中Web前端界面的静态资源，包括JavaScript脚本（如Kelpy-Codos.js和custom.js）、样式表（custom.css）、字体文件（simsun.ttc）、图标（favicon.ico）及其他图像资源（如avatar.png），为界面交互与展示提供必要的资源支持。

其中，文件Kelpy-Codos.js的主要功能是为网页中的 <pre> 代码块动态添加“复制”按钮，用户单击按钮即可将代码块内容复制到剪贴板。该脚本通过MutationObserver监听DOM变化，确保在新添加的代码块中也能自动插入复制按钮，并提供图标提示复制状态的反馈。

```
(function () {
  "use strict";

  function addCopyButton(pre) {
    var code = pre.querySelector("code");
    if (!code) {
      return; // 如果没有找到 <code> 元素，则不添加按钮
    }
    var firstChild = code.firstChild;
    if (!firstChild) {
      return; // 如果 <code> 元素没有子节点，则不添加按钮
    }
```

```
    var button = document.createElement("button");
    button.textContent = "\uD83D\uDCCE";   // 使用 📎 符号作为“复制”按钮的文本
    button.style.position = "relative";
    button.style.float = "right";
    button.style.fontSize = "1em";                   // 可选：调整按钮大小
    button.style.background = "none";                // 可选：去掉背景颜色
    button.style.border = "none";                    // 可选：去掉边框
    button.style.cursor = "pointer";                 // 可选：显示指针样式
    button.addEventListener("click", function () {
      var range = document.createRange();
      range.selectNodeContents(code);
      range.setStartBefore(firstChild);              // 将范围设置为第一个子节点之前
      var selection = window.getSelection();
      selection.removeAllRanges();
      selection.addRange(range);

      try {
        var success = document.execCommand("copy");
        if (success) {
          button.textContent = "\u2714";
          setTimeout(function () {
            button.textContent = "\uD83D\uDCCE"; // 恢复按钮为“复制”
          }, 2000);
        } else {
          button.textContent = "\u2716";
        }
      } catch (e) {
        console.error(e);
        button.textContent = "\u2716";
      }

      selection.removeAllRanges();
    });
    code.insertBefore(button, firstChild); // 将按钮插入第一个子元素之前
  }

  function handleNewElements(mutationsList, observer) {
    for (var mutation of mutationsList) {
      if (mutation.type === "childList") {
        for (var node of mutation.addedNodes) {
          if (node.nodeName === "PRE") {
            addCopyButton(node);
          }
        }
      }
    }
```

```
    }

    var observer = new MutationObserver(handleNewElements);
    observer.observe(document.documentElement, {
      childList: true,
      subtree: true,
    });

    document.querySelectorAll("pre").forEach(addCopyButton);
})();
```

6.6.6 模型推理

文件inference.py的主要功能是为DeepSeek-VL2模型提供推理生成的逻辑，包括加载模型、转换对话为模型输入格式、定义停止条件，以及进行多模态推理生成文本的核心过程。代码通过调用generate函数支持流式输出文本，结合图像输入和多种生成参数控制文本生成。

```
def load_model(model_path, dtype=torch.bfloat16):
    # 加载 DeepSeek-VL2 处理器、分词器和模型
    vl_chat_processor = DeepseekVLV2Processor.from_pretrained(model_path)
    tokenizer = vl_chat_processor.tokenizer

    vl_gpt: DeepseekVLV2ForCausalLM = AutoModelForCausalLM.from_pretrained(
        model_path, trust_remote_code=True, torch_dtype=dtype
    )
    vl_gpt = vl_gpt.cuda().eval()
    return tokenizer, vl_gpt, vl_chat_processor


def convert_conversation_to_prompts(conversation: Conversation):
    # 将对话转换为模型需要的提示格式
    conv_prompts = []
    last_image = None
    messages = conversation.messages
    for i in range(0, len(messages), 2):
        if isinstance(messages[i][1], tuple):
            text, images = messages[i][1]
            last_image = images[-1]
        else:
            text, images = messages[i][1], []

        prompt = {
            "role": messages[i][0],
            "content": text,
```

```
            "images": images
        }
        response = {"role": messages[i + 1][0], "content": messages[i + 1][1]}
        conv_prompts.extend([prompt, response])

    return conv_prompts, last_image


class StoppingCriteriaSub(StoppingCriteria):
    # 定义自定义停止条件类，当模型生成特定序列时停止生成
    def __init__(self, stops=[], encounters=1):
        super().__init__()
        self.stops = [stop.to("cuda") for stop in stops]

    def __call__(
        self, input_ids: torch.LongTensor, scores: torch.FloatTensor, **kwargs
    ):
        for stop in self.stops:
            if input_ids.shape[-1] < len(stop):
                continue
            if torch.all((stop == input_ids[0][-len(stop):])).item():
                return True
        return False


@torch.inference_mode()
def deepseek_generate(
    conversations: list,
    vl_gpt: torch.nn.Module,
    vl_chat_processor: DeepseekVLV2Processor,
    tokenizer: transformers.PreTrainedTokenizer,
    stop_words: list,
    max_length: int = 256,
    temperature: float = 1.0,
    top_p: float = 1.0,
    repetition_penalty: float = 1.1,
    chunk_size: int = -1
):
    # 用于处理多模态模型推理并生成文本
    pil_images = []
    for message in conversations:
        if "images" not in message:
            continue
        pil_images.extend(message["images"])

    prepare_inputs = vl_chat_processor.__call__(
```

```
        conversations=conversations,
        images=pil_images,
        inference_mode=True,
        force_batchify=True,
        system_prompt=""
    ).to(vl_gpt.device)

    return generate(
        vl_gpt,
        tokenizer,
        prepare_inputs,
        max_gen_len=max_length,
        temperature=temperature,
        repetition_penalty=repetition_penalty,
        top_p=top_p,
        stop_words=stop_words,
        chunk_size=chunk_size
    )


@torch.inference_mode()
def generate(
    vl_gpt,
    tokenizer,
    prepare_inputs,
    max_gen_len: int = 256,
    temperature: float = 0,
    repetition_penalty=1.1,
    top_p: float = 0.95,
    stop_words: List[str] = [],
    chunk_size: int = -1
):
    """ 以流式输出方式从多模态模型中生成文本。"""
    streamer = TextIteratorStreamer(tokenizer, skip_prompt=True)

    # 将停止词转换为 ID 列表
    stop_words_ids = [
        torch.tensor(tokenizer.encode(stop_word)) for stop_word in stop_words
    ]
    stopping_criteria = StoppingCriteriaList(
        [StoppingCriteriaSub(stops=stop_words_ids)]
    )

    if chunk_size != -1:
        # 增量预填充模式
        inputs_embeds, past_key_values = vl_gpt.incremental_prefilling(
```

```
            input_ids=prepare_inputs.input_ids,
            images=prepare_inputs.images,
            images_seq_mask=prepare_inputs.images_seq_mask,
            images_spatial_crop=prepare_inputs.images_spatial_crop,
            attention_mask=prepare_inputs.attention_mask,
            chunk_size=chunk_size
        )
    else:
        inputs_embeds = vl_gpt.prepare_inputs_embeds(**prepare_inputs)
        past_key_values = None

    generation_config = dict(
        inputs_embeds=inputs_embeds,
        input_ids=prepare_inputs.input_ids,
        images=prepare_inputs.images,
        images_seq_mask=prepare_inputs.images_seq_mask,
        images_spatial_crop=prepare_inputs.images_spatial_crop,
        attention_mask=prepare_inputs.attention_mask,
        past_key_values=past_key_values,
        pad_token_id=tokenizer.eos_token_id,
        bos_token_id=tokenizer.bos_token_id,
        eos_token_id=tokenizer.eos_token_id,
        max_new_tokens=max_gen_len,
        do_sample=True,
        use_cache=True,
        streamer=streamer,
        stopping_criteria=stopping_criteria,
    )

    if temperature > 0:
        generation_config.update(
            {
                "do_sample": True,
                "top_p": top_p,
                "temperature": temperature,
                "repetition_penalty": repetition_penalty,
            }
        )
    else:
        generation_config["do_sample"] = False

    # 使用线程异步生成文本
    thread = Thread(target=vl_gpt.generate, kwargs=generation_config)
    thread.start()

    yield from streamer
```

上述代码实现了 DeepSeek-VL2 模型的推理功能，其核心功能如下所示。

- 模型加载 (load_model)：加载 DeepSeek-VL2 模型及对应的处理器和分词器，支持在 GPU 上进行推理。
- 会话转换 (convert_conversation_to_prompts)：将用户和模型之间的交互会话转换为适合模型输入的格式，包括文本与图像信息。
- 文本生成 (deepseek_generate)：根据输入的多模态内容（文本和图像），调用模型生成文本输出，支持流式生成与多种超参数调节（如温度和重复惩罚）。
- 流式输出 (generate)：在推理过程中实现了流式生成，通过 TextIteratorStreamer 实现逐步输出文本，以提高交互性。
- 停止条件控制 (StoppingCriteria)：实现自定义停止条件，以支持在满足特定词汇条件时终止文本生成。

总之，文件 inference.py 实现了 DeepSeek-VL2 模型的核心推理功能，涵盖了从数据预处理到多模态生成输出的全过程。

6.7 图文对话推理

在 DeepSeek-VL2 项目根目录下，也有一个推理文件 inference.py，其主要功能是通过 DeepSeek-VL2 模型进行图文对话推理。它包括模型加载、会话输入处理、多模态生成和结果展示等功能。其中，通过对输入图像及文本的处理，结合模型生成新的对话输出，并支持视觉目标定位等高级功能。

```
def load_pil_images(conversations: List[Dict[str, str]]) -> List[PIL.Image.
    Image]:
    """
    加载对话中的图像路径并转换为 PIL 图像。

    参数：
        conversations (List[Dict[str, str]]): 包含消息的对话列表。
        例如：
            [
                {
                    "role": "User",
                    "content": "<image>\n提取该图像中的所有信息并以Markdown格式输出。",
                    "images": ["./examples/table_datasets.png"]
                },
                {"role": "Assistant", "content": ""},
            ]
    返回：
        pil_images (List[PIL.Image.Image]): PIL 图像的列表。
```

```
    """
    pil_images = []
    for message in conversations:
        if "images" not in message:
            continue
        for image_path in message["images"]:
            pil_img = PIL.Image.open(image_path)
            pil_img = pil_img.convert("RGB")
            pil_images.append(pil_img)
    return pil_images

def main(args):
    dtype = torch.bfloat16
    # 指定模型路径
    model_path = args.model_path
    vl_chat_processor: DeepseekVLV2Processor = DeepseekVLV2Processor.from_
                                               pretrained(model_path)
    tokenizer = vl_chat_processor.tokenizer
    vl_gpt: DeepseekVLV2ForCausalLM = AutoModelForCausalLM.from_pretrained(
        model_path,
        trust_remote_code=True,
        torch_dtype=dtype
    )
    vl_gpt = vl_gpt.cuda().eval()

    # 多图像对话示例
    # 请注意 <|grounding|> 标记专门用于基础图像描述功能。普通对话不需要该标记
    conversation = [
        {
            "role": "<|User|>",
            "content": "<image>\n<image>\n<|grounding|>在第一张图像中，一个对象
             被红色矩形标记。请在第二张图像中定位相同类别的对象。",
            "images": [
                "images/incontext_visual_grounding_1.jpeg",
                "images/icl_vg_2.jpeg"
            ],
        },
        {"role": "<|Assistant|>", "content": ""},
    ]

    # 加载图像并准备模型输入
    pil_images = load_pil_images(conversation)
    print(f"len(pil_images) = {len(pil_images)}")

    prepare_inputs = vl_chat_processor.__call__(
        conversations=conversation,
```

```
        images=pil_images,
        force_batchify=True,
        system_prompt=""
    ).to(vl_gpt.device, dtype=dtype)

    with torch.no_grad():
        if args.chunk_size == -1:
            inputs_embeds = vl_gpt.prepare_inputs_embeds(**prepare_inputs)
            past_key_values = None
        else:
            # 使用增量填充模式，当 GPU 内存为 40G 时适用于 vl2-small 模型
            inputs_embeds, past_key_values = vl_gpt.incremental_prefilling(
                input_ids=prepare_inputs.input_ids,
                images=prepare_inputs.images,
                images_seq_mask=prepare_inputs.images_seq_mask,
                images_spatial_crop=prepare_inputs.images_spatial_crop,
                attention_mask=prepare_inputs.attention_mask,
                chunk_size=args.chunk_size
            )
        # 运行模型获取响应
        outputs = vl_gpt.generate(
            inputs_embeds=inputs_embeds,
            input_ids=prepare_inputs.input_ids,
            images=prepare_inputs.images,
            images_seq_mask=prepare_inputs.images_seq_mask,
            images_spatial_crop=prepare_inputs.images_spatial_crop,
            attention_mask=prepare_inputs.attention_mask,
            past_key_values=past_key_values,
            pad_token_id=tokenizer.eos_token_id,
            bos_token_id=tokenizer.bos_token_id,
            eos_token_id=tokenizer.eos_token_id,
            max_new_tokens=512,
            do_sample=True,
            temperature=0.4,
            top_p=0.9,
            repetition_penalty=1.1,
            use_cache=True,
        )

        # 解码输出并打印结果
        answer = tokenizer.decode(outputs[0][len(prepare_inputs.input_
                ids[0]):].cpu().tolist(), skip_special_tokens=False)
        print(f"{prepare_inputs['sft_format'][0]}", answer)

        # 可视化输出中的标注信息
        vg_image = parse_ref_bbox(answer, image=pil_images[-1])
```

```
        if vg_image is not None:
            vg_image.save("./vg.jpg", format="JPEG", quality=85)

if __name__ == "__main__":
    # 解析命令行参数
    parser = ArgumentParser()
    parser.add_argument("--model_path", type=str, required=True,
                        default="deepseek-ai/deepseek-vl2",
                        help="模型名称或本地路径")
    parser.add_argument("--chunk_size", type=int, default=-1,
                        help="模型增量填充的块大小。当 GPU 为 40G 时适用于 vl2-small
                              模型。"
                             "默认为 -1，即不使用增量填充。")
    args = parser.parse_args()
    main(args)
```

注意

主目录中的文件 inference.py 更偏向于离线实验和本地推理测试；而 serve 目录中的文件 inference.py 是为 Web 部署和在线服务推理设计的，具备流式输出和更高的交互灵活性。

6.8 Web 测试

DeepSeek-VL2 提供了模型的 Web 体验功能，通过文件 web_demo.py，用户可以快速搭建一个基于 Web 的交互界面，用于体验和测试 DeepSeek-VL2 模型。这个 Web 界面允许用户与模型进行实时对话，并观察模型对文本和图像输入的响应。

6.8.1 Web 前端实现

文件 web_demo.py 提供了一个基于 Gradio 的 Web 前端界面，通过该界面可以与 DeepSeek-VL2 模型进行多模态对话和推理交互。主要实现了模型加载、对话生成、参数设置、图片上传与展示、预测和重新生成的功能。用户通过界面输入文本和图像，系统能够调用 DeepSeek-VL2 模型生成对应的输出。

文件 web_demo.py 的具体实现流程如下所示。

（1）下列代码定义了一个名为 MODELS 的列表，其中包含了 DeepSeek-VL2 模型的多种变体名称。这些变体包括不同大小的模型（如 tiny、small 和标准版），以及它们在 Hugging Face 模型库中的完整路径。这个列表的作用是提供一个集中管理的模型名称集合，方便在代码中动态选择和加载不同的模型版本。通过这种方式，用户可以根据需要轻松切换不同的模型，而无须修改代码逻辑。

```
MODELS = [
    "DeepSeek-VL2-tiny",
    "DeepSeek-VL2-small",
    "DeepSeek-VL2",

    "deepseek-ai/deepseek-vl2-tiny",
    "deepseek-ai/deepseek-vl2-small",
    "deepseek-ai/deepseek-vl2",
]
```

（2）下列两行代码定义了两个全局变量，其中DEPLOY_MODELS是一个空字典，用于存储已部署的模型；IMAGE_TOKEN被赋值为字符串 "<image>"，它可能是一个占位符，用于表示与图像相关的特殊标记或符号。

```
DEPLOY_MODELS = dict()
IMAGE_TOKEN = "<image>"
```

（3）下列代码定义了一个名为examples_list的列表，包含多个示例，主要用于视觉引导和视觉问答任务。每个元素是一个列表，其中包含一组图像路径和与之相关的文本描述或指令。

```
examples_list = [
    # 视觉引导 - 1
    [
        ["images/visual_grounding_1.jpeg"],
        "<|ref|>The giraffe at the back.<|/ref|>",
    ],

    # 视觉引导 - 2
    [
        ["images/visual_grounding_2.jpg"],
        "找到<|ref|>淡定姐<|/ref|>",
    ],

    # 视觉引导 - 3
    [
        ["images/visual_grounding_3.png"],
        "Find all the <|ref|>Watermelon slices<|/ref|>",
    ],

    # 引导对话
    [
        ["images/grounding_conversation_1.jpeg"],
        "<|grounding|>I want to throw out the trash now, what should I do?",
    ],
```

```
    # 上下文视觉引导
    [
        [
            "images/incontext_visual_grounding_1.jpeg",
            "images/icl_vg_2.jpeg"
        ],
        "<|grounding|>In the first image, an object within the red rectangle
         is marked. Locate the object of the same category in the second
         image."
    ],

    # 视觉问答
    [
        ["images/vqa_1.jpg"],
        "Describe each stage of this image in detail",
    ],

    # 多图像
    [
        [
            "images/multi_image_1.jpeg",
            "images/multi_image_2.jpeg",
            "images/multi_image_3.jpeg"
        ],
        "能帮我用这几个食材做一道菜吗？",
    ]

]
```

（4）下列代码的功能是根据提供的模型名称model_name加载并返回模型。如果模型已经加载（在DEPLOY_MODELS中存在），则直接返回该模型的信息；如果没有则加载模型并将其存储在DEPLOY_MODELS中，以便后续使用。该函数支持通过args.local_path来指定本地路径加载模型，或者使用默认的模型名称。加载过程中会显示相应的提示信息。

```
def fetch_model(model_name: str, dtype=torch.bfloat16):
    global args, DEPLOY_MODELS

    if args.local_path:
        model_path = args.local_path
    else:
        model_path = model_name

    if model_name in DEPLOY_MODELS:
        model_info = DEPLOY_MODELS[model_name]
        print(f"{model_name} has been loaded.")
```

```
    else:
        print(f"{model_name} is loading...")
        DEPLOY_MODELS[model_name] = load_model(model_path, dtype=dtype)
        print(f"Load {model_name} successfully...")
        model_info = DEPLOY_MODELS[model_name]

    return model_info
```

（5）下列代码的功能是生成一个包含历史对话的提示（prompt），它根据提供的文本、图片、历史对话信息及预设的最大长度（max_length）来构建对话。如果提供了图片，则会调整文本中的图片标记（<image>）与图片数量的匹配情况，确保生成的提示可以被有效处理。该函数会在生成的提示长度超过最大限制时，尝试通过移除较早的对话历史来缩短提示。若最终无法生成合适的提示，则返回None。

```
def generate_prompt_with_history(
    text, images, history, vl_chat_processor, tokenizer, max_length=2048
):
    """
    为 deepseek 应用生成包含历史的提示。

    参数：
        text (str): 文本提示。
        images (list[PIL.Image.Image]): 图片提示。
        history (list): 之前对话的历史记录。
        tokenizer: 用于编码提示的分词器。
        max_length (int): 提示的最大长度。

    返回：
        tuple: 包含生成的提示、图片列表、对话和对话副本的元组。如果在 max_length 限制内
无法生成提示，则返回 None。
    """
    global IMAGE_TOKEN

    sft_format = "deepseek"
    user_role_ind = 0
    bot_role_ind = 1

    # 初始化对话
    conversation = vl_chat_processor.new_chat_template()

    if history:
        conversation.messages = history

    if images is not None and len(images) > 0:
```

```
    num_image_tags = text.count(IMAGE_TOKEN)
    num_images = len(images)

    if num_images > num_image_tags:
        pad_image_tags = num_images - num_image_tags
        image_tokens = "\n".join([IMAGE_TOKEN] * pad_image_tags)

        # 在文本提示后添加 <image> 标签
        text = image_tokens + "\n" + text
    elif num_images < num_image_tags:
        remove_image_tags = num_image_tags - num_images
        text = text.replace(IMAGE_TOKEN, "", remove_image_tags)

    # print(f"prompt = {text}, len(images) = {len(images)}")
    text = (text, images)

conversation.append_message(conversation.roles[user_role_ind], text)
conversation.append_message(conversation.roles[bot_role_ind], "")

# 创建对话副本，以避免在 UI 中截断历史记录
conversation_copy = conversation.copy()
logger.info("=" * 80)
logger.info(get_prompt(conversation))

rounds = len(conversation.messages) // 2

for _ in range(rounds):
    current_prompt = get_prompt(conversation)
    current_prompt = (
        current_prompt.replace("</s>", "")
        if sft_format == "deepseek"
        else current_prompt
    )

    if torch.tensor(tokenizer.encode(current_prompt)).size(-1) <= max_
                length:
        return conversation_copy

    if len(conversation.messages) % 2 != 0:
        gr.Error("用户和助手之间的消息没有配对。")
        return

    try:
        for _ in range(2):  # 连续弹出两条消息
            conversation.messages.pop(0)
    except IndexError:
```

```
            gr.Error("输入文本处理失败，无法在此回合做出响应。")
            return None

    gr.Error("在max_length限制内无法生成提示。")
    return None
```

（6）下列代码的功能是将一个对话（conv）转换为Gradio聊天机器人格式，它遍历对话中的每条消息，并根据消息的角色（用户或助手）及消息内容进行处理。如果消息中包含图片，则会将图片转为Base64编码的字符串，并将其嵌入消息中。最终，函数返回一个适用于Gradio聊天框架的格式，供其展示对话。

```
def to_gradio_chatbot(conv):
    """将对话转换为Gradio聊天机器人格式。"""
    ret = []
    for i, (role, msg) in enumerate(conv.messages[conv.offset:]):
        if i % 2 == 0:
            if type(msg) is tuple:
                msg, images = msg

                if isinstance(images, list):
                    for j, image in enumerate(images):
                        if isinstance(image, str):
                            with open(image, "rb") as f:
                                data = f.read()
                            img_b64_str = base64.b64encode(data).decode()
                            image_str = (f'<img src="data:image/
                                            png;base64,{img_b64_str}" '
                                          f'alt="用户上传图片" style="max-
                                            width: 300px; height: auto;" />')
                        else:
                            image_str = pil_to_base64(image, f"用户上传图片_
                                         {j}", max_size=800, min_size=400)

                        # 替换消息中的<image>标签
                        msg = msg.replace(IMAGE_TOKEN, image_str, 1)

                else:
                    pass

            ret.append([msg, None])
        else:
            ret[-1][-1] = msg
    return ret
```

（7）下列代码的功能是将对话（conv）转换为Gradio历史记录的状态，它通过返回对话中的所有消息（从偏移量开始），来创建一个适用于Gradio历史记录格式的输出。

```
def to_gradio_history(conv):
    return conv.messages[conv.offset:]
```

（8）下列代码的功能是生成一个用于生成的提示（prompt）。根据对话对象（conv）中的设置，构建并返回一个提示字符串。如果使用的是DeepSeek风格的分隔符（SeparatorStyle.DeepSeek），那么，它会根据消息的角色和内容拼接提示，并加上分隔符；如果没有特定的风格，则会调用get_prompt方法获取默认的提示。

```
def get_prompt(conv) -> str:
    """ 获取生成的提示。"""
    system_prompt = conv.system_template.format(system_message=
                                                conv.system_message)
    if conv.sep_style == SeparatorStyle.DeepSeek:
        seps = [conv.sep, conv.sep2]
        if system_prompt == "" or system_prompt is None:
            ret = ""
        else:
            ret = system_prompt + seps[0]
        for i, (role, message) in enumerate(conv.messages):
            if message:
                if type(message) is tuple:  # 多模态消息
                    message, _ = message
                ret += role + ": " + message + seps[i % 2]
            else:
                ret += role + ":"
        return ret
    else:
        return conv.get_prompt()
```

（9）下列代码的功能是将输入的文本和图片进行转换，并返回多个值，包括更新后的文本框和按钮的状态。具体来说，将输入的文本和图像传递并返回一个元组，其中包括输入文本、输入图像、清空文本框、清空图像框，以及将按钮设置为可见。

```
def transfer_input(input_text, input_images):
    print("transferring input text and input image")

    return (
        input_text,
        input_images,
        gr.update(value=""),      # 清空文本框
        gr.update(value=None),    # 清空图像框
        gr.Button(visible=True)   # 设置按钮为可见
    )
```

（10）下列代码定义了一个用于预测用户输入的响应函数predict()。其会根据用户提供的文本、图像及选择的模型进行处理，并生成相应的聊天机器人输出。首先加载并处理图像、生成包含历史记录的提示，其次通过调用生成模型生成聊天机器人的响应。会不断更新生成的内容，并在过程中将结果传递给Gradio界面，以便及时显示对话信息。

```
@wrap_gen_fn
def predict(
    text,
    images,
    chatbot,
    history,
    top_p,
    temperature,
    repetition_penalty,
    max_length_tokens,
    max_context_length_tokens,
    model_select_dropdown,
):

    print("running the prediction function")
    try:
        tokenizer, vl_gpt, vl_chat_processor = fetch_model(model_select_dropdown)

        if text == "":
            yield chatbot, history, "Empty context."
            return
    except KeyError:
        yield [[text, "No Model Found"]], [], "No Model Found"
        return

    if images is None:
        images = []

    pil_images = []
    for img_or_file in images:
        try:
            # load as pil image
            if isinstance(images, Image.Image):
                pil_images.append(img_or_file)
            else:
                image = Image.open(img_or_file.name).convert("RGB")
                pil_images.append(image)
        except Exception as e:
            print(f"Error loading image: {e}")
```

```
conversation = generate_prompt_with_history(
    text,
    pil_images,
    history,
    vl_chat_processor,
    tokenizer,
    max_length=max_context_length_tokens,
)
all_conv, last_image = convert_conversation_to_prompts(conversation)

stop_words = conversation.stop_str
gradio_chatbot_output = to_gradio_chatbot(conversation)

full_response = ""
with torch.no_grad():
    for x in deepseek_generate(
        conversations=all_conv,
        vl_gpt=vl_gpt,
        vl_chat_processor=vl_chat_processor,
        tokenizer=tokenizer,
        stop_words=stop_words,
        max_length=max_length_tokens,
        temperature=temperature,
        repetition_penalty=repetition_penalty,
        top_p=top_p,
        chunk_size=args.chunk_size
    ):
        full_response += x
        response = strip_stop_words(full_response, stop_words)
        conversation.update_last_message(response)
        gradio_chatbot_output[-1][1] = response

        # sys.stdout.write(x)
        # sys.stdout.flush()

        yield gradio_chatbot_output, to_gradio_history(conversation),
        "Generating..."

if last_image is not None:
    vg_image = parse_ref_bbox(response, last_image)
    if vg_image is not None:
        vg_base64 = pil_to_base64(vg_image, f"vg", max_size=800,
                                  min_size=400)
        gradio_chatbot_output[-1][1] += vg_base64
        yield gradio_chatbot_output,
        to_gradio_history(conversation), "Generating..."
```

```
    print("flushed result to gradio")
    torch.cuda.empty_cache()

    if is_variable_assigned("x"):
        print(f"{model_select_dropdown}:\n{text}\n{'-' * 80}\n{x}\n{'=' * 80}")
        print(
            f"temperature: {temperature}, "
            f"top_p: {top_p}, "
            f"repetition_penalty: {repetition_penalty}, "
            f"max_length_tokens: {max_length_tokens}"
        )

    yield gradio_chatbot_output, to_gradio_history(conversation), "Generate: Success"
```

（11）下列代码定义了一个名为retry()的函数，用于重新预测聊天机器人的响应。当聊天记录为空时，函数retry()会返回一个提示消息；如果聊天记录中有内容，则会删除最后的历史记录并将其重置为上一个有效的输入，然后调用predict()函数重新生成响应。

```
# @wrap_gen_fn
def retry(
    text,
    images,
    chatbot,
    history,
    top_p,
    temperature,
    repetition_penalty,
    max_length_tokens,
    max_context_length_tokens,
    model_select_dropdown,
):
    if len(history) == 0:
        yield (chatbot, history, "Empty context")
        return

    chatbot.pop()
    history.pop()
    text = history.pop()[-1]
    if type(text) is tuple:
        text, image = text

    yield from predict(
        text,
        images,
```

```
        chatbot,
        history,
        top_p,
        temperature,
        repetition_penalty,
        max_length_tokens,
        max_context_length_tokens,
        model_select_dropdown,
        args.chunk_size
    )
```

（12）下列代码的功能是接收一组文件，并返回每个文件的路径。具体来说，它会遍历所有文件，提取每个文件的路径，并将路径保存到一个列表中，最后返回该列表，以便进行图片预览。

```
def preview_images(files):
    if files is None:
        return []

    image_paths = []
    for file in files:
        # 使用 file.name 获取文件路径
        # image = Image.open(file.name)
        image_paths.append(file.name)
    return image_paths  # 返回所有图片路径，用于预览
```

（13）下列代码构建了一个Gradio应用函数build_demo()，主要的功能是为聊天机器人创建图形用户界面（GUI），包括用户输入、图片上传、聊天记录显示、模型选择、参数设置等功能。用户可以输入文本、上传图片，调整模型生成的参数，并与聊天机器人交互。通过Gradio组件，用户可以查看生成的聊天记录、图片预览，设置生成文本的参数并执行操作，如发送消息、重试生成、清空对话等。

```
def build_demo(args):
    # fetch model
    if not args.lazy_load:
        fetch_model(args.model_name)

    with open("deepseek_vl2/serve/assets/custom.css", "r", encoding="utf-8")
        as f:
        customCSS = f.read()

    with gr.Blocks(theme=gr.themes.Soft()) as demo:
        history = gr.State([])
        input_text = gr.State()
        input_images = gr.State()
```

```
        with gr.Row():
            gr.HTML(title)
            status_display = gr.Markdown("Success", elem_id="status_display")
        gr.Markdown(description_top)

        with gr.Row(equal_height=True):
            with gr.Column(scale=4):
                with gr.Row():
                    chatbot = gr.Chatbot(
                        elem_id="deepseek_chatbot",
                        show_share_button=True,
                        bubble_full_width=False,
                        height=600,
                    )
                with gr.Row():
                    with gr.Column(scale=4):
                        text_box = gr.Textbox(
                            show_label=False, placeholder="Enter text",
                            container=False
                        )
                    with gr.Column(
                        min_width=70,
                    ):
                        submitBtn = gr.Button("Send")
                    with gr.Column(
                        min_width=70,
                    ):
                        cancelBtn = gr.Button("Stop")
                with gr.Row():
                    emptyBtn = gr.Button(
                        "🧹 New Conversation",
                    )
                    retryBtn = gr.Button("🔄 Regenerate")
                    delLastBtn = gr.Button("🗑️ Remove Last Turn")

            with gr.Column():
                upload_images = gr.Files(file_types=["image"], show_
                                    label=True)
                gallery = gr.Gallery(columns=[3], height="200px", show_
                                    label=True)

                upload_images.change(preview_images, inputs=upload_images,
                                    outputs=gallery)

                with gr.Tab(label="Parameter Setting") as parameter_row:
                    top_p = gr.Slider(
```

```
            minimum=-0,
            maximum=1.0,
            value=0.9,
            step=0.05,
            interactive=True,
            label="Top-p",
        )
        temperature = gr.Slider(
            minimum=0,
            maximum=1.0,
            value=0.1,
            step=0.1,
            interactive=True,
            label="Temperature",
        )
        repetition_penalty = gr.Slider(
            minimum=0.0,
            maximum=2.0,
            value=1.1,
            step=0.1,
            interactive=True,
            label="Repetition penalty",
        )
        max_length_tokens = gr.Slider(
            minimum=0,
            maximum=4096,
            value=2048,
            step=8,
            interactive=True,
            label="Max Generation Tokens",
        )
        max_context_length_tokens = gr.Slider(
            minimum=0,
            maximum=8192,
            value=4096,
            step=128,
            interactive=True,
            label="Max History Tokens",
        )
        model_select_dropdown = gr.Dropdown(
            label="Select Models",
            choices=[args.model_name],
            multiselect=False,
            value=args.model_name,
            interactive=True,
        )
```

```
            # show images, but not visible
            show_images = gr.HTML(visible=False)
            # show_images = gr.Image(type="pil", interactive=False,
                                     visible=False)

    def format_examples(examples_list):
        examples = []
        for images, texts in examples_list:
            examples.append([images, display_example(images), texts])

        return examples

    gr.Examples(
        examples=format_examples(examples_list),
        inputs=[upload_images, show_images, text_box],
    )

    gr.Markdown(description)

    input_widgets = [
        input_text,
        input_images,
        chatbot,
        history,
        top_p,
        temperature,
        repetition_penalty,
        max_length_tokens,
        max_context_length_tokens,
        model_select_dropdown,
    ]
    output_widgets = [chatbot, history, status_display]

    transfer_input_args = dict(
        fn=transfer_input,
        inputs=[text_box, upload_images],
        outputs=[input_text, input_images, text_box, upload_images, submitBtn],
        show_progress=True,
    )

    predict_args = dict(
        fn=predict,
        inputs=input_widgets,
        outputs=output_widgets,
        show_progress=True,
```

```
        )

        retry_args = dict(
            fn=retry,
            inputs=input_widgets,
            outputs=output_widgets,
            show_progress=True,
        )

        reset_args = dict(
            fn=reset_textbox, inputs=[], outputs=[text_box, status_display]
        )

        predict_events = [
            text_box.submit(**transfer_input_args).then(**predict_args),
            submitBtn.click(**transfer_input_args).then(**predict_args),
        ]

        emptyBtn.click(reset_state, outputs=output_widgets, show_progress=True)
        emptyBtn.click(**reset_args)
        retryBtn.click(**retry_args)

        delLastBtn.click(
            delete_last_conversation,
            [chatbot, history],
            output_widgets,
            show_progress=True,
        )

        cancelBtn.click(cancel_outputing, [], [status_display],
                        cancels=predict_events)

    return demo
```

（14）下列代码是应用的主入口。首先，通过解析命令行参数来配置应用的启动选项，例如模型名称、IP地址、端口、根路径等。其次，调用函数build_demo()创建Gradio应用，并设置其标题。最后，通过调用demo.queue() 和demo.launch()来启动Gradio应用，并设置一些运行参数，例如并发请求数、最大事件队列等。

```
if __name__ == "__main__":
    parser = ArgumentParser()
    parser.add_argument("--model_name", type=str, required=True,
                        choices=MODELS, help="model name")
    parser.add_argument("--local_path", type=str, default="",
                        help="huggingface ckpt, optional")
```

```
    parser.add_argument("--ip", type=str, default="0.0.0.0", help="ip address")
    parser.add_argument("--port", type=int, default=37913, help="port number")
    parser.add_argument("--root_path", type=str, default="", help="root path")
    parser.add_argument("--lazy_load", action='store_true')
    parser.add_argument("--chunk_size", type=int, default=-1,
                        help="chunk size for the model for prefiiling. "
                             "When using 40G gpu for vl2-small, set a chunk_
                              size for incremental_prefilling."
                             "Otherwise, default value is -1, which means we
                              do not use incremental_prefilling.")
    args = parser.parse_args()

    demo = build_demo(args)
    demo.title = "DeepSeek-VL2 Chatbot"

    reload_javascript()
    demo.queue(concurrency_count=CONCURRENT_COUNT, max_size=MAX_EVENTS).launch(
        # share=False,
        share=True,
        favicon_path="deepseek_vl2/serve/assets/favicon.ico",
        inbrowser=False,
        server_name=args.ip,
        server_port=args.port,
        root_path=args.root_path
    )
```

6.8.2 启动 Web 测试

在启动Web测试之前需要准备好模型，各个模型被保存在Huggingface上，具体信息请参考本章6.2节中的内容。

1. 使用 deepseek-vl2-tiny 模型

deepseek-vl2-tiny模型总共有 3.37B参数，激活 1B参数，可以在单个GPU（< 40GB）上运行。使用deepseek-vl2-tiny模型启动Web测试的命令如下：

```
CUDA_VISIBLE_DEVICES=2 python web_demo.py \
--model_name "deepseek-ai/deepseek-vl2-tiny"  \
--port 37914
```

对上述参数的具体说明如下：

- CUDA_VISIBLE_DEVICES：指定要使用的GPU设备编号。例如，CUDA_VISIBLE_DEVICES=2 表示使用编号为 2 的GPU。
- --model_name：指定要加载的模型名称。例如，deepseek-ai/deepseek-vl2-tiny。

● --port：指定Web服务器的端口号。例如，37914。

2. 使用 deepseek-vl2-small 模型

deepseek-vl2-small模型总共有 16.1B参数，激活 2.4B参数。

（1）如果在A100 40GB GPU上运行：需要设置 --chunk_size 512 以节省内存，但可能会比较慢。使用deepseek-vl2-small模型启动Web测试的命令如下：

```
CUDA_VISIBLE_DEVICES=2 python web_demo.py \
--model_name "deepseek-ai/deepseek-vl2-small"  \
--port 37914 \
--chunk_size 512
```

可选参数--chunk_size用于设置增量预填充的块大小，以节省内存。注意，仅在GPU内存有限时需要设置。

（2）如果在 > 40GB的GPU上运行：可以忽略 --chunk_size参数，以获得更快的响应。使用deepseek-vl2-small模型启动Web测试的命令如下：

```
CUDA_VISIBLE_DEVICES=2 python web_demo.py \
--model_name "deepseek-ai/deepseek-vl2-small"  \
--port 37914
```

3. 使用 deepseek-vl2 模型

deepseek-vl2 模型总共有 7.5B参数，激活 4.2B参数。使用deepseek-vl2 模型启动Web测试的命令如下：

```
CUDA_VISIBLE_DEVICES=2 python web_demo.py \
--model_name "deepseek-ai/deepseek-vl2"  \
--port 37914
```

4. 访问 Web 界面

运行上述命令后，Web服务器将启动。你可以通过以下地址访问Web界面：

```
http://localhost:37914
```

如果你使用的是远程服务器，可以将localhost替换为服务器的IP地址。

下列代码展示了使用deepseek-vl2-small模型启动Web测试方法。

```
# 安装依赖
pip install -e .[gradio]

# 运行 Web Demo
CUDA_VISIBLE_DEVICES=2 python web_demo.py \
--model_name "deepseek-ai/deepseek-vl2-small"  \
--port 37914 \
```

```
--chunk_size 512
```

通过这些步骤，可以启动DeepSeek-VL2 的Web测试界面，并通过浏览器与模型进行交互，效果如图6-4所示。

图6-4　Web测试界面

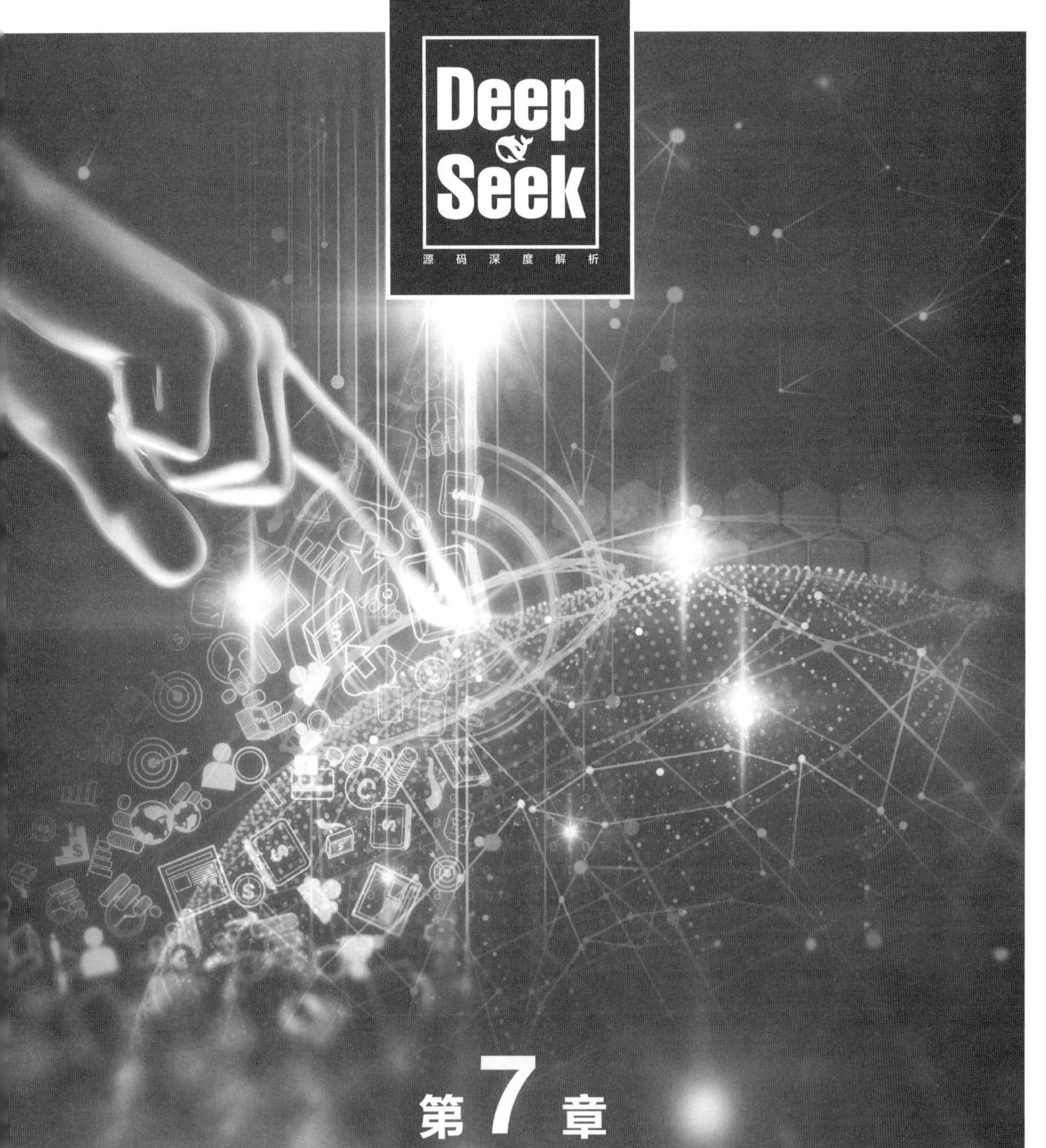

第7章 DeepSeek-R1 推理大模型

DeepSeek-R1是DeepSeek AI团队推出的第一代推理模型，包括DeepSeek-R1-Zero和DeepSeek-R1。通过大规模强化学习训练，这些模型不依赖监督微调进行初步训练，在多个推理相关基准测试中表现出色。DeepSeek-R1-Zero展示了强大的推理能力，能够自然地涌现出许多强大而有趣的推理行为。DeepSeek-R1在解决可读性和语言混合问题的同时，进一步提升了推理性能。

7.1 背景介绍

近年来，随着人工智能技术的飞速发展，大语言模型不断迭代与进化，逐步缩小了与人工通用智能之间的差距。在这一进程中，推理能力的提升成为LLMs发展的重要方向之一。传统上，推理能力的提升主要依赖监督微调，这种方法虽然有效，但需要大量的监督数据，数据收集成本高且耗时。

为了突破这一瓶颈，DeepSeek R1的研究团队尝试采用纯强化学习来提升大语言模型的推理能力，不再依赖SFT作为初步步骤。他们以DeepSeek-V3-Base为基础模型，采用组相对策略优化（Group Relative Policy Optimization，GRPO）作为强化学习框架，通过大规模的强化学习训练，使模型在推理任务中展现出强大的自我进化能力。例如，在AIME 2024测试中，DeepSeek-R1-Zero的单样本通过率从15.6%飙升至71.0%，通过多数投票策略甚至能达到86.7%，与OpenAI-o1-0912的水平相当。

然而，DeepSeek-R1-Zero在推理过程中也暴露出一些问题，如可读性差、语言混用等。为了解决这些问题并进一步提升推理性能，研究团队引入了DeepSeek-R1，该模型在训练初期加入了少量冷启动数据，并采用了多阶段训练流程。最终，DeepSeek-R1在多个推理基准测试中达到了与OpenAI-o1-1217相当的性能。

此外，DeepSeek团队还探索了将DeepSeek-R1的推理能力通过蒸馏技术传递给更小的模型，成功地提升了小型密集模型的推理性能。这一成果不仅为LLMs的推理能力提升提供了新的思路，也为未来的研究和应用开辟了更广阔的空间。

7.2 项目介绍

DeepSeek-R1 项目由DeepSeek AI团队推出，旨在通过强化学习显著提升LLMs的推理能力。近年来，大语言模型在不断演进过程中逐渐逼近AGI的水平，而推理能力的提升被证明不仅能够提高数学、编程、科学推理等任务的准确性，还能更好地符合社会价值和用户偏好，同时降低训练计算资源的需求。

7.2.1 模型演进

在DeepSeek-R1项目中涉及两个模型——DeepSeek-R1-Zero和DeepSeek-R1，具体说明如下。

1. DeepSeek-R1-Zero

这是项目中的第一代推理模型，采用纯强化学习训练，没有经过任何监督微调的冷启动阶段。通过大规模的强化学习训练，模型在AIME 2024 等推理基准上表现出惊人的性能提升。例如，pass@1 分数从 15.6% 提升至 71.0%，在多数投票机制下甚至达到 86.7%，与OpenAI的o1-0912 模型表现相当。该模型在训练过程中自然涌现出很多有趣而强大的推理行为，但也存在可读性差和语言混杂等问题。

2. DeepSeek-R1

为了解决DeepSeek-R1-Zero的不足，DeepSeek团队引入了少量冷启动数据和多阶段训练流程。在强化学习之前，先用数千个链式推理（Chain-of-Thought，CoT）数据对基础模型（DeepSeek-V3-Base）进行微调，再利用强化学习（基于GRPO算法）进一步强化推理能力。训练过程还结合了拒绝采样生成的监督微调数据，以确保输出既具备高性能的推理能力，又能保持清晰和用户友好的格式。最终，DeepSeek-R1 在多项任务上表现与OpenAI-o1-1217 相当。

7.2.2 训练方案

在提升语言模型性能时，往往依赖大量监督数据进行训练，这需要花费大量时间和人工成本来收集和标注数据。相较之下，本项目提出了一种全新的思路：即便不利用监督微调作为冷启动，仅通过大规模的强化学习方法，也能显著提升模型的推理能力。通过这种方法，模型在没有预先灌输大量人工示例的情况下，依靠自我探索与奖励机制，逐步发现并掌握有效的推理策略，从而在多项推理任务中取得了令人瞩目的成绩。

DeepSeek团队发现，在强化学习前引入少量高质量的冷启动数据，能够进一步促进模型性能的提升。具体而言，DeepSeek设计了如下三种训练方案：

- DeepSeek-R1-Zero：该方法直接在基础模型上进行强化学习训练，全程不依赖任何监督微调数据。这一策略充分验证了纯强化学习方法在激发模型自我进化方面的巨大潜力。
- DeepSeek-R1：为了克服纯强化学习在可读性和语言一致性等方面存在的不足，该方法首先利用链式推理的示例数据对基础模型进行微调，从而形成一个更为“温和”的冷启动检查点；其次，通过强化学习进一步优化，提升模型在复杂推理任务中的表现。
- 推理能力蒸馏到小型模型：在获得性能优异的DeepSeek-R1 后，研究者利用蒸馏技术将其推理能力迁移至体积更小的密集模型中。这样，即便是在计算资源有限的情况下，也能获得高效且具有强大推理能力的模型。

这种多阶段、混合策略的训练流程，不仅突破了传统依赖海量监督数据的局限，还为大规模语言模型训练提供了一条更为经济高效的新途径，有助于在降低训练成本的同时实现卓越的推理性能。

本项目采用GRPO（Group Relative Policy Optimization，组相关策略优化）来进行强化学习训练，该方法通过从旧策略中采样一组输出，并利用规则设计的奖励（包括准确性奖励和格式奖励）来优化模型。奖励机制确保模型不仅在任务解答上正确，同时将推理过程清晰地嵌入特定标签（如 <think> 和 </think>）中。为了有效引导强化学习过程，在项目文档中介绍了如下两种奖励方案。

- 准确性奖励：评估模型响应的正确性，例如在数学问题中要求答案格式符合预定义标准。
- 格式奖励：要求模型将推理过程以固定格式输出，确保易读性和结构化表达。

通过设计固定的训练模板（用户提示—模型生成推理过程和答案），DeepSeek-R1-Zero在强化学习过程中展现了"顿悟时刻"：模型能够在中间版本中自主学会重新评估初始方法并延长思考时间，从而处理更复杂的问题。

7.2.3 蒸馏小型模型

为了降低模型体量并便于部署，DeepSeek团队进一步探索了从DeepSeek-R1 到更小密集模型的蒸馏技术。利用Qwen和Llama作为基础，通过直接微调 800K个推理样本，得到了包括 1.5B、7B、8B、14B、32B、70B在内的六个密集模型。这些蒸馏模型在多项推理基准测试中表现出色，部分甚至超过了现有的其他开源大模型。

下面对DeepSeek团队得到的评估结果进行详细分析，将分别从数学推理、问答推理、代码推理等多个角度比较不同模型的性能表现。

1. 数学与逻辑推理能力（AIME 2024 pass@1 和 MATH-500 pass@1 指标）

- 从AIME 2024 的表现来看，GPT-4o-0513（9.3%）和Claude-3.5-Sonnet（16.0%）的成绩明显低于其他模型，说明这两款模型在解决较高难度数学题目时存在较大挑战。
- o1-mini、DeepSeek-R1-Distill系列模型和QwQ-32B-Preview则表现更为出色，尤其是随着模型容量的提升（如DeepSeek-R1-Distill-Qwen从 1.5B到 32B），AIME pass@1 分数从28.9%提升至 72.6%，而MATH-500 pass@1 也相应提升至接近 94%。这表明通过强化学习和蒸馏技术，较大模型能够更好地捕捉和利用推理模式，从而显著提高数学和逻辑推理能力。

2. 问答推理能力（GPQA Diamond pass@1 指标）

- 在该指标上，Claude-3.5-Sonnet得分为 65.0%，略高于GPT-4o-0513 的 49.9%，显示出一定的问答推理优势。
- DeepSeek-R1-Distill系列中，随着模型规模的增加，得分也有逐步提升，从Qwen-1.5B的 33.8% 到Qwen-32B的 62.1%，Llama-70B达到 65.2%，这说明较大容量的模型在处理事实问答或复杂推理问题时具备更强的能力。

3. 代码推理与工程任务能力（LiveCodeBench pass@1 与 CodeForces rating 指标）

- 在LiveCodeBench指标上，o1-mini取得 53.8%，而DeepSeek-R1-Distill系列中的Qwen-32B和Llama-70B分别为 57.2% 和 57.5%，显示出这些模型在代码生成或编程任务上的竞争力。
- 在CodeForces rating指标上，o1-mini以 1820 的高分领先，表明其在编程竞赛任务中表现尤为突出；而DeepSeek-R1-Distill模型的评分在 954 至 1691，虽然略逊于o1-mini，但随着模型容量的增加（如Qwen-32B达到 1691），性能也显著提升。

4. 模型规模与性能关系

从整体趋势来看，蒸馏得到的DeepSeek-R1-Distill模型呈现出明显的规模效应：

- 较小模型（如Qwen-1.5B）的各项指标相对较低；
- 随着模型参数量增加（从 7B、14B到 32B），无论是在数学、问答还是代码任务上，性能都有明显提升。

● 同时，基于不同基础模型（Qwen与Llama）的蒸馏结果在各项指标上各有优势，但总体趋势均表明大模型在推理任务上具有更高的表现潜力。

7.2.4 开源信息介绍

目前（作者写本书时），在DeepSeek-R1的GitHub仓库，开源内容主要是一个详细的PDF文档DeepSeek_R1.pdf，这份文档全面介绍了项目的背景、模型架构、训练方法（包括纯强化学习、冷启动策略、推理导向的强化学习、拒绝采样和监督微调等）及蒸馏技术等核心内容。

文档中提到，DeepSeek AI已经开源了两大系列模型。

● 推理模型：包括通过纯强化学习训练得到的DeepSeek-R1-Zero和经过冷启动数据微调后的DeepSeek-R1。

● 蒸馏模型：利用DeepSeek-R1生成的800K推理样本，对基于Qwen和Llama平台的小型密集模型进行微调，涵盖了从1.5B到70B不等的多个参数规模版本。

目前仓库中仅提供这份PDF文档，尚未见到实际的代码或模型权重。这表明DeepSeek团队仅先公开了详细的研究成果和方法论，为社区提供一个完整的技术参考，后续版本中可能会陆续发布相应的代码和预训练模型。

虽然仓库中没有提供具体的训练代码，但确实开源了模型。根据仓库信息，DeepSeek-R1-Zero和DeepSeek-R1 两个模型都基于DeepSeek-V3-Base进行训练，拥有671B的总参数和37B的激活参数，同时支持长达128K的上下文长度，并可通过Hugging Face下载。

1. DeepSeek-R1-Zero 模型

DeepSeek-R1-Zero是DeepSeek AI开源的第一代推理模型，其训练过程完全基于大规模强化学习，没有使用监督微调作为初始步骤。该模型在数学、代码和逻辑推理等任务上展现了卓越的性能，并在训练过程中自然涌现出多种强大而有趣的推理行为。DeepSeek-R1-Zero模型在HuggingFace发布，如图7-1所示，大家可以从Hugging Face上获取并部署。

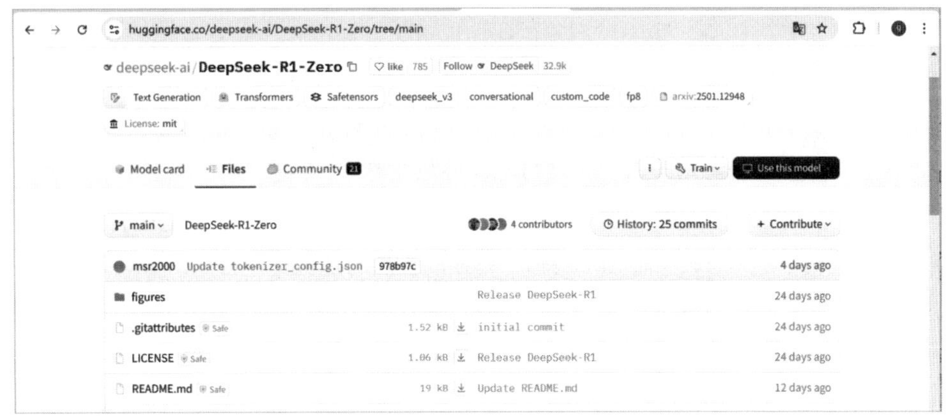

图7-1 DeepSeek-R1-Zero模型

2. DeepSeek-R1 模型

虽然DeepSeek-R1-Zero模型展示出了强大的推理能力，但是也暴露出了重复、可读性差和语言混杂等问题。为了解决这些挑战并进一步提升性能，DeepSeek团队发布了DeepSeek-R1模型，在进行强化学习训练前引入了少量冷启动数据，从而实现了在数学、编程和逻辑推理等任务上与OpenAI-o1 相媲美的效果。DeepSeek-R1模型在Hugging Face发布，如图7-2所示，大家可以从Hugging Face上获取并部署。

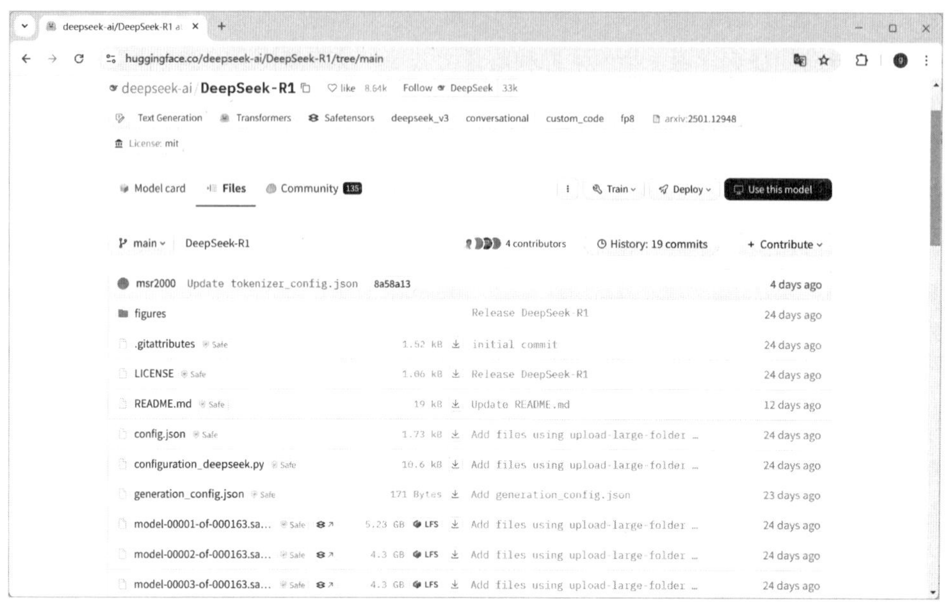

图7-2　DeepSeek-R1模型

我们可以直接从Hugging Face上获取这些预训练好的模型权重，而无须重新训练模型。

7.2.5 结论

本项目的评估结果充分证明了通过大规模强化学习和蒸馏策略提升模型推理能力的有效性，且模型规模对性能具有重要影响。较大模型不仅能捕捉更多推理模式，而且在多个复杂任务中均能取得显著提升。

- 强化学习与蒸馏效果明显：DeepSeek-R1-Distill系列模型在多个推理任务中表现优异，特别是在数学推理上，较大容量模型表现甚至超过了部分商业模型。
- 商业模型与开源模型对比：GPT-4o-0513 和Claude-3.5-Sonnet在某些指标上存在不足，而经过蒸馏的开源模型（尤其是较大模型）展示了更强的推理与任务适应能力。
- 应用场景多样：虽然在编程竞赛任务上o1-mini评分最高，但DeepSeek-R1-Distill模型在数学、问答、代码生成等任务上均表现出较高的综合能力，表明其在多场景推理任务中具有良好的实用性。

7.3 DeepSeek-R1-Zero 训练方案

在传统的推理任务研究中，很多方法往往依赖大量监督数据来提升模型性能，而这种数据的采集和标注过程通常既耗时又成本高昂。为了解决这一问题，本研究探讨了如何在完全没有监督数据的条件下，通过纯粹的强化学习过程来实现大型语言模型的自我进化和推理能力的显著提升。为此，我们首先设计了一种名为DeepSeek-R1-Zero的方案，其主要思想是直接在未经监督微调的基础模型上进行强化学习训练，从而让模型在无须预先依赖人工示例的情况下，自动挖掘并优化其内部推理策略。

7.3.1 强化学习算法

在具体实现上，为了降低强化学习训练的计算开销并提高效率，我们引入了GRPO算法。GRPO算法与传统强化学习方法相比，有如下所示的两个关键特点。

1. 不使用独立的评论家模型

常规的策略梯度方法通常需要一个与策略模型规模相匹配的评论家模型来评估当前策略的表现，而GRPO巧妙地放弃了这一做法。它通过从当前策略中采样一组候选输出，并利用这些输出的组内分布统计（如标准差和均值）来估计一个基线，从而直接为策略模型提供反馈。这种方法既节省了计算资源，又避免了构建和训练额外评论家模型的复杂性。

2. 设计采样函数与目标函数

具体来说，对于每个问题（记为 i），GRPO从旧策略（$\pi\theta_{old}$）中采样出一组候选输出 $\{a_i^1, a_i^2, \cdots, a_i^A\}$。接下来，通过最大化下面的目标函数来对当前策略 $\pi\theta$ 进行优化：

- 对每个候选输出 a_i，根据其在旧策略下的概率和当前策略下的概率之比，采用剪切策略来控制更新幅度，确保新策略不会偏离旧策略太远。
- 为了鼓励模型探索更优解，目标函数中引入了优势（A_i）的概念，即利用当前组内输出的标准差（std）与均值（mean）的差值来衡量每个候选解的优劣。
- 目标函数中还加入了KL散度项，用以约束新旧策略之间的差异，防止策略更新过程中发生过大变化。具体超参数 ε 和 β 分别控制剪切范围和KL惩罚的力度。

优势 A_i 的计算公式如下：

$$A_i = \text{std}(\{a_i^1, a_i^2, \cdots, a_i^A\}) / (a_i - \text{mean}(\{a_i^1, a_i^2, \cdots, a_i^A\}))$$

这意味着，每一组候选输出的离散程度（标准差）与其平均水平之间的关系，会影响策略的更新方向，从而使模型在没有任何监督信息的情况下，通过内部比较不断改进自身的推理策略。

通过这种方式，DeepSeek-R1-Zero能够在强化学习过程中逐步“觉醒”出强大的推理能力，并在多个推理任务中展示出令人兴奋的性能提升。研究团队通过这一方法证明，即使完全摒弃依赖监督数据的传统训练方式，大型语言模型仍然能够在纯强化学习的驱动下实现自我进化，为后续更高效、低成本的推理模型训练提供了有力的理论依据与实践支持。

7.3.2 奖励建模

在强化学习训练过程中，奖励信号扮演着至关重要的角色，它不仅为模型提供了方向性的反馈，而且直接决定了模型优化的目标和路径。为了训练DeepSeek-R1-Zero，我们设计了一套基于规则的奖励系统，这套系统主要分为两大类奖励机制，以确保模型在推理任务中既能正确回答问题，又能遵循特定的格式输出，从而便于后续的评估与验证。

1. 准确性奖励

准确性奖励的核心在于衡量模型生成答案的正确性。在数学问题中，如果模型被要求将最终答案以特定格式（如在方框内）呈现，这样不仅便于规则化检查，也能确保答案可以被程序或规则自动验证其正确性。类似的，对于LeetCode编程题目，我们可以利用编译器对模型生成的代码进行预设测试，通过测试用例的反馈来评估代码是否符合要求并正确运行。准确性奖励模型通过这种基于规则的自动化验证方式，精确地衡量模型回答问题的正确率，为模型的进一步优化提供了明确而可靠的训练信号。

2. 格式奖励

除了答案的正确性，输出的格式同样至关重要。为了确保模型不仅能给出正确的答案，还能生成结构清晰、便于理解的推理过程，我们设计了格式奖励模型。这一奖励机制要求模型在回答问题时，将其思考过程严格地嵌入指定的标签中。例如，要求将推理过程封装在 <think> 和 </think> 标签之间。这样做的目的有两方面：一方面，它能帮助评估系统或人类审阅者快速捕捉到模型的推理轨迹；另一方面，通过强制固定的输出格式，还能降低输出内容中的噪声，提升整体可读性和一致性。

在DeepSeek-R1-Zero的开发过程中，我们特意没有引入基于神经网络的结果或过程奖励模型。原因在于，尽管神经奖励模型在某些场景下可能提供细粒度的反馈，但在大规模强化学习训练中，它们容易遭遇奖励黑客攻击，也就是模型可能通过学习漏洞规避真实的训练目标，从而获得虚假的奖励。此外，训练和维护神经奖励模型本身需要大量额外的训练资源，并且会显著增加整个训练流程的复杂性。为此，我们选择了基于规则的奖励方式，以确保奖励信号的稳定性和可靠性，从而在大规模强化学习过程中更好地指导模型的自我进化。

总之，通过采用准确性奖励与格式奖励这两种基于规则的奖励机制，我们能够为DeepSeek-R1-Zero提供既准确又直观的反馈信号，确保模型在纯强化学习训练中既提升推理能力，也保持了输出的规范性和易读性，从而为后续的性能优化打下坚实的基础。

7.3.3 训练模板

为了训练DeepSeek-R1-Zero，DeepSeek团队精心设计了一种简单而明确的训练模板，该模板为基础模型提供了清晰的指令，引导其在生成答案时遵循特定的格式。具体来说，该模板要求模型在回答用户问题时，先详细描述其内部的推理过程，然后再给出最终的答案。这样的设计有助于我们在训练过程中系统地观察和记录模型的思考轨迹，从而更好地分析其自我进化和推理策略的形成。

在训练模板中规定了严格的结构格式，具体格式如下：

```
    用户输入提示信息，
    模型输出时必须先生成包含完整推理过程的部分（用 <think> 和 </think> 标签包裹），接着
再生成包含最终答案的部分（用 <answer> 和 </answer> 标签标识）。
```

这种设计有两个主要目的。

- 规范输出格式：通过固定的输出结构，确保所有生成内容具有一致的格式，便于后续的自动化评估和人类审查。
- 避免内容偏差：模板的简洁设计刻意避免引入额外的指令，例如要求模型反思或解释特定解决策略，从而防止因人为预设的偏好而影响模型自然发展出的推理过程。这样，我们就能更加客观地观察模型在纯强化学习过程中如何自发地构建和优化其内部推理机制。

总之，这一训练模板不仅为模型提供了明确的生成指南，也为研究人员观察模型在强化学习过程中自然演化推理能力提供了宝贵的数据支持。

7.3.4 DeepSeek-R1-Zero 的性能

在之前的研究中，模型性能的提升通常依赖大量的监督数据，这些数据的收集既耗时又费力。而 DeepSeek 的研究表明，即使在没有监督微调作为初始步骤的情况下，通过大规模的强化学习，模型的推理能力也能显著提高。此外，加入少量的初始数据，可以进一步提升模型性能。DeepSeek 团队提出了以下方法。

- DeepSeek-R1-Zero：直接在基础模型上应用强化学习，不依赖任何监督微调数据。
- DeepSeek-R1：从经过长链推理示例微调的检查点开始应用强化学习。
- 将 DeepSeek-R1 的推理能力蒸馏到小型密集模型：通过知识蒸馏技术，将大型模型的推理能力传递到较小的模型中，以提高其性能和效率。

这些方法的实施，展示了在不同程度的监督微调和强化学习结合下，模型推理能力的提升，为未来的研究提供了新的思路和方向。

7.3.5 DeepSeek-R1-Zero 的自我进化过程

这里的自我进化过程，是指强化学习在没有任何额外监督数据支持的情况下，驱动 DeepSeek-R1-Zero 模型自主提升推理能力的全过程。这一过程不仅直观地证明了纯强化学习方法的有效性，更为我们揭示了模型内部自发优化机制的奥秘。

通过直接在基础模型上实施强化学习，能够实时、细致地监控模型在面对复杂推理任务时的逐步演化情况。由于没有依赖传统监督微调阶段的预先引导，模型在训练过程中完全依靠奖励信号的

反馈来不断调整和改进其内部策略。这种纯粹的自我驱动进化，为研究者提供了一个难得的窗口，可以观察到模型在面临不同难度和复杂度问题时，其思维过程和策略如何自然地从简单向复杂转变。

如图7-3所示，训练过程中模型的“思考时间”呈现出明显且持续的提升趋势。

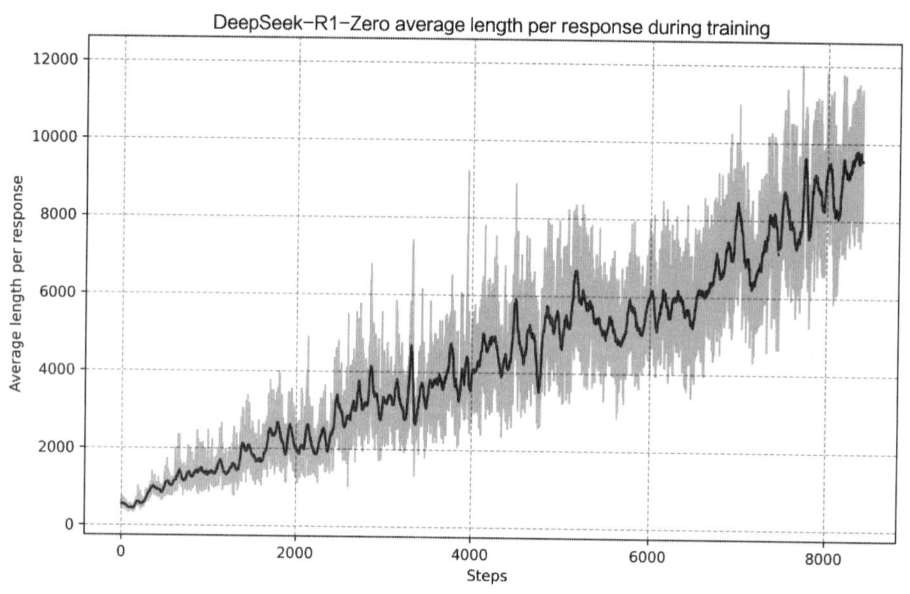

图7-3　训练过程可视化

这里的“思考时间”可以理解为模型在生成答案前所使用的推理步骤或计算量。起初，模型可能只生成数百个推理标记来完成简单任务；然而，随着训练的深入，它逐渐学会在面对更具挑战性的问题时，主动扩展推理计算，甚至可以生成数千个推理标记，以便对问题进行更全面、细致的解析。这样的扩展不仅说明了模型在处理复杂任务时具有灵活性和适应性，也证明了其内部策略在不断优化和迭代。

更引人注目的现象在于，当模型开始重新评估其初始解题策略时，它会主动为难题分配更多的思考时间。这种现象，可以看作模型在“顿悟”时刻的体现——一种突然突破、策略升级的瞬间。在这一时刻，模型不再局限于早期所采用的简单方法，而是根据过去的反馈和奖励信息，发现了更为高效的推理路径，从而大幅提升了最终答案的准确性和合理性。

总的来说，DeepSeek-R1-Zero的自我进化过程为我们展示了以下几点重要启示。

- 自我优化能力：模型能够在纯强化学习的驱动下，自主调整和优化其推理过程，无须额外的监督数据干预，这证明了强化学习在激发模型内在潜力方面的强大力量。
- 适应性和灵活性：通过延长推理计算和增加思考步骤，模型可以更深入地探索问题解决策略，从而更有效地处理复杂推理任务。这种适应性为面对多变任务提供了强有力的支持。
- 意外的复杂行为涌现：模型在训练过程中自然涌现出的“顿悟时刻”，不仅标志着性能上的跃升，更体现了强化学习方法能够催生出意想不到的高级行为模式，为未来智能系统的自主进化指明了方向。

这种自发的发展和策略优化，为我们理解和构建未来更自主、更智能的系统提供了宝贵的理论依据和实践经验。通过深入研究这种自我进化的机制，我们有望在不远的将来设计出能够在极少人类干预的情况下，自主学习并解决复杂任务的高级智能模型。

7.3.6 在 DeepSeek-R1-Zero 的“顿悟时刻”

在DeepSeek-R1-Zero的训练过程中，我们观察到一个极具启发性且意义深远的现象——“顿悟时刻”。这一现象不仅展示了纯强化学习的巨大潜力，还为我们理解模型如何自主进化提供了宝贵的线索。

1. “顿悟时刻”的定义与表现

所谓“顿悟时刻”，指的是在训练的中间阶段，模型突然表现出一种明显超越以往状态的能力跃迁。这时，DeepSeek-R1-Zero不再依赖单一的初始解题策略，而是开始重新评估自身的推理路径，主动为复杂问题分配更多的思考资源和时间，从而大幅提高了解题的准确率和效率。比如，在AIME 2024的基准测试中，模型的平均pass@1分数从最初的15.6%迅速跃升到71.0%，这不仅是数值上的提升，更反映出模型在内部策略上的根本性改进。

2. 机制与内在逻辑

这一现象的背后，是强化学习算法在大规模数据环境下对模型自我调整能力的激发。通过设计合理的奖励机制，我们并没有预先告知模型如何处理复杂问题，而是通过不断提供正确的激励信号，让模型在试错过程中自主发现更高效的解题方法。具体来说，模型在每一次生成输出后，会根据预设的准确性和格式奖励获得反馈；随着训练的深入，它逐渐学会在遇到更难问题时，延长推理过程，并利用额外的“思考时间”进行更充分的信息整合和策略调整。这种自我反思和优化的过程正是“顿悟时刻”的核心所在。

3. 意义与启示

“顿悟时刻”不仅证明了纯强化学习方法在不依赖大量监督微调数据的情况下也能激发出模型的内在潜力，还为未来智能系统的研发指明了方向，能够给我们如下启示。

- 自主进化的可能性：模型可以通过适当的激励和反馈，自主发现并优化解决问题的策略，而不必完全依赖人类专家预设的规则。
- 自适应与灵活性：在面对多样化和复杂问题时，模型能够灵活地调整自身行为，从而在不同任务间保持较高的通用性和适应性。
- 未来研究的契机：这一现象为研究人员提供了一个全新的视角，即通过观察和解析“顿悟时刻”，我们或许可以更深入地理解智能系统内部的决策机制，为设计更为高效和自我完善的AI系统奠定理论基础。

总之，DeepSeek-R1-Zero的“顿悟时刻”不仅标志着模型在推理能力上实现了质的飞跃，也展示了强化学习方法在大规模语言模型训练中的独特优势。通过这种自主进化的过程，模型在不依赖外部监督的情况下，逐步形成了更高效的问题解决策略，为构建未来更自主、更智能的系统提供了宝贵的实验范例和理论依据。这种现象无疑激励着我们在强化学习和自我进化机制的探索中迈出更大胆的步伐，期待未来能涌现出更多突破性的研究成果。

注意

尽管DeepSeek-R1-Zero在推理能力和自主发展方面展现了显著的优势，但它也面临一些挑战。具体而言，模型在生成内容的可读性和语言混合方面存在问题。为了解决这些问题，并使推理过程更易于理解和共享，我们开发了DeepSeek-R1。该模型通过在初始阶段引入人类友好的监督数据，结合强化学习，旨在提升模型生成内容的可读性和一致性。这种方法不仅改善了模型的输出质量，还增强了其在不同任务和语言环境下的适应能力。

7.4 DeepSeek-R1 训练方案

在强化学习领域，DeepSeek-R1-Zero的出色表现引起了广泛关注。它通过创新的方法实现了一系列令人瞩目的成果，为后续研究提供了宝贵的思路和方向。然而，随着研究的不断深入，一些新的问题也逐渐浮现出来。首先，如果在训练过程中引入少量但质量极高的数据作为冷启动，是否能够进一步提升模型的推理性能，或者加快模型的收敛速度？这是一个极具挑战性的问题，因为数据的质量和数量在强化学习中起着至关重要的作用。其次，如何训练出一个真正用户友好的模型？这个模型不仅要能够生成清晰、连贯的链式推理，还要具备强大的通用能力，能够在多种不同的场景和任务中表现出色。为了解决这些复杂而关键的问题，DeepSeek团队经过深入研究和反复试验，精心设计了一个包含四个阶段的训练流程。在本节内容中，将详细介绍这个训练流程的每个阶段及其目标信息。

7.4.1 冷启动

为了避免从基础模型开始进行强化学习训练时可能出现的初期推理不稳定性问题，DeepSeek-R1采用了一种不同的方法。我们精心构建并收集了一小部分链式数据，以微调模型，作为初始的强化学习策略模型。这种策略旨在为模型提供一个稳定的起点，确保训练过程的平稳进行。

在数据收集过程中，DeepSeek探索了多种方法。

- 少样本提示：使用包含长链式思维的示例，提供少量样本来提示模型生成类似的推理过程。
- 直接提示：直接提示模型生成包含反思和验证的详细答案，鼓励模型在回答问题时进行深度思考和自我检查。
- 输出收集：收集DeepSeek-R1-Zero生成的可读格式输出，筛选出高质量的响应，作为训练数据。
- 人类后处理：通过人类标注者对模型生成的响应进行后处理，完善结果，确保数据的准确性和可读性。

通过上述方法收集了数千条冷启动数据，用于微调 DeepSeek-V3-Base，作为强化学习训练的起点。与 DeepSeek-R1-Zero 相比，冷启动数据具有以下优势。

- 可读性：DeepSeek-R1-Zero 的一个主要限制是其生成的内容通常不适合阅读，响应可能混合多种语言或缺乏突出显示答案的格式。在创建 DeepSeek-R1 的冷启动数据时，我们设计了一个可读的模式，包括在每个响应末尾添加总结，并过滤掉对读者不友好的响应。具体而言，我们定义了输出格式为 |special_token|<reasoning_process>|special_token|<summary>，其中 <reasoning_process> 是查询的链式推理，<summary> 用于总结推理结果。
- 潜力：通过精心设计带有人类先验知识的冷启动数据模式，我们观察到与 DeepSeek-R1-Zero 相比，模型性能有了显著提升。我们相信，迭代训练是提升推理模型性能的有效方法。

总之，采用冷启动数据为 DeepSeek-R1 提供了一个稳定且有效的训练起点，显著提升了模型的可读性和推理能力。这种方法展示了结合人类知识与模型自我学习的潜力，为未来推理模型的训练提供了宝贵的经验。

7.4.2 推理导向的强化学习

在完成冷启动数据的微调之后，对 DeepSeek-V3-Base 模型实施了与 DeepSeek-R1-Zero 相同的大型强化学习训练。这一阶段的核心目标是显著提升模型在推理密集型任务中的推理能力。推理密集型任务通常涉及明确定义的问题及清晰的解决方案，例如编码任务、数学问题、科学实验分析及逻辑推理等，这些任务对模型的逻辑思维能力和问题解决能力提出了极高的要求。

在模型训练的链式推理过程中经常出现语言混合的情况，尤其是在强化学习提示涉及多种语言时。这种语言混合可能会导致模型的输出变得混乱，影响推理的准确性和可读性。为了减少这一问题，在强化学习训练中引入了一种新的奖励机制——语言一致性奖励。具体来说，语言一致性奖励是通过计算链式推理中目标语言单词的比例来实现的。例如，如果目标语言是英语，那么模型在生成推理过程时，使用英语单词的比例越高，获得的语言一致性奖励就越高。

然而，在进行消融实验时发现这种对齐机制虽然能够有效提升语言一致性，但也会导致模型性能略有下降。尽管如此，我们仍然认为这种奖励机制是必要的。因为它使模型的输出与人类的偏好更加一致，从而提高了输出的可读性。毕竟，模型的最终目标是服务于人类用户，输出清晰、连贯且符合人类语言习惯的结果至关重要。

为了综合考虑推理任务的准确性和语言一致性，将推理任务的准确性奖励和语言一致性奖励直接相加，形成了最终的奖励函数。然后，在微调后的模型上应用强化学习训练，持续优化模型的参数，直到模型在推理任务上达到收敛。这一阶段的训练不仅提升了模型的推理能力，还使其输出更加符合人类的语言习惯，为后续的应用奠定了坚实的基础。

7.4.3 拒绝采样和监督微调

在推理导向的强化学习训练收敛后，接下来利用由此产生的检查点为下一轮训练收集监督微调数据。这一阶段的目标是进一步优化模型，使其不仅在推理任务上表现出色，还能在其他通用任务中表现良好。

1. 推理数据的生成与筛选

DeepSeek团队策划了一系列推理提示，并通过拒绝采样（Rejection Sampling）从上述强化学习训练的检查点中生成推理轨迹。在之前的推理导向强化学习阶段，主要关注的是可以使用基于规则的奖励进行评估的数据。然而，在这一阶段进一步扩展了数据集，加入了使用生成式奖励模型评估的数据。具体来说，将真实值和模型预测输入DeepSeek-V3进行判断，以评估模型输出的质量。

此外，由于模型在某些情况下会生成混乱且难以阅读的输出，对生成的数据进行了严格的筛选操作。过滤掉了包含混合语言、长段落和代码块的链式推理，以确保数据的高质量和可读性。对于每个推理提示，采样了多个响应，并仅保留正确的响应。通过这种方式，总共收集了大约60万个与推理相关的训练样本。

2. 非推理数据的生成与整合

除了推理数据，还收集了非推理数据，包括写作、事实问答、自我认知和翻译等任务。对于这些非推理任务，沿用了DeepSeek-V3的整体流程，并复用了DeepSeek-V3的部分监督微调数据。具体来说，这一流程包括以下几个关键步骤：

（1）数据生成与整合。

- 对于某些非推理任务，通过提示调用DeepSeek-V3生成潜在的思维链，以便在回答问题之前提供更详细的推理过程。
- 对于更简单的查询（如“你好”），不会提供推理链作为回应，因为这些任务不需要复杂的推理。

（2）数据筛选与优化。

对生成的数据进行了筛选，确保数据的质量和可读性。例如，过滤掉了混合语言、长段落和代码块的输出。最终，收集了大约20万个与推理无关的训练样本。

3. 监督微调

使用上述策划的数据集（约80万个样本，包括推理和非推理数据）对DeepSeek-V3-Base进行了两个epoch的微调。这一过程不仅进一步优化了模型在推理任务上的表现，还增强了其在其他通用任务中的能力。

7.4.4 全场景强化学习

在第二阶段的强化学习中，目标是让模型在推理能力不断提升的同时，更加符合人类的偏好，确保其生成的内容既有用又安全无害。为此，设计了一种全场景的强化学习策略，该策略整合了多种奖励信号，并使用了多样化的提示分布，以覆盖不同任务和场景。

具体来说，这一阶段的训练主要分为两大类数据：推理数据和通用数据。

- 推理数据：依照之前在DeepSeek-R1-Zero中提出的方法，继续使用基于规则的奖励机制。

这种奖励机制主要应用于数学、编码及逻辑推理等需要精确解题步骤的任务，通过对模型生成的推理过程进行严格评估，确保其能够在解决复杂问题时展示出清晰、有条理的思考路径。

● 通用数据：引入了奖励模型，以捕捉那些更为复杂且细微的人类偏好。本项目将基于DeepSeek-V3流程，采用了类似的偏好对和训练提示分布，从而让模型不仅在专业推理任务中表现优异，同时也能在写作、对话、角色扮演等通用任务中提供符合用户期望的回答。

为了确保模型生成的内容既实用又无害，在奖励设计上进行了细致区分：

● 有用性奖励：在这一部分，主要关注生成内容的最终总结，确保回答能切实满足用户需求和实际应用场景，从而提升整体响应的相关性和实用性，同时尽量减少对底层推理细节的干扰。

● 无害性奖励：为防止模型生成潜在的有害、偏见或风险内容，我们对整个响应（包括推理过程和最终总结）进行全面评估，确保输出既符合安全标准，又能有效避免可能引发的负面效应。

最终，通过将这两类奖励信号与多样化的数据分布相结合，我们成功地训练出了一款在推理任务上表现出色，同时在有用性和无害性方面也达到了较高标准的模型。这一全场景强化学习策略不仅提升了模型的推理能力，也使其在面对多种任务时能更好地满足用户需求，并确保输出内容符合安全和伦理要求。

7.5 蒸馏处理

为了使更高效的小型模型具备像DeepSeek-R1这样的强大推理能力，我们探索了一种基于知识蒸馏的方法。具体来说，我们直接使用DeepSeek-R1生成的80万个高质量训练样本对一系列开源模型进行微调。这些开源模型包括Qwen和Llama等。通过这种简单的蒸馏方法，我们发现小型模型的推理能力得到了显著提升。

7.5.1 基础模型的选择与蒸馏过程

在DeepSeek-R1项目中，DeepSeek开发团队选择的基础模型包括Qwen2.5-Math-1.5B、Qwen2.5-Math-7B、Qwen2.5-14B、Qwen2.5-32B、Llama-3.1-8B和Llama-3.3-70B-Instruct。在这些模型中，特别选择了Llama-3.3，因为它在推理能力上略优于Llama-3.1。这些基础模型涵盖了不同规模和能力，使我们能够全面评估蒸馏方法的有效性。

DeepSeek提供了如表7-1所示的蒸馏模型（Distill Models），这些模型基于DeepSeek-R1 的推理能力，通过知识蒸馏技术将推理能力迁移到较小规模的基础模型中，从而提升这些小型模型的推理性能。

表7-1 蒸馏模型

模型名称	基础模型	模型名称	基础模型
DeepSeek-R1-Distill-Qwen-1.5B	Qwen2.5-Math-1.5B	DeepSeek-R1-Distill-Qwen-14B	Qwen2.5-14B

续表

模型名称	基础模型	模型名称	基础模型
DeepSeek-R1-Distill-Qwen-7B	Qwen2.5-Math-7B	DeepSeek-R1-Distill-Qwen-32B	Qwen2.5-32B
DeepSeek-R1-Distill-Llama-8B	Llama-3.1-8B	DeepSeek-R1-Distill-Llama-70B	Llama-3.3-70B-Instruct

上述蒸馏模型在Hugging Face平台公开发布，例如DeepSeek-R1-Distill-Qwen-1.5B在Hugging Face平台的信息如图7-4所示，大家可以选择需要的模型部署到本地或云服务器中进行测试。

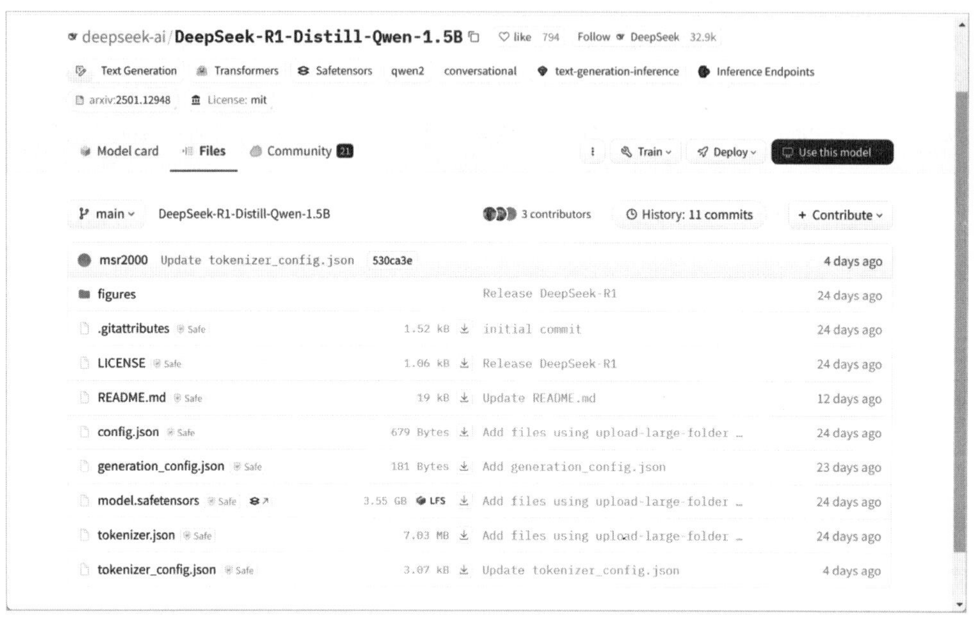

图7-4 DeepSeek-R1-Distill-Qwen-1.5B模型

在蒸馏过程中，DeepSeek-R1仅应用了监督微调，而没有包括强化学习阶段。尽管强化学习可以显著提升模型性能，但在这里，我们的主要目标是展示知识蒸馏技术本身的强大有效性。通过使用DeepSeek-R1生成的高质量数据对这些小型模型进行微调，成功地将DeepSeek-R1的推理能力“蒸馏”到了这些模型中。

7.5.2 模型蒸馏的技术原理

DeepSeek-R1提出的模型蒸馏技术是一种将大型教师模型中获得的高级推理能力迁移到较小、效率更高的学生模型的方法，这种方法的优势在于能够以较低的计算成本和资源消耗，将大型模型的推理能力迁移到小型模型上。这对于资源有限的用户和应用场景来说具有重要意义，因为它使得小型模型能够在推理任务中表现出色，同时保持高效的运行速度。

1. 技术原理

- 知识迁移与软标签：传统的知识蒸馏方法主要依靠教师模型生成的“软标签”作为训练目标，

这些软标签不仅包含了标准的正确答案信息，还蕴含了教师模型对样本的置信度分布和内部推理模式。DeepSeek-R1经过大规模强化学习训练后，掌握了复杂的推理策略，其输出中包含详细的链式推理过程。通过将这些高质量、细粒度的推理数据作为训练样本，学生模型可以学习到教师模型在解决复杂问题时的思考路径和策略，而不仅仅是答案本身。

- 监督微调作为蒸馏手段：在蒸馏过程中，DeepSeek-R1直接使用其生成的约80万条推理样本对开源的学生模型（如Qwen和Llama系列）进行监督微调。相比于强化学习，监督微调的过程更加稳定和高效，因为它利用的是现成的高质量标签数据。这样，学生模型能够在较短时间内捕捉到教师模型传递的推理模式和知识。
- 模型容量与性能传递：蒸馏技术利用了教师模型在大模型容量下形成的强大推理能力，并将这一能力迁移到参数规模较小的模型中。尽管学生模型的规模较小，但由于接受了高质量推理数据的指导，它们在推理任务（如数学、编码、逻辑推理等）上能达到与教师模型相当的性能。这种方法有效地打破了大模型与小模型之间性能差距的瓶颈，为资源受限的应用场景提供了实用的解决方案。

2. 实现方法

（1）数据生成与收集。

- 教师数据生成：首先，利用DeepSeek-R1模型生成大规模的推理样本，这些样本不仅包含最终答案，还详细记录了推理过程。生成的数据通常经过精心设计，确保每个样本都具有清晰的链式思维和总结部分。
- 数据清洗与过滤：为了确保高质量的数据输入学生模型中，研究团队对生成的样本进行了筛选和过滤，剔除混杂语言、格式混乱或不符合预期的低质量输出，最终构成约80万条高质量数据集。

（2）监督微调训练。

- 目标模型选择：研究人员选择了多个开源学生模型，如Qwen2.5-Math系列（1.5B、7B、14B、32B）和Llama系列（Llama-3.1-8B、Llama-3.3-70B-Instruct）。这些模型作为基础，通过蒸馏数据进行微调。
- 训练过程：在训练过程中，学生模型通过对比教师模型生成的推理过程和最终答案，学习到如何生成类似的链式推理输出。训练过程中采用标准的监督学习损失函数，如交叉熵损失，通过多轮迭代使得学生模型逐步逼近教师模型的表现。
- 强化学习阶段的排除：在蒸馏过程中，研究团队仅使用监督微调，而不引入额外的强化学习训练。这既简化了训练流程，又将强化学习阶段的复杂性和高资源消耗留给教师模型，重点展示了蒸馏技术本身的有效性。

（3）模型评估与性能提升。

- 性能验证：经过蒸馏训练后，小型模型在多项推理基准测试中表现显著提升，证明了教师模型的推理模式成功迁移。具体的评估指标涵盖数学题目、编码任务及逻辑推理等多个领域，显示出蒸馏后的模型在准确性和推理深度上均有优异表现。
- 应用场景扩展：由于蒸馏后的模型体积较小、计算效率高，因此，其在实际应用中更加灵活，能够满足在资源受限的环境下快速部署的需求。

总之，DeepSeek-R1 提出的模型蒸馏技术，通过将大规模、经过强化学习训练获得强大推理能力的教师模型的输出作为监督信号，成功地将复杂的推理模式迁移到小型学生模型中。其技术原理依赖知识迁移和软标签机制，而实现方法则基于大规模数据生成、精细数据清洗及标准的监督微调训练。该技术不仅显著提升了小型模型在推理任务上的性能，还为低资源环境下的高效推理模型提供了可行方案，同时为后续的模型优化和应用推广奠定了坚实基础。